INTRODUCTION TO
ENGINEERING
MECHANICS

Schlenker / McKern

THE JACARANDA PRESS

First published 1976 by
JACARANDA WILEY LTD
33 Park Road, Milton, Qld 4064

Offices also in Sydney and Melbourne

Typeset by Hartland & Hyde Typesetting Pty Ltd, Sydney

© Jacaranda Wiley Ltd 1976

National Library of Australia
Cataloguing-in-Publication data

Schlenker, Barry Royce, 1940–.
    Introduction to engineering mechanics.

    Index.
    First published, Sydney: Wiley, 1976.
    ISBN 0 7016 1224 X.

    1. Mechanics, Applied. I. McKern, Donald, 1934–.
    joint author. II. Title.

620.1

To — Jenny, Margo, Wendy, Melinda,
Theresa, Sally, Michael and Christie —
who all have their moments.

Produced in Australia

Contents

Preface

Students today are becoming increasingly aware of the importance of industry and technology and at the same time increasingly critical of some of its effects on our environment and related lifestyle. The professional person who is often either directly or indirectly involved in finding solutions to many current controversies and problems is the engineer, whether he is working in the civil, mechanical, aeronautical, marine or electrical fields. Consequently, it was thought necessary to develop a course at senior high school level in which some aspects of engineering and its impact upon society could be explored.

Such a course was introduced into New South Wales high schools in 1966; it was known as Industrial Arts. To satisfy the changing requirements of modern education a new course called Engineering Science was introduced in 1975. It has much broader aims and objectives than the original Industrial Arts course and is designed to appeal to a wider cross-section of students. It includes many units of work centred around engineering mechanics and materials science—two principal areas of knowledge used by most engineers.

The purpose of our book is to explain some of the basic principles of engineering mechanics in as simple a manner as possible and without recourse to more advanced mathematics. Graphical solutions, while apparently providing less accuracy, are used extensively throughout the text as they are usually simpler and often reveal a clear understanding of the concepts involved. Practice is needed to achieve reasonable accuracy with graphical techniques and the use of drawing office equipment and procedures is recommended.

The concepts of friction and stress are introduced early in the text so that, wherever possible, realistic problems could be used. In this way *frictionless surfaces* and *weightless bodies* are avoided wherever possible.

The SI system of units, unlike earlier systems, provides a clear distinction between mass and weight. Thus the acceleration of free-fall, g, is now used in determining weight-force and needs to be used in statics as well as in dynamics. Values of g vary considerably from place to place and, in order to simplify calculations and place greater emphasis upon basic principles, g has often been approximated to 10 m/s^2, particularly in the early chapters.

We believe that, while it may be important in many instances to obtain an exact answer to a particular problem, it is *always important* to appreciate the nature of the problem even if an exact answer to it is not possible under certain circumstances or at a certain, given level of knowledge.

Students who use this book and who are interested in the work of engineers will be able to analyse simple problems, perhaps involving the sizing and proportioning of components or the effects of forces on structures or on bodies in motion. In the text, emphasis is not placed upon the derivation and use of formulas, but rather upon the gaining of a real understanding of the laws and principles of mechanics. To this end our book begins with a study of Newton's laws and their implications and all sample problems are solved from first principles rather than by using derived formulas which are, in themselves, often rather meaningless.

At the same time, our book has been designed to generate questions in the minds of its readers, questions that are sometimes left unanswered in order to encourage students to carry out further reading and research.

This book will be successful if it causes two things to happen in the minds of the students who use it. First, they will gain some understanding of the implications of Newtonian mechanics for engineering, and second, they will develop the ability to examine critically the products of engineering within the limits of this knowledge. The final chapter was written with the latter objective in mind.

If our students also develop an interest which leads to further study in one of many fields of engineering, then this will be an added bonus.

Sydney   DON McKERN
November 1975   BARRY SCHLENKER

ACKNOWLEDGEMENTS

We wish to acknowledge with gratitude the help and cooperation of the following people and organisations.

Photographs and technical information have been supplied by The Broken Hill Proprietary Coy Ltd (photographs on pp. 16, 85, 86 and 374);   Caterpillar Tractors;   New South Wales Department of Main Roads (photograph on p. 140);   The Electricity Commission of New South Wales (photograph on p. 136);   NASA (photographs on pp. 20, 194, 221 and 276);   The Public Transport Commission of New South Wales (photograph on p. 22, Railways Dept);   QANTAS Airways Ltd;   Rex Aviation;   The Library of New South Wales;   The Metric Conversion Board;   Professor Stan Hall;   Dr George Haggarty;   and Sir Isaac Newton. Acknowledgement is made to the McGraw-Hill Book Co. for permission to use problems similar to Review Problems 3/43, 3/69, 3/70 and 3/71 from Beer and Johnson: *Mechanics for Engineers*, 2nd ed., 1960.

Our thanks go to the many industrial arts teachers who contributed, advised, criticised and assisted. In particular, we are indebted to Ian Poppitt, science teacher, Ron Balderston, Peter Hough, and Denis Whitfield, all industrial arts teachers, for helping with the solutions to the hundreds of review problems.

A special tribute goes to our wives, Helen and Bridget, not only for many hours spent in typing and re-typing of the manuscript, but also for putting up with us and helping to keep us sane through the very difficult times which this production has engendered.

Finally, we wish to express our appreciation to the staff of our publisher for the editorial assistance given during the planning and preparation of this book.

USEFUL INFORMATION

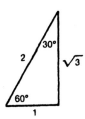

$\sin 30° = \frac{1}{2}$ (0.5) $\sin 60° = \frac{\sqrt{3}}{2}$ (0.866)

$\cos 30° = \frac{\sqrt{3}}{2}$ (0.866) $\cos 60° = \frac{1}{2}$ (0.5)

$\tan 30° = \frac{1}{\sqrt{3}}$ (0.577) $\tan 60° = \sqrt{3}$ (1.732)

$\sin 45° = \frac{1}{\sqrt{2}}$ (0.707)

$\cos 45° = \frac{1}{\sqrt{2}}$ (0.707)

$\tan 45° = 1$

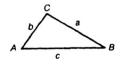

Sine rule: $\dfrac{a}{\sin A} = \dfrac{b}{\sin B} = \dfrac{c}{\sin C}$

Cosine rule: $c^2 = a^2 + b^2 - 2ab \cos C$

Coefficients of Friction

Surfaces in Contact	μ_s (static)	μ_k (kinetic)	μ_R (rolling)
Steel on steel (dry)	0.6	0.4	0.006
Steel on steel (greasy)	0.1	0.05	
Steel on brass (dry)	0.5	0.4	
Steel on brass (greasy)	0.1	0.07	
Brake lining on cast iron	0.4	0.3	
Rubber tyres on smooth road	0.9	0.8	0.02

MODULUS of ELASTICITY $E_{steel} \approx 200$ GPa $E_{concrete} \approx 14$ GPa $E_{wood} \approx 10$ GPa

Approximate Densities of Some Common Materials
Water = 1 (1 tonne/cubic metre)

Softwood	0.5	Aluminium	2.7
Hardwood	0.8	Titanium	3.0
Ice	0.9	Cast iron	7.2
Oil	0.9	Steel	7.8
Earth	1.5	Lead	11.4
Concrete	2.4	Mercury	13.6
Glass	2.6	Gold	19.3

A recent development in Australian Standard, AS 1000, is preference for r as the symbol for revolution, instead of rev.

Thus, r/s, r/s², and r/min are favoured over rev/s, rev/s², and rev/min (or rpm).

FORMULAS

STATICS

$$M = Fd; \qquad\qquad \mu = \frac{F_t}{N} = \tan\phi \qquad (F_t = \mu N)$$

If a body is in equilibrium then: $\Sigma F = 0 \ [\Sigma Fx = 0; \ \Sigma F_y = 0;]; \ \Sigma M = 0$

DYNAMICS

Linear	Circular/Rotational
$v = u + at$	$\omega = \omega_0 + \alpha t$
$s = ut + \frac{1}{2}at^2$	$\theta = \omega_0 t = \frac{1}{2}\alpha t^2$
$v^2 = u^2 + 2as$	$\omega^2 = \omega_0^2 + 2\alpha\theta$
$F = ma$	$T = I\alpha$
	$F_c = mv^2/r = m\omega^2 r$
$I = Ft = m(v_2 - v_1)$	
$M = mv$	$M = I\omega$
$KE = \frac{1}{2}mv^2$	$KE = \frac{1}{2}I\omega^2$
$PE = mgh$	
$U = Fs$	$U = T\theta$
$P = U/t$	
	$I = mk^2$

$$s = r\theta$$
$$v = r\omega$$
$$a = r\alpha$$

MACHINES

$$MA = L/E \qquad\qquad VR = d_E/d_L \qquad\qquad \eta = \frac{\text{output}}{\text{input}} = \frac{MA}{VR}$$

STRENGTH OF MATERIALS

$$\sigma = F/A \qquad \epsilon = e/L \qquad\qquad\qquad \frac{\sigma}{y} = \frac{M}{I} = \frac{E}{r}$$

$$E = \sigma/\epsilon = \frac{PL}{Ae} \qquad\qquad\qquad I(\text{rect.}) = \frac{bd^3}{12}$$

$$SE(\text{per unit volume}) = \sigma^2/2E \qquad\qquad I(\text{round}) = \frac{\pi d^4}{64}$$

1

Introduction

Engineers plan, design and supervise the construction of buildings, machines, components and the many different types of structures which form the basis of our technology. Each component, machine, building or other structure is designed to fulfil a definite purpose or to carry out a task or operation considered essential by those who are responsible for planning the development of our society.

The first step in the solution of any engineering problem is the *identification of the problem to be solved*. This may be as simple as selecting a suitable timber beam to hold up part of the floor of a building or as complex as building a space vehicle to carry men to distant planets.

Once the problem has been identified, *several different types of solutions may be conceived*. For example, in designing a suitable river crossing, the alternatives available may include a ferry, a tunnel, several different types of bridges (such as a simple arch, a multiple span or a suspension bridge) or a combination of a bridge and a tunnel.

When the preliminary alternative designs are available, usually in sketch form, *a preliminary feasibility study* is conducted. This takes into account such factors as environmental impact (where applicable), the engineering difficulties involved, availability of required materials and anticipated costs. Often one perfectly satisfactory solution is rejected at this stage simply because an alternative solution will be just as satisfactory but will cost less.

ANALYSIS

When one solution (or design) has been selected from the alternatives available, *a detailed analysis* is undertaken. This involves:

(i) *A mechanical analysis* which identifies and calculates the design loads (the force analysis), determines the relative sizes of members and components that will be used, and also allows the calculation of relative movements and deformations of individual elements or members of the design when they are subjected to expected service loads.

(ii) *A materials analysis,* which involves the selection of suitable materials for each element or component of the design, usually on the basis of such factors as strength properties, stability in service or durability, availability, cost and anticipated future maintenance to ensure continuing satisfactory service. Sometimes the need for the development of new materials with specific properties arises during this part of the analysis, a good example of this being the space programme to land a man on the moon.

In order that the mechanical analysis can be successfully carried out, it is obvious that the engineer must possess a thorough knowledge of the principles and techniques of *engineering mechanics,* particularly as these apply to his particular problem.

ENGINEERING MECHANICS

Engineering mechanics is the branch of applied science which deals with the effects of forces on an object or body and the resultant motion, if any, that occurs. In other words, mechanics describes and predicts the conditions of rest or motion of a body subjected to a given force or system of forces. At an elementary level, engineering mechanics can be divided into three main sections.

(i)   Force analysis, in which the emphasis is placed upon identifying those particular forces acting on a given body at a given time. The concepts of *resultant force, equilibrant force* (or balancing force) and *reactive force* are important in the study of these systems.

(ii) Statics, which deals with the effects of forces on bodies that remain at rest. In this respect, the concept of *equilibrium* becomes most important.

(iii) Dynamics, which is concerned with forces and the resultant movements (motions) of bodies.

It is often convenient in the initial stages of a mechanical analysis of a particular problem to assume that the object or body is *rigid*; that is, that the body does not deform at all under the applied loads. However, all bodies do deform when loaded. As these deformations are usually very small in well-designed structures they generally do not affect considerations of equilibrium or resultant motion. Such deformations, however, are very important as far as resistance to failure is concerned and must therefore become important when the *strength of materials* is being considered.

The study of engineering mechanics can be traced back to the time of Aristotle (384–322 B.C.), but the first satisfactory expression of its fundamental principles occurred in the sixteenth century when (Sir) Isaac Newton* proposed his three now-famous laws of motion. While Einstein's theory of relativity has helped point out some of its limitations, Newtonian mechanics remains valid as the basis of engineering mechanics today. Only four basic concepts are used in Newtonian mechanics, these being *space, time, mass* and *force.*

The concept of *space* is essential if the position of any given point is to be expressed in relation to a known point called the *origin,* Cartesian coordinates being one

---

*A study of Newton's life and work is given in the Appendix.

mathematical method of expressing the position of a point relative to a system of three mutually perpendicular axes. However, the position of an event cannot be defined unless the *time* of the event is stated as well.

Mass is a measure of the *inertia* of a body and is directly related to the amount of matter or substance present. Two bodies of equal mass in similar positions in relation to the centre of the earth are equally attracted to the earth and thus possess the same weight.* Also, two bodies of equal mass behave in the same way when acted upon by equal forces.

A *force* may be described as the action of one body on another and is simply a push or pull in a given direction. Forces can be direct, such as the force with which a sledgehammer hits a chisel, or indirect, such as the earth's gravitational pull on an orbiting satellite.

Forces are readily experienced as bodily sensations. For instance, your hand experiences a downward *pull* if you are holding a heavy mass in it. If you stretch a spring between your hands, you can feel your hands being *pulled* together, and you thus assume that the same spring stretched between any two supports likewise tends to pull them together. If you are sitting in a car that is accelerating you *feel* a force on your body in the direction of that acceleration. Force, therefore, is really part of our everyday experience. (Forces are dealt with in detail in Chapter 2.)

NEWTON'S LAWS

The statements of Newton's three laws of motion given below are not exactly as stated in his original works but are expressed as simply as possible in modern terminology.

First Law

A body will remain at rest or in uniform motion in a straight line unless it is compelled to change this state by forces impressed upon it.

This is sometimes known as the *inertia law* since the states of rest or of uniform motion in a straight line are both inertial states. Thus, to alter the inertial state of a body, an external force must act upon it, thereby causing an acceleration. It follows that, if there is no unbalanced force acting upon a body, it is in an *equilibrium* state, that is, no acceleration is occurring.

Second Law

A body acted upon by an external unbalanced force will accelerate in proportion to the magnitude of this force in the direction in which this force acts.

This law is the basis of dynamics and can be expressed mathematically as
$$F \propto a \qquad (1)$$
However, Newton also showed that, when a force causes the acceleration of a body, the magnitude of the acceleration produced is proportional to the mass of that body
$$a \propto m \qquad (2)$$

---

*The concept of weight or weight-force is fully discussed in Chapter 2.

Thus, by combining (1) and (2) we obtain the most useful mathematical expression of Newton's second law, which is

$$F \propto ma$$

or
$$F = kma$$

where k is a constant whose value depends upon the types of units selected for the values of force, mass and acceleration. If coherent units are selected, k becomes 1, and the mathematical expression becomes $F = ma$. This law is the basis of dynamics.

Third Law
To every action (or force) there is an equal and opposite reaction (or force).

Known as the *action-reaction law*, this law reminds us that forces always exist in pairs. For example, if a girl stands on a ladder, the girl pushes down on the ladder with a force equal to her weight-force, while the ladder pushes up against the feet of the girl with an equal and opposite *reaction* force. Thus the girl is said to be in equilibrium.

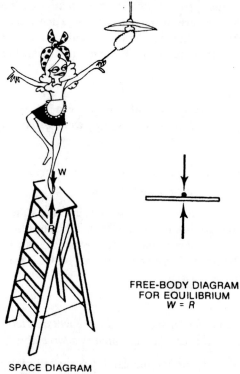

FREE-BODY DIAGRAM
FOR EQUILIBRIUM
W = R

SPACE DIAGRAM

Figure 1.1

Newton explained this law thus:

Whatever draws or presses another is as much drawn or pressed by that other. If you press a stone with your finger, the finger is also pressed by the stone. If a horse draws a stone tied to a rope, the horse (if I may say so) will be equally drawn back towards the rope; for the distended rope, by the same endeavour to relax or unbend itself, will draw the horse as much towards the stone as it does the stone towards the horse, and will obstruct the progress of the one as much as it advances that of the other. If a body impinge upon another, and by its force change the motion of the other, that body also (because of the equality of the mutual pressure) will undergo an equal change, in its own motion, towards the contrary part. The changes made by these actions are equal, not in the velocities but in the motions of the bodies; that is to say, if the bodies are not hindered by any other impediments.

Newton emphasised several times in his *Principia Mathematica* that the above three laws are based upon the results of experiments performed by himself and others. Taken collectively, they form the basis of engineering mechanics and their ramifications are explored more fully in subsequent chapters.

UNITS

The text of this book is written entirely in SI (*Système International d'Unités*, i.e. the International System of Units) which is the coherent metric system of units adopted by the Eleventh General Conference on Weights and Measures in 1960 and subsequently modified by successive conferences.*

There has long been a need to standardise various measures from one country to another in order to simplify national and international trade. This could only be done by rationalising the vast numbers of often similar units available. For example, various measures of weight (mass)† often used for international trade include the long ton (Britain) which is 2240 lb or 1016.047 kg, the short ton (U.S.A.) which is 2000 lb or 907.184 kg, and the tonne (or the metric ton) which is 1000 kg. Similarly with volumes, we have had the litre (1 cubic decimetre or 1000 cubic centimetres) in European countries, the Imperial gallon (4.546 litres), the U.S. dry gallon for solids (4.404 88 litres), the U.S. liquid gallon (3.785 litres) as well as the cubic metre, the cubic foot, and the cubic yard ($27 \text{ ft}^3 = 1 \text{ yd}^3 = 0.764 \text{ } 555 \text{ m}^3$). Some highly specialised units have also been used, such as the *barrel* for oil production ($1 \text{ barrel} = 0.158 \text{ } 987 \text{ m}^3$).

The advantages of standardising all measurement using the SI system of units can be summed up in the following three points.

(i) Since the SI system has only seven easily defined *base units*, standardisation can be readily accomplished.

(ii) Unlike the older Imperial system, there is a direct relationship between the units of length, volume, mass and force (i.e. the SI system is *coherent*). This will become apparent in the later text.

(iii) All multiples of *base units* are in powers of ten, therefore we can use our ordinary decimal (i.e. tens) system of counting and all calculations are subsequently simplified. ‡

---

*For a complete statement of SI units, refer to AS 1000—1974, *The International System of Units (SI) and its Application*, Standards Association of Australia, Sydney, 1974.
†A crucial distinction, fully explained in subsequent chapters.
‡For a fascinating short study of the history of measurement, the following is recommended reading: R.M.E. Diamant, *Understanding SI Metrication*, Angus and Robertson, Sydney, 1970.

For example, compare
Imperial: 12 inches = 1 foot;   3 feet = 1 yard;   1760 yards = 1 mile.
SI: 1000 (10^3) mm = 1 metre;   1000 (10^3) m = 1 kilometre.
    The following *base units* of the SI system are those from which all other units are derived; for completeness all seven are given in Table 1.1, but for this introductory course in engineering mechanics you need only a working knowledge of the base units for *length* (metre), *mass* (kilogram) and *time* (second).

Table 1.1. Base Units of SI

Quantity	Name of unit	Symbol	Definition of unit
Length	metre	m	The *metre* is the length equal to 1 650 763.73 wavelengths in vacuum of the radiation corresponding to the transition between the energy levels $2p_{10}$ and $5d_5$ of the krypton-86 atom.
Mass	kilogram	kg	The *kilogram* is the International kilogram, being the mass of the cylinder that is deposited in the International Bureau of Weights and Measures and was declared by the First General Conference on Weights and Measures held in Paris in the year 1889 to be the International Prototype Kilogram.
Time	second	s	The *second* is the duration of 9 192 631 770 periods of the radiation corresponding to the transition between the two hyperfine levels of the ground state of the caesium-133 atom.
Electric current	ampere	A	The *ampere* is the unvarying electric current that, when flowing in each of two parallel straight conductors of infinite length of negligible cross-section separated by a distance of one metre from each other in free space, produces between those conductors a force equal to 0.000 000 2 newton per metre length of conductor.
Thermodynamic temperature	kelvin	K	The *kelvin*, a unit of thermodynamic temperature, is a temperature equal to the fraction 1/273.16 of the temperature of the triple point of water.
Luminous intensity	candela	cd	The *candela* is the luminous intensity in the perpendicular direction of a surface of 1/600 000 of one square metre of a full radiator at the temperature of solidification of platinum under a pressure of 101 325 newtons per square metre.
Amount of substance	mole	mol	The *mole* is an amount of substance of a system which contains as many elementary units as there are carbon atoms in 0.012 kg (exactly) of carbon-12.

    As well as these seven base units, *two supplementary units* are used, these being the *radian* for the measurement of plane angles and the *steradian* for the measurement of solid angle. However, in this book the *degree* (360° in a circle) which is a *permitted SI unit*, is used for the measurement of all plane angles, while the *radian* (2π radians = 360°) is used only in some areas of dynamics.
    All other SI units are *derived* from base units and Table 1.2 sets out those base units, derived units and permitted units commonly used in engineering mechanics. A feature of this book is that when a new unit is introduced into the text, its derivation from the base units is explained.

Table 1.2. Units Commonly Used in Engineering Mechanics

Quantity	SI unit and symbol	Multiples or sub-multiples of SI unit	Other permitted units or quantities that can be used	Remarks
Time	second (s)	kilosecond (ks), millisecond (ms)	day (d) (1 d = 24 h), hour (h) (1 h = 60 minutes), minute (min) (1 min = 60 s)	Use of the second (s) is preferred in derived units, *i.e.* km/s and *not* km/min. However, km/h may be used for non-technical work.
Mass	kilogram (kg)	gram (10^{-3} kg), milligram (10^{-6} kg)	tonne (t), 1 t = 10^3 kg	The names kilokilogram or megagram are not used for 10^3 kg (1000 kg).
Density (mass density)	kilogram per cubic metre (kg/m³)		tonne per cubic metre (t/m³)	
Momentum	kilogram metre per second (kg m/s)		tonne metre per second (t m/s)	
Length	metre (m)	kilometre (km), millimetre (mm)	centimetre (cm)	The centimetre is not permitted for technical work and is NOT used in the text.
Area	square metre (m²)	square kilometre (km²), square millimetre (mm²)	square centimetre (cm²)	The unit cm² is not used in this book since it is a non-technical unit.
Volume	cubic metre (m³)	cubic decimetre or litre (dm³ or *l*), cubic millimetre (mm³)	1 litre = 10^{-3} m³, 1 millilitre = 10^{-6} m³ = 1 cm³	The name *litre* (*l*) is used only for liquid volumes and is the same as a cubic decimetre. (1 dm³ = 10^{-3} m³).
Force	newton (N) (1 N = 1 kg m/s²)	meganewton (MN), kilonewton (kN)		(i) A newton is the force which when applied to a body having a mass of 1 kilogram causes an acceleration of 1 metre per second per second. (ii) Units such as kgf are not permitted under any circumstances.
Moment of force	newton metre (N m)	meganewton metre (MN m), kilonewton metre (kN m), newton millimetre (N mm)		Although dimensionally the same as the unit of work (the joule), moment of a force is measured in Nm or one of its multiples or sub-multiples.
Pressure and stress	pascal (Pa) 1 Pa = 1 N/m²	gigapascal (GPa), megapascal (MPa), kilopascal (kPa), 1 MPa = 1 N/mm²	bar (1 bar = 10^5 Pa) atmosphere (1 atm = 101 325 Pa)	A pascal is the pressure or stress arising when a force of 1 newton is applied uniformly over an area of 1 square metre. The *bar* is used primarily in meteorology.
Work and energy	joule (J)	gigajoule (GJ), megajoule (MJ), kilojoule (kJ)	kilowatt hour 1 kW h = 3.6×10^6 J = 3.6 MJ (used in electrical industry only)	The joule is the work done or the energy expended when a force of 1 newton moves its point of application a distance of 1 metre in the direction of that force. (1 J = 1 N m.)
Power	watt (W)	megawatt (MW), kilowatt (kW)		The watt is the power used when work is done or energy expended at the rate of 1 joule per second. (1 W = 1 J/s.) (Horsepower is not a permitted unit and thus disappears from usage).

Unit Prefixes

While the use of the actual SI units (either *base* or *derived*) is preferred, the prefixes given in Table 1.3 are in common usage. The use of other multiples or sub-multiples is to be avoided wherever possible. A few examples of the use of multiples and sub-multiples of base units are set out below:

$$12\,000\,m = 12\,km$$
$$0.00394\,m = 3.94\,mm$$
$$1\,400\,000\,N = 1.4\,MN$$

Note that a space rather than a comma (which now becomes equivalent to a decimal point) separates groups of numbers,

that is   3,406,502 (Imperial) is written as 3 406 502 (SI)

and   3.406 is the same as 3,406.

Table 1.3. SI Prefixes

Prefix	Factor by which the unit is multiplied	Symbol
tera	10^{12} (1 000 000 000 000)	T
giga	10^{9} (1 000 000 000)	G
mega	10^{6} (1 000 000)	M
kilo	10^{3} (1 000)	k
milli	10^{-3} (0.001)	m
micro	10^{-6} (0.000 001)	μ
nano	10^{-9} (0.000 000 001)	n
pico	10^{-12} (0.000 000 000 001)	p
femto	10^{-15}	f
atto	10^{-18}	a

AND SHE'S 90—60—90 DAD!

Figure 1.2 It is important that one begins to "think metric".

The major problem you will find when using SI units is to develop a concept of size relationships; it is thus most important to "think metric" and to get used to metric sizes in isolation. For this reason, no exercises in conversion from other systems of units to SI or vice versa are given and *this kind of exercise is to be avoided.*

SCALAR AND VECTOR QUANTITIES

In general there are two groups of quantities that are encountered in mechanics problems; these are *scalar* quantities and *vector* quantities.

Scalar quantities can be defined completely by a number. For instance, the fundamental quantities of length (or space) L, mass, m, and time, t, are scalars as are the multiples of length like area (L^2) and volume (L^3). The number of washers in a jar is a scalar quantity since a number such as 100 would fully describe the quantity present. A few examples of scalar quantities are listed below:

10 kilograms	(10 kg)	7 square metres	(7 m^2)
5 metres	(5 m)	2 cubic metres	(2 m^3)
3 seconds	(3 s)	1 litre	(1 *l*)

Other quantities, however, need more description than magnitude alone. These are known as *vector quantities* and include such things as force, displacement, velocity and acceleration. *Vector quantities possess both magnitude and direction* and can be represented by a line drawn to scale in a certain direction with an arrowhead indicating its *sense.* This line is called a *vector.*

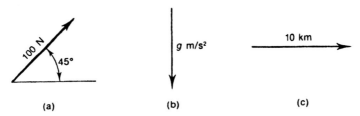

Figure 1.3 Graphical representation of vectors. When this is done as part of the solution of a problem, the scale used should be clearly stated.

Thus the vector in Figure 1.3 (a) represents a force of 100 newtons acting upwards to the right at 45° to the horizontal (or in a NE direction); the vector in Figure 1.3 (b) the acceleration, g, due to free fall; and the vector in Figure 1.3 (c) a 10-kilometre displacement in an easterly direction.

Other examples of vectors are:

a force of 100 N acting vertically upwards

a displacement of 5 metres south

a velocity of 7 metres per second (7 m/s) west

an acceleration of 10 metres per second per second (10 m/s^2) vertically down

Scalars are added algebraically (for example, the distances of 5 m + 7 m are simply added to give 12 metres); however, vectors must be added geometrically because their line of action and direction are just as important as their magnitude. When vectors are added, the solution is known as the *resultant.* This is the one vector which could replace all the others without changing the overall effect. For instance, if a man walked 1000 metres east from *A* to *B* and then turned and walked 1000 metres north from *B* to *C*, he would arrive at the same place as if he had walked 1414 metres north-east (i.e. from *A* directly to *C*). This is shown in Figure 1.4.

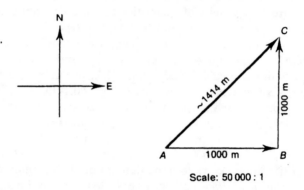

Scale: 50 000 : 1

Figure 1.4

If the vector diagram *ABC* is drawn to a suitable scale, the resultant vector *AC* will be to the same scale, and the magnitude of this resultant vector can easily be obtained by measuring the length of the vector (line) *AC*.

This method works equally well for other vector quantities besides displacement. For example, a force has magnitude, line of action, direction, and point of application and all of these factors can be represented by a single line, such as the line *AB* in Figure 1.5. The length of the line *AB* represents the magnitude of the force (3 kN) to a suitable scale and the angle θ of the line *AB* gives the direction of the force.

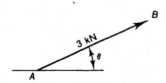

Figure 1.5 *AB* is a vector representing a force of 3 kN inclined at the angle θ to the horizontal. The arrowhead indicates the *sense* of this force; that is, its direction along the line of action *AB*.

When two forces of 5 newtons each are acting on a body, it does not necessarily mean that there is a total of 10 newtons tending to move the body. If the two forces have the same line of action but opposite directions, they will cancel each other out.

The resultant force is thus zero and the body is in equilibrium. This is shown in Figure 1.6.

Figure 1.6 The body is in equilibrium since the two opposing 5 N forces cancel each other out.

When the lines of action of the two forces are at an angle to each other, the body will tend to move along a line somewhere between these two directions. Such a situation is shown in Figure 1.7 (a). When the two forces *AB* and *BC* are drawn to scale and added together as shown in Figure 1.7 (b) the full description of the resultant force *AC* can be read from the diagram.

The method shown in Figure 1.7 (b) is known as the *triangle of vectors* and is a quick, simple method of determining the resultant of two forces or other vectors acting at a given angle to each other. Note the similarity of adding the two distances in Figure 1.4 and adding the two forces in Figure 1.7 *All vector quantities can be added in a like manner.*

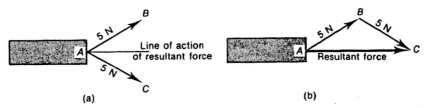

(a)   (b)

Figure 1.7 Two forces can be added in a manner similar to the adding of distances (as shown in Figure 1.4).

Another method used to determine the resultant of two forces (or other vectors) is to turn the original force (or vector) diagram into a parallelogram, the diagonal of which becomes the resultant force (or vector). Newton first expounded this method for the addition of forces and other vector quantities and an example of this is shown in Figure 1.8.

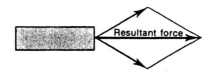

Figure 1.8 Vectors, including forces, can also be added using the parallelogram method. However, this is often not as convenient as the triangle of forces shown in Figure 1.7.

The parallelogram and triangle of vectors will be dealt with in much greater detail in Chapter 3, "Concurrent Forces".

ACCURACY

Referring back to the diagram shown in Figure 1.4 it is obvious that the magnitude of the resultant displacement can be obtained graphically by the direct measurement of the resultant R, or analytically from

$$R = 1000 \sin 45° \text{ m}$$
$$\therefore R = 1000 \sqrt{2} \text{ m}$$
$$\therefore R = 1414.213\,56 \text{ m}$$

Of course, it is very difficult to measure the distance a man walks to within one hundred-thousandth of a metre, thus the excessive accuracy of the answer is meaningless and the accuracy of the graphical solution is quite adequate. In some cases the analytical method is to be preferred to the graphical, either because of a need for greater accuracy than the graphical method will produce, or just because it is quicker. If our man had walked 300 metres east and then 400 metres north we know from Pythagoras that the resultant displacement would be 500 metres at a bearing whose tangent is 3/4. (From tables $\theta = 36°52'$. Refer to Figure 1.9.)

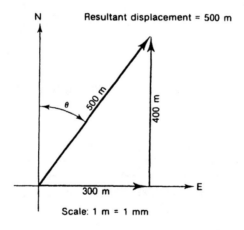

Figure 1.9

A proper appreciation of the order of accuracy with which engineering calculations can be done is necessary in order that the student does not bog down in meaningless detail or become over-casual. Slide rule accuracy, usually involving three significant figures, is considered satisfactory for the majority of engineering calculations. The decimal point should be found by a rough longhand approximation which also prevents a major slide rule error. The final check is on the validity of the obtained answer: is it reasonable and not obviously ridiculous? It is useless to have achieved exact mathematical accuracy when it is based on the wrong concept. Whether it be in a consulting engineer's office or in a school examination, having the proper concepts and making the correct approach to the problem is much more important than getting the exact answer. In the drawing office a structural design will be checked and re-checked by many people (and machines), while in the school examination the examiner is much more interested in the candidate's knowledge of mechanics than in his mathematical ability.

ORDER OF MAGNITUDE

An important characteristic of an answer to any practical problem is its *order of magnitude*. This involves the correct location of the decimal point or, when using the expontential system, the use of the correct power of 10. *If the order of magnitude of an answer is incorrect, the answer itself is useless* and costly errors may result.

Consider the two examples given below.

(i) 0.0368×184

0.0368 is approximately 0.04 which is 4×10^{-2}
184 is approximately 200 which is 2×10^{2}
Thus, using exponents only,
order of magnitude $= 10^{-2} \times 10^{2} = 10^{0}$
and, using approximate significant figures only,
the answer is: $4 \times 2 = 8$
The significant figure of the answer is about 8 and the order of magnitude is 10^{0}. From a slide rule, the significant figures obtained are 677. Thus the answer, using the slide rule, is obviously 6.77.

(ii) $\dfrac{760 \times 1000 \times 85.6}{0.0027 \times 0.0198}$

$$\text{Order of magnitude} = \frac{10^{3} \times 10^{3} \times 10^{2}}{(2.5 \times 10^{-3}) \times (2 \times 10^{-2})}$$
$$= \frac{10^{8}}{5 \times 10^{-5}}$$
$$= 2 \times 10^{12}$$

The slide rule gives 122 as the significant figures, thus the slide rule answer is obviously 1.22×10^{12}. Note that, even though the significant figure obtained from the slide rule was considerably less than 2, the order of magnitude (10^{12}) was correct; obtaining this order of magnitude was the only purpose of the above approximate calculation.

REVIEW PROBLEMS

1/1
Explain the difference between mass and weight.

1/2
All stationary objects are in equilibrium.
(a) Explain the meaning of the term "equilibrium" as used in this sense.
(b) Describe a situation in which a moving object is in equilibrium.

1/3
Briefly explain the meaning of the term *inertia* and its significance to Newton's second law.

1/4
Express the following quantities in their simplest forms by converting them to multiples or sub-multiples of SI units, either base or derived:

0.001 2 metres   56 000 kilograms
12 600 grams   0.2 centimetres
1 minute 6 seconds 10^7 newtons

1/5
Express the following quantities in terms of base SI units using the exponential system:

1 tonne   6 millimetres
2 kilometres   4 hours
1 meganewton

1/6
Classify the following quantities as vectors or scalars:

50 kilograms   9.8 metres per second^2
70 cubic metres   7 metres
30 newtons   7 metres south-east
1 second

1/7
Draw vectors to represent forces of 5 kN vertically up, 100 N horizontally to the left, 50 N down to the right at 30° to the horizontal.

1/8
Determine both graphically and analytically the resultant displacement of a helicopter which flies 100 km west and then 200 km north.

1/9

A pilot on a cross-country navigational exercise has been instructed to fly 100 km north to point A, then 200 km west to point B, then 300 km south to point C and return to base. Using both graphical and analytical methods, determine:

(a) How far is he from base when he is over point C?

(b) What is the bearing he will need to set on his compass in order to fly from point C to base M?

1/10

Determine the order of magnitude of the following calculations using approximation techniques and then obtain the significant figures for each calculation.

(a) $0.794 \times 386 \times 184$;   (b) $\dfrac{980 \times 110 \times 8}{0.002 \times 127}$

2

Force

A force is a push or a pull on a body.

A *force* is a push or a pull on a body. We cannot see a force, we see only its effect on the object upon which the force acts. Many forces are seen in operation because they require physical contact, and we see the cause and effect—a man pushing or pulling on a door—but there are also unseen forces which do not require any physical contact, such as electrostatic forces, magnetic forces, and gravitational forces.

It is the responsibility of the engineer to design structures ranging from such huge monoliths as a dam wall to the commonplace crane-hook, with the certain knowledge that they will not fail in service. In each case tremendous forces are at work trying to destroy the wall, or the hook, but if the engineer has done his work properly, no failure will occur.

Figure 2.1 Many tonnes of scrap steel are lifted by these magnetic cranes in the BHP scrap yard.

Figure 2.2   Gravity destroyed this bridge in the course of erection. Could the engineering calculations and procedures have been inadequate?

A force will produce movement of a body which is free to move, and if the force is applied continuously for a given length of time, the body will accelerate. From Newton's second law, this acceleration is directly proportional to the magnitude of the force ($a \propto F$), that is, a bigger force gives a bigger acceleration and vice versa.

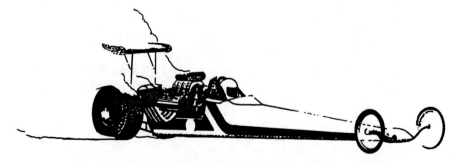

Figure 2.3   A big force can give a big acceleration (if the mass is small).

The acceleration is also inversely proportional to the magnitude of the mass of the body ($a \propto 1/m$), that is, a bigger mass gets a smaller acceleration.

Figure 2.4 A big mass gets a small acceleration.

Combining these two relationships, we obtain

$$a \propto \frac{F}{m}$$

or

$$F = kma$$

By choosing a suitable unit of force the constant k becomes 1 and the relationship therefore becomes

$$a = \frac{F}{m}$$

or $\qquad\qquad F = ma \qquad$ (Refer to Newton's second law, Chapter 1.)

This unit of force is known as the *newton* (N) and is that force which, when applied to a body with a mass of 1 kg, gives it an acceleration of 1 metre per second per second (1 m/s^2).

The newton is not a very big force (about as big as the force required to lift Sir Isaac Newton's original apple) and for engineering purposes multiples of it, such as the *kilonewton* (kN = 1000 N), the *meganewton* (MN = 1 000 000 N) and the *giganewton* (GN = 1 000 000 000 N) are commonly used.

GRAVITY

If a body is placed within the earth's gravitational field, a force of mutual attraction occurs between the body and the earth. This is a gravitational force and, unless the body is restricted by opposing forces, will cause an acceleration known as *the acceleration of free fall* or *the acceleration due to gravity* (g).

Because the earth's gravitational force acts not only on the mass of the body, but is also caused by that mass attracting the earth, it follows that the magnitude of the force will be directly proportional to the magnitude of the mass ($F \propto m$). (See Chapter 10 for a more detailed discussion of Newton's law of gravitational attraction.)

It has already been shown that the force producing motion is equal to the product of the mass of the body and the resulting acceleration ($F = ma$) but, as the gravitational force is proportional to the mass of the body on which it acts ($F \propto m$), we can see that a must have a constant value, quite independent of the mass of the body.

Figure 2.5 The earth's gravitational force causes an acceleration known as "the acceleration of free fall" or "the acceleration due to gravity" (*g*).

Figure 2.6 During his moon walk, Astronaut David Scott dropped a hammer and a feather in the vacuum on the moon. They landed together.

This means that objects of different masses, when allowed to fall freely, have the same acceleration. This acceleration, g, has a value of approximately 9.8 m/s^2 on earth.*

Galileo experimented by dropping stones from the Leaning Tower of Pisa in attempts to prove this, but he did not have the benefit of a vacuum. The Apollo 15 Astronaut, David Scott, gave us a more spectacular proof in the vacuum on the moon, when he dropped, simultaneously, a hammer and a feather. They landed together.

WEIGHT-FORCE

As has already been stated, it is the gravitational pull acting on the mass of a body which gives us that force known as the *weight* of the body.

From Newton's second law, $F = ma$, the weight-force is found from $W = mg$, where m is the mass of the body and g the acceleration of free fall.

Consider a man of mass 100 kilograms standing on his bathroom scales. The force in newtons that the man exerts on the platform of the scales is found from

$$W = mg$$
$$\approx 100 \times 9.8$$
$$\therefore W \approx 980 \text{ N}$$

where:
$m = 100$ kg
$g \approx 9.8$ m/s^2

thus the man exerts a force of approximately 980 newtons on the platform of the scales; this force is his *weight*, or, more properly, his *weight-force* or *gravity force*.

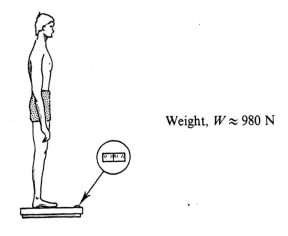

Weight, $W \approx 980$ N

Figure 2.7

When a body, such as the bridge shown in Figure 2.8, is at rest relative to the surface of the earth, the *sum of the vertical supporting forces* (the visible support and the invisible air buoyancy, etc.) is *exactly equal and opposite to the weight-force of the body*. This must be exactly so, since if an unbalanced vertical force remained, it would give rise to a vertical acceleration. It is more usual to say: "The body applies a force on its support equal to its weight-force" than to say: "The support exerts an upward force on the body". However, this is just a matter of convenience and convention.

---

*The moon, being smaller than the earth, has a smaller gravitational force, so the acceleration due to gravity is correspondingly less, approximately 1.67 m/s^2.

Figure 2.8 The weight-force exerted by the bridge is exactly resisted by the reactions at the supports.

The calculation of the force the body exerts upon its supports depends upon a knowledge of, and the use of, the acceleration of free fall, g, at that location. The calculations of forces acting in this manner must be very frequently considered since they enter into all problems of statics in which bodies, which must always have mass, are at rest on some form of support structure. Prior to the use of SI units, the acceleration of free fall, g, did not appear in engineer's calculations for statics problems, in which it rightly plays a part. Instead g, either as an acceleration or as a numeric, appeared only in dynamics problems, in many of which it plays no logical part. With the use of SI units this anomaly is removed.

The use of the coherent unit of force, the newton, will result in a clear appreciation of the difference between mass and force, especially weight-force, which has so often been obscure in the past. Problems in statics, of bodies at rest on some form of structure, will require more calculation to solve (since masses must be multiplied by g to obtain gravity forces or weight-forces) but problems in dynamics will be very much easier to deal with. These advantages of SI units, plus the many other advantages of SI as a whole, are considerable but may only become fully apparent to new generations of students receiving all of their education in the new units.

EQUILIBRIUM OF BODIES

A body is said to be *in equilibrium* if it is at rest (or moving with uniform velocity in a straight line). In this situation all forces acting must balance each other, cancelling all tendency to change. An object in a vice, for instance, is pressed equally by the fixed and movable jaws and the weight-force of the object is opposed by the friction between the object and the vice jaws.

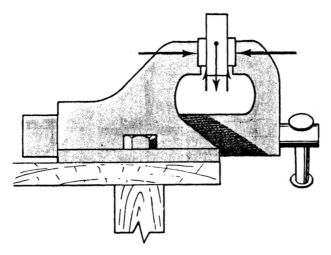

Figure 2.9 Large forces are produced by the vice but the object presses equally hard on each vice jaw. These forces are all in equilibrium.

A boat at anchor is pressing down on water, and the water is pressing equally hard up on the boat. The current is pushing the boat downstream and the line attached to the mooring is pulling equally hard upstream. These four forces are said to be in equilibrium and the boat does not move, even though the forces may be large.

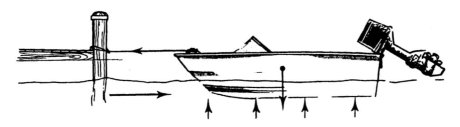

Figure 2.10 Many forces are present here but they are in equilibrium.

The man of mass 100 kilograms standing on his bathroom scales was also *in equilibrium* since he pushed down on the platform of the scales with a force of 980 N and *the platform pushed back with an equal and opposite force* (see Figure 2.11 (a)). Also, the scales themselves exerted a force of 980 N vertically downwards on the bathroom floor and *the floor pushed back with an equal and opposite force* (see Figure 2.11 (b)).

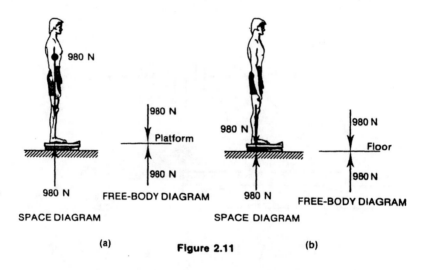

(a)　　　　　**Figure 2.11**　　　　　(b)

The forces exerted upward by the platform of the scales and the bathroom floor are called *reaction forces* and serve to illustrate Newton's third law: *To every action there is an equal and opposite reaction* (refer back to Chapter 1).

If these reaction forces were not present, that is, if the floor was not strong enough, the man would accelerate downwards through the floor under the influence of his own weight-force, which is the only active force present.

Figure 2.12

USING *g* IN CALCULATIONS

The gravitational attraction of the earth upon any mass varies inversely with the square of the distance of that mass from the centre of the earth. But the earth is not truly spherical. Due to millions of years of rotating, it has become distended towards the equator, and flattened at the poles. The poles are therefore closer to the centre of the earth and the acceleration due to gravity is greater there than it is near the equator. Similarly, *g* at the top of a mountain or in a high-flying aeroplane is significantly less. The variations of *g* for some of the more important locations are shown below.

The Moon	1.67
Sydney (Australia)	9.796 830
Washington (U.S.A.)	9.800 984
Denver (U.S.A.)	9.796 122
London (Great Britain)	9.811 960
Sevres (France)	9.809 408
Leningrad (U.S.S.R.)	9.819 308
Cape Kennedy (U.S.A.)	9.797 320
International Standard	9.806 650

A proper appreciation of the order of accuracy required in the use of *g* in calculations is essential if the results are to be meaningful.

In dynamics problems involving precise measurement of distance, mass, and time, it may be necessary to use an equally precise value for *g* so that the accuracy of the result is consistent with the accuracy of the information. For example, if the calculation of the thrust required to launch a rocket towards the moon had a 2% error, the astronauts would be understandably upset when they found themselves lost in space. In this case the computers would be fed with the most accurate values for local *g* available.

On the other hand, the engineer, when deciding the size of the bolt to use to support a structural steel beam, may use $g = 10 \text{ m/s}^2$ in his calculations, knowing that the 2% "safe" error he incurs by doing so will be absorbed by an arbitrarily chosen "safety factor" (which could be as high as 10 : 1). Achieving a theoretical bolt diameter of 23.278 046 mm is quite meaningless. He will specify 24 mm bolts anyway, and probably 4 (or 6) of them (factor of safety 4 or 6).

To enable a quicker grasp of the fundamental concepts of mechanics, the value of g in some of the examples in this book has often been "rounded up" from the awkward approximate values of 9.8 or 9.81 to an even 10 when calculating weight-force. The minor inaccuracies associated with this 2% error are more than compensated for by the resulting simplification of the calculation.

It should be emphasised that this order of accuracy is perfectly suitable for a vast range of engineering problems and is probably more accurate than some of the other information with which the engineer is supplied. However, an understanding of which value of *g* should be used for any particular problem is essential in order to produce worth-while results. To this end, the value of *g* used in each problem is stated when it is considered to be significant.

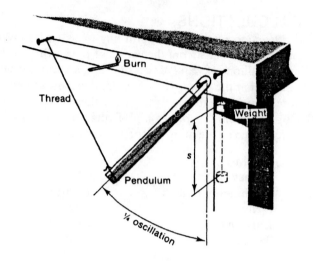

Figure 2.13

The local value for *g* can be assessed by a simple laboratory experiment. Support a pendulum at an angle with a weight so positioned that, if they are released simultaneously, the pendulum will strike the falling weight at the instant the pendulum becomes vertical (see Figure 2.13). A simple method of ensuring that the pendulum and the weight are released at the same instant is to burn the supporting thread. Measure the distance the weight falls from its original supported position to the mark it makes on the pendulum. Calculate the time taken for this quarter of an oscillation of the pendulum by timing several full oscillations. This is also the time taken for the measured fall of the weight. From the formula $s = ut + \frac{1}{2}at^2$ (see Chapter 9) derive a formula for *g* in terms of *s* and *t*

$$s = \tfrac{1}{8}gt^2 \quad \text{so} \quad g = \frac{2s}{t^2}$$

Substitute your experimental values for *s* in metres, and *t* in seconds in this formula and calculate *g*.

What restrictions should you impose on the validity of your result? How many significant figures could you justifiably quote? Design a more accurate experiment.

MASS AND FORCE

The mass of an object is the measure of its inertia, and is related to the amount of material it contains. It is measured in kilograms (kg). In order to determine the mass of an object, we can apply a known force and measure the acceleration produced, since $m = F/a$. This is the method used to determine the mass of sub-atomic particles such as the electron.

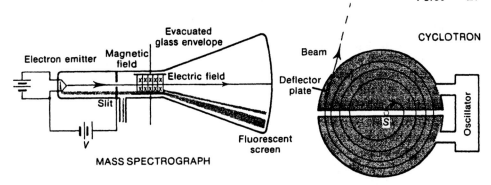

Figure 2.14 The mass spectrograph applies a known force to particles giving them an acceleration which is measurable, and thus enabling their mass to be calculated (*F = ma*).

However, acceleration of potatoes along the counter to determine their mass is not the method favoured by the local greengrocer (for obvious reasons!) He places the potatoes in the tray of a spring balance where they are subjected to a force (their own weight-force) which will produce a known acceleration (*g*) and then he measures the force required to resist the gravitational pull acting on the potatoes. (The grocer probably does not realise this.) This resisting force supplied by the spring is, of course, equal to the weight-force of the potatoes. The scale of the spring balance is graduated to read in mass units since potatoes, like many other common items, are sold by *mass*.*

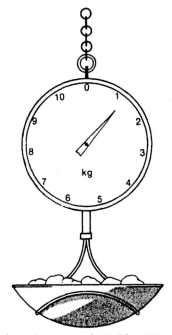

Figure 2.15 The grocer's scales are graduated in mass units for convenience.

---

*It is interesting to note that some countries (for example, South Africa) have abandoned the use of the terms *weight* and *weighing*; they call their scales *mass metres* and the operation becomes *massing*.

Another method of determining the mass of an object is to compare it with a known mass. This will involve a balance of some sort using a beam instead of a spring. However, we are still comparing the relative gravitational forces exerted by the masses in order to discover the magnitude of the mass. As $W = mg$ and g is a constant, $F \propto m$ and we find it more useful to read the scales in mass units (kg) instead of force units (newtons). This gives a more convenient result, as mass is directly related to the weight-force it produces. If the weight-force itself is required, it is a simple matter to multiply the mass shown on the scales in kilograms by 9.8 (or ≈ 10) to get the weight-force in newtons.

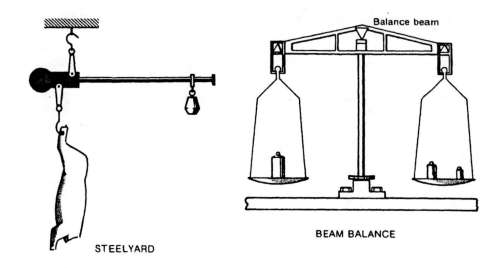

Figure 2.16 The steelyard and the beam balance use different methods to compare masses.

CENTRE OF MASS

All masses have a finite volume, but to simplify calculations it is usual to assume that the effect of the mass is either distributed over a specified area (a uniformly distributed load or UDL) or is concentrated at a single point.* This point is known as the *centre of mass* (or *centre of gravity*) of the object, and is *the point at which the mass could be assumed to be concentrated* without altering the overall effect of its inertia or the weight-force it produces. For example, the centre of mass of a spherical ball would be at the centre of the sphere and its weight-force acts through this centre. The centres of mass of a plate, cube and prism are at the intersection of the diagonals of these solids, and their weight-forces act through these centres (see Figure 2.18).

---

*UDLs are discussed in Chapter 17 ("Beams and Bending").

Figure 2.17 The centre of mass of the ball can be considered to be at the centre of the sphere.

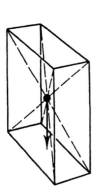
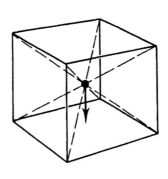

Figure 2.18

CHARACTERISTICS OF A FORCE

When considering forces other than weight-force, we need to specify more than just the magnitude in order to describe the force completely. (A given force may, for instance, be horizontal, and could be acting to the left or to the right.) In all cases, four major characteristics must be specified. They are:

        (a) magnitude,
        (b) line of action,   ⎫
        (c) direction (or sense), ⎬ or attitude
                      ⎭
        (d) point of application,   or location

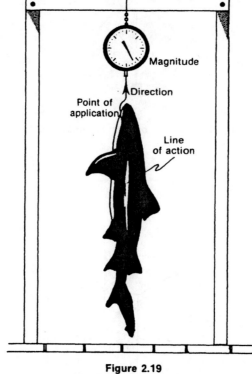

Figure 2.19

(a) The *magnitude* is the size of the force, expressed in newtons or multiples thereof.
(b) The *line of action* is the attitude of the force to a known reference line. It may be expressed as *30° to the vertical*, *45° to the horizontal*, or *60° to the beam AB* and so on.

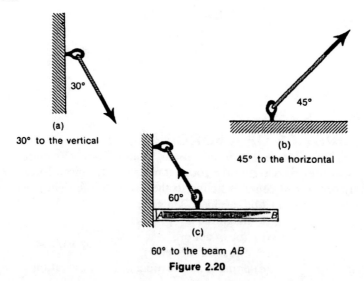

Figure 2.20

(c) The *direction* (sometimes called *sense*) is the direction along the line in which the force is acting. We may say horizontally *to the right*, vertically *down*, *due north*, and so on.

(d) The *point of application* (the location) is the point on the body through which the force acts; it could be described as *four metres to the left of the support*, or *through the centre of mass*, and so on. The term *attitude* can be used instead of *line of action and sense* and *location* or *position* instead of *point of application*.

The position of the point of application of the force is important only as far as the net effect on the body is concerned. Its position along the line of action is unimportant. For example, a locomotive *pushing* a carriage up the line has the same effect as a locomotive *pulling* a carriage up a line. This is known as the *principle of transmissibility* of a force.

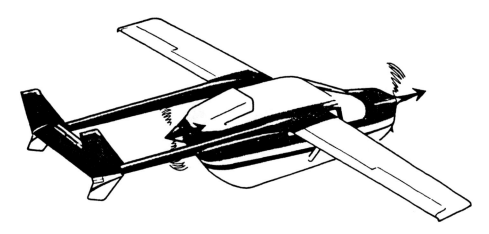

Figure 2.21 The Cessna 337 Skymaster demonstrates the principle of transmissibility of a force. It can fly on either engine.

The Cessna 337 Skymaster is an interesting example of the principle of the transmissibility of a force. It has an engine at both ends of the fuselage. The forward engine pulls, and the rear engine pushes. The plane can fly with either engine operating, because each can provide enough force to propel the plane through the air. With both engines in operation, the force (or thrust) is more than doubled, and the performance is proportionally improved.

PRESSURE AND STRESS

Not all forces act through or upon a single point. In many cases they act on an area of specific size. If the area is significant, the resultant *force per unit area* is termed a *pressure* or *stress* and is described as a force per unit area (e.g. N/m^2, N/mm^2, etc.).

Pressure and stress have the same units but there is a subtle distinction in their use. Pressure is used to describe the forces per unit area that gases and liquids (i.e. fluids) exert on the walls of their containers, or the upward pressure of the ground on a foundation. Pressure is always positive and infers compression; for example, the compressed air in a football, balloon, or a car tyre, or the pressure of a load on its support causing compression of that support.

Stress refers more to the forces per unit cross-sectional area induced in a solid body by a load or force and can be *tensile, compressive* or *shear stress*. For example:

(1) the tensile stress in a cable, a chain or a bolt subjected to a direct tensile load;
(2) the compressive stress in a column, a pillar, or a support carrying a direct compressive load;
(3) the shear stress (slicing) in a bolt, a rivet, or a beam carrying a shear load.

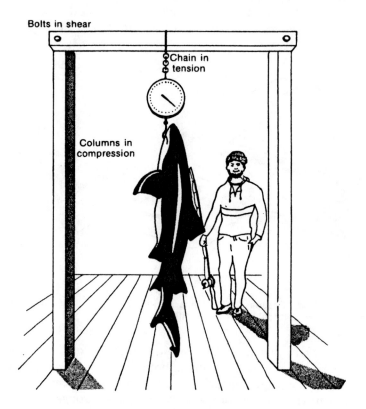

Figure 2.22 All three basic stresses are shown here: the chain is in tension; the columns are in compression; the bolts are in shear.

The SI unit of pressure (or stress) has been named after the famous physicist, Blaise Pascal—and one *pascal* is the pressure caused by the force of one newton acting on an area of one square metre ($1\,Pa = 1\,N/m^2$). Because the newton is such a small force, it follows that the pascal is a very small pressure, and multiples of it, such as the kilopascal (kPa) and the megapascal (MPa) are used in engineering. One megapascal is also equal to the force of 1 newton acting on an area of 1 square millimetre.

$$1\,\text{N/mm}^2 = \frac{1\,\text{N}}{1\,\text{mm}^2}$$

$$= \frac{1\,\text{N}}{1 \times 10^{-6}\,\text{m}^2}$$

$$= \frac{1 \times 10^6\,\text{N}}{1\,\text{m}^2}$$

$$= 10^6\,\text{N/m}^2$$

$$= 1\,\text{MN/m}^2 \quad (\text{since } 10^6\,\text{N} = 1\,\text{MN})$$

$$\therefore\, 1\,\text{N/mm}^2 = 1\,\text{MPa} \quad (\text{since } 1\,\text{N/m}^2 = 1\,\text{Pa})$$

Since most calculations involve areas in square millimetres it is most useful to remember that $1\,\text{N/mm}^2 = 1\,\text{MN/m}^2 = 1\,\text{MPa}$.

Even though the pascal is the accepted SI unit for pressure or stress it is more descriptive to use N/m^2, or N/mm^2 and MN/m^2 for megapascals. This is done in the text, particularly at intermediate stages of calculations.

Here are some approximate values of mass, force, pressure and stress which you may find useful in assessing relative magnitudes.

An apple has a mass of about 100 g and requires a force of about 1 N to lift it.

A large loaf of bread has a mass of 1 kg and needs a force of about 10 N to lift it.

The average school suitcase full of books etc. has a mass of 10 kg and needs a force of 100 N to lift it.

Atmospheric air pressure is approximately 100 kPa.

Air pressure in the tyres of the average car is about 200 kPa.

Water pressure available at the household tap is about 600 kPa.

Maximum yield stress in 250 grade mild steel plate is 250 MPa.

Maximum yield stress in 400 grade mild steel plate is 400 MPa.

SAMPLE PROBLEM 2/1

Describe fully the forces shown in the diagrams (a), (b), (c) and (d).

Solution:

(a) Force of 10 kN, vertically down, at point A.

(b) Force of 2 MN, up to the right, at 30° to the horizontal through A.

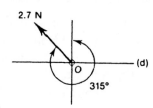

(c) Force of 5 N, horizontally to the right, 10 mm above A.

(d) Force of 2.7 N, NW (or at a bearing of 315°) through O. ■

Problem 2/1

SAMPLE PROBLEM 2/2

What is the weight-force exerted by a mass of 51 kg? (Use $g = 9.8$ m/s².)

Solution:

$$F = ma \ (= mg)$$
$$= 51 \times 9.8$$
$$\approx 500 \text{ N}$$

■

SAMPLE PROBLEM 2/3

Two equal masses of 10.2 kg each are suspended by a light rope over a pulley, which is hanging on a chain. The pulley has a mass of 0.51 kg. (a) What is the tension in the rope? (b) What is the tension in the chain? (Use $g = 9.8$ m/s².)

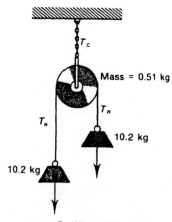

Problem 2/3

Analysis:
Consider the forces on one mass only. Gravity is attempting to pull the mass down, and the rope is preventing this by exerting an equal and opposite force upwards. Therefore the force supplied by the rope (the tension in the rope) is equal to the weight-force of the mass.

Solution:

$$F = ma \ (= mg)$$
$$= 10.2 \times 9.8$$
$$\approx 100 \, \text{N}$$

The tension in the rope is 100 N.
The chain is supporting both masses and the pulley. The total weight-force exerted by this combined mass will be

$$F = (10.2 + 10.2 + 0.51) \times g$$
$$= 20.91 \times 9.8$$
$$\approx 205 \, \text{N}$$

The tension in the chain is 205 N.   ■

SAMPLE PROBLEM 2/4

Mass = 60 kg

If a shoe has an area of 15 000 mm² in contact with the floor, what is the average pressure exerted on the floor by a boy with a mass of 60 kg standing normally? (Use $g = 10 \, \text{m/s}^2$.)

Solution:
Weight-force exerted by the boy

$$F = ma \ (= mg)$$
$$= 60 \times 10$$
$$= 600 \, \text{N}$$

Problem 2/4

Area of contact of two shoes is $15\,000 \times 2 = 30\,000 \, \text{mm}^2$.
Average pressure (assuming uniform distribution of load) is the force per unit area.

$$\text{Pressure} = \frac{600}{30\,000} = 0.02 \, \text{N/mm}^2 = 0.02 \, \text{MPa}$$
$$= 20 \, \text{kPa}$$

■

REVIEW PROBLEMS
(Use g = 10 m/s² unless otherwise specified.)

2/5
What is the SI unit of force? What other units can be used? Why is it incorrect to use the old gravitational unit, kg f?

2/6
Describe two effects of a force. Explain the relationship between mass and weight-force. How can the mass of a body change? How can the weight-force of a body change?

2/7
What is the weight-force produced by masses of 1 kg; 10 kg; 10.2 kg; 27.3 kg; 2000 kg? (Use $g = 9.8$ m/s² first; then use $g = 10$ m/s².)

2/8
Suggest values for g suitable for use in each of the following calculations. Justify your answer.
  (i) Time taken for a pencil to fall from the desk to the floor.
 (ii) Time taken for a stone to fall down a deep mine shaft.
(iii) Thickness of a suspended concrete floor slab.
(iv) Tension in a crane cable.
 (v) Time of rocket burn for a satellite launch.

2/9
A bag containing 10 apples is hung from a spring balance which then shows a reading of 1 kg. What force is the hook supporting?

Problem 2/9

2/10
Completely describe the force acting on each anchor bolt holding the ends of a suspended cable if the tension in the cable is 5 kN.

20°

Problem 2/10

2/11
A child hangs at rest from a rope. Determine the approximate mass of the child if the tension in the rope is 300 newtons. What is the magnitude of the force tending to withdraw the eyebolt from the overhead beam?

Problem 2/11

2/12
What is the magnitude of the force exerted by the floor on each leg of a table of mass 50 kg? What are such forces termed?

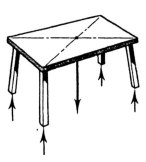

2/13
Determine the tension in the cable supporting the spring balance. The spring balance itself has a mass of 2 kg and it is showing a reading of 20 kg.

Problem 2/12

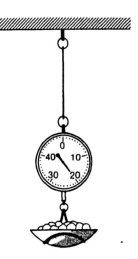

2/14
A shopper tries out his new bathroom scales in the elevator while leaving the shop. He finds that the reading on the scale varies considerably. Why? If the shopper has a mass of 80 kg, suggest likely readings during three stages of the descent:
(a) while leaving the top floor;
(b) halfway down;
(c) nearing the ground floor.

Problem 2/13

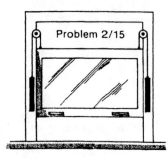

2/15

The window shown is open and in equilibrium.
What is the tension in the sash cords if each sash
balance has a mass of 5 kg?

2/16

Determine the force in the chain supporting the
steelyard. The balance weight has a mass of 3 kg
and the beam a mass of 2 kg. The steelyard is
horizontal when the balance weight is positioned
on the 20 kg graduation.

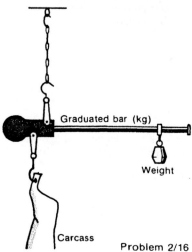

Problem 2/16

2/17

The tower crane is 100 m high and is supporting a
load of 5 tonnes at ground level. Determine the
approximate tension in the cable
 (i) near the hook, and
(ii) at the winding drum,
if the cable has a mass of 2 kg/m.

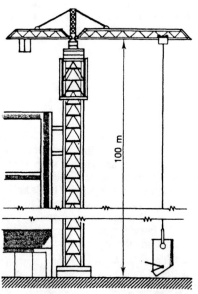

Problem 2/17

2/18

Billiard tables have an extremely flat surface, made of 75-mm thick, polished slate which is covered with green cloth. Determine the approximate force in each leg of a billiard table, given that the slate has a mass of 750 kg and a full-size table is 4 m × 2 m and has eight legs.

If the maximum point loading the floor can support is 1 kN, should a slate table be installed, or will it be necessary to choose a lighter table with a chipboard top?

2/19

Each 10-person elevator in a 30-storey building has a mass of 1500 kg when empty, and is supported by six cables, each with a mass of 1 kg/m. The counterweight has a mass of 2000 kg.

(a) Determine the maximum static tension in each cable.

(b) Under what set of conditions will this occur?

(c) An elevator car with three persons aboard is stationary at the tenth floor. Determine the approximate tension in each cable: (i) near the elevator; (ii) near the counterweight; (iii) near the winding drum.

(d) Is the value for tension obtained in (a) suitable for all conditions of operation? Justify your answer.

Floor separation in the building can be taken as 3.5 m and 85 kg as the mass of an average adult.

2/20

Determine the approximate size of the square concrete foundation base required to support the steel column of a building. The calculated load on the column is 12 tonnes and field and laboratory tests have indicated a suitable bearing pressure for the ground on which the base is to rest would be 100 kPa (0.1 MPa).

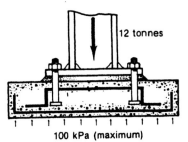

100 kPa (maximum)

Problem 2/20

2/21

You are sitting on a drawing room stool which has a mass of 4 kg. The total area of the feet of the stool in contact with the floor is 1600 mm².

(i) What is the pressure exerted by each leg on the floor?

(ii) What is the stress produced in the floor under each leg?

If you now tilt the stool back, the area of contact with the floor is reduced to about 80 mm².

(iii) What pressure does each leg now produce on the floor?

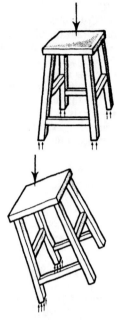

Problem 2/21

2/22

A car is raised on a single column hoist for service. The car has a mass of 1 tonne and the hoist has a mass of 0.5 tonne.

(i) Determine the compressive force in the base of the column of the hoist.

(ii) What air pressure will be required to just support the car if the column has a diameter of 250 mm?

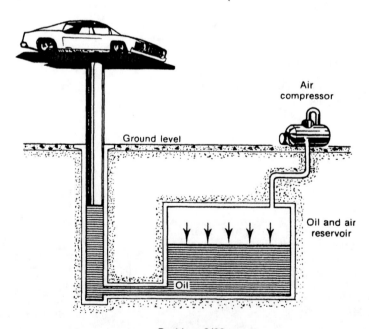

Problem 2/22

3

Concurrent Forces

$$\Sigma F_x = 0$$
$$\Sigma F_y = 0$$

In many instances force systems operating on an object consist of a number of separate forces applied independently. For example, if two tugs are pushing an ocean liner into its berth as shown in Figure 3.1, each tug by virtue of the thrust of its screw exerts a separate, independent force on the hull of the liner.

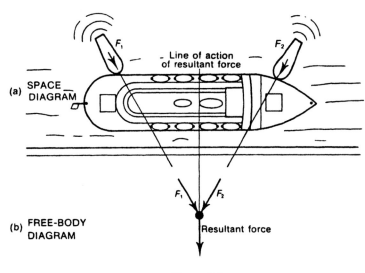

Figure 3.1

Similarly, lamps are suspended over pedestrian crossings by cables attached to poles on either side of the roadway in order to illuminate the crossing at night. A typical situation is shown in Figure 3.2. In this situation the weight-force of the lamp is balanced by the *tensions* (T_1 and T_2) induced in the supporting cables and the lamp is held in equilibrium; that is, it does not fall.

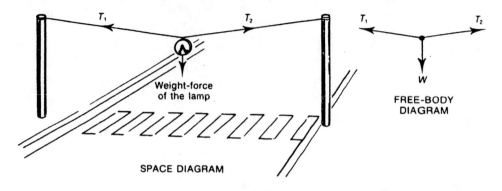

Figure 3.2

A third example is a weightlifter who has lifted the barbell above his head as shown in Figure 3.3. Since the bar is in equilibrium, the total weight-force of the bar and weights must exactly equal the upward (reactive) forces present in the arms of the man.

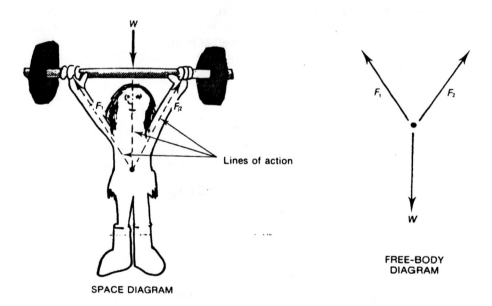

Figure 3.3

In *each* of these three situations it is obvious that all of the forces present are in the same plane (i.e. are *coplanar*) and their lines of action, when projected if necessary, pass through the same point in that plane. That is, *in each situation a system of coplanar concurrent forces is present.*

Consider the two tugs pushing the ocean liner into its berth. The liner will move along the *line of action* of the resultant force which is found by *adding* the two forces, F_1 and F_2, provided by the tugs. To find the magnitude and direction of this *resultant force*, both the magnitudes and the directions of these two forces must be considered (refer back to "Scalar and Vector Quantities", Chapter 1). If a free-body diagram is drawn of this situation, the active forces become apparent as does the line of action of the resultant force (refer to Figure 3.1 (b)).

RESULTANTS OF TWO CONCURRENT FORCES

Resultant forces may be determined by vector addition, and this is done either graphically or analytically. However, no matter which method is used, the first step is to *analyse the problem and determine the active vectors (forces) present.*

For example, consider the situation of the reluctant donkey shown in Figure 3.4 (a).

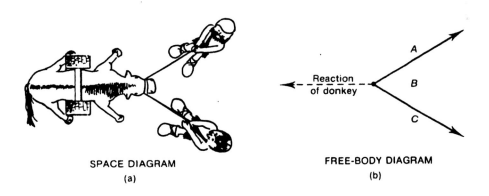

SPACE DIAGRAM
(a)

FREE-BODY DIAGRAM
(b)

Figure 3.4 If no movement occurs, these three forces must be in equilibrium.

The two men pulling on the bridle ropes are trying to move the donkey forward while the donkey obviously does not want to go. The two active forces are thus the *tensions* in the bridle ropes.

A free-body diagram of this situation appears in Figure 3.4 (b). The magnitude and direction of the resultant force acting on the donkey can now be found by one of several methods.

GRAPHICAL SOLUTION

Draw the free-body diagram (as shown in Figure 3.4) (b)) so that each of the forces (vectors) are represented graphically by lines as shown. The *magnitude* of each vector is represented by drawing each line to a suitable scale (such as 1 N = 10 mm). The angles at which the lines are drawn are parallel to the lines of action of the forces they represent. An arrowhead on each line shows the direction in which that force acts.

A letter is placed on either side of each force in order to refer to it by name. This is known as *Bow's notation* and is of the greatest importance, particularly in complex force analysis. Thus, the two forces become *AB* and *BC*.

First Method: Parallelogram of Forces

Project lines parallel to each vector so that the diagram is turned into a parallelogram. The diagonal *AC* of this parallelogram represents the *resultant force*, that is, it is the force which could replace the original forces *AB* and *BC* without changing their overall effect on the body upon which they act.

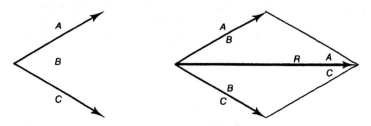

Figure 3.5   The parallelogram of forces.

Second Method: Triangle of Forces

It is apparent that both halves of the previous parallelogram are triangles containing lines representing each vector and their resultant. It is therefore simpler to draw only half the original parallelogram, the resultant *R* of the two forces *AB* and *BC* being represented by the line *AC*, which joins the starting point *A* to the terminal point *C* of the triangle.

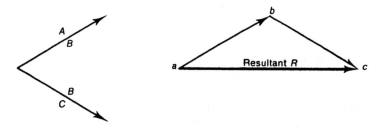

Figure 3.6   The triangle of forces.

Now reconsider the situation of the two men trying to move the reluctant donkey (Figure 3.4). If the resultant force *AC* is equal to the resisting force that the donkey is applying, then the donkey is in equilibrium and this resisting force is termed the *equilibrant force*. It is thus obvious that an *equilibrant force* is equal in magnitude but opposite in direction to the resultant of a system of forces.

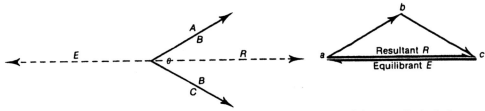

Figure 3.7 The equilibrant and resultant forces are equal in magnitude but opposite in direction.

Analytical Solution

Once the triangle of forces has been drawn, as in Figures 3.6 or 3.7, it is relatively simple to calculate the value of the resultant AC using one of several methods. Since the sides AB and BC and the angle $\widehat{ABC}$ are all known, from the *cosine rule*, for the triangle ABC,

$$AC^2 = AB^2 + BC^2 - 2.AB.BC.\cos\widehat{ABC}$$

Note that $\widehat{ABC}$ is the *supplement* of the angle θ between the vectors AB and BC in the force diagram.

RESULTANT OF THREE OR MORE CONCURRENT FORCES

Now consider the eye-bolt shown in Figure 3.8 to which three cords are attached. Two of the cords are passed over pulleys as shown and the weights W_1, W_2 and W_3 are attached to the ends of the cords. Thus the forces (tensions) F_1, F_2 and F_3 are present in the cords. Figure 3.8 (b) represents the free-body diagram of this situation.

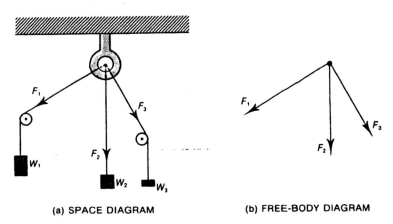

(a) SPACE DIAGRAM   (b) FREE-BODY DIAGRAM

Figure 3.8

The resultant force on the eye-bolt can be readily determined graphically using an extension of the "triangle of forces" method known as a *polygon of forces*.

Again, draw the force diagram (free-body diagram) to scale so that the magnitude and direction of each force is graphically represented by a vector. Refer to Figure 3.9 (a).

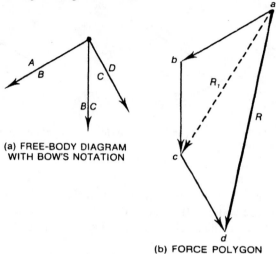

(a) FREE-BODY DIAGRAM
WITH BOW'S NOTATION

(b) FORCE POLYGON

Figure 3.9 The polygon of forces.

Now label each force using Bow's notation as before. Combine the forces AB and BC in a triangle of forces so that their resultant AC is obtained (refer to Figure 3.9 (b)). Add this resultant AC to the force CD in another triangle of forces to obtain the resultant R of the three-force system (again refer to Figure 3.9 (b)).

From Figure 3.9 (b) it is obvious that the drawing of the intermediate resultant AC is unnecessary. The polygon of forces $ABCD$ can thus be drawn if each of the vectors AB, BC and CD are placed "head-to-tail" to each other, the join AD of the polygon being the *resultant* force on the metal eye-bolt.

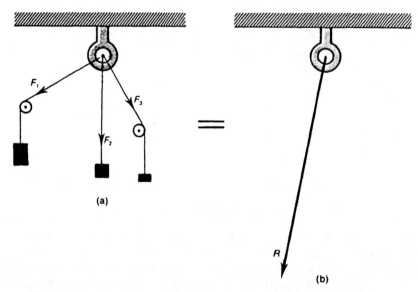

(a)

(b)

Figure 3.10 The resultant force, R, replaces the original force system.

The original force system shown in Figure 3.10 (a) is *equal* to the single force *R* shown in Figure 3.10 (b). Thus, the resultant *replaces* the original force system.

The "polygon of forces" construction can be applied to coplanar force systems having three or more forces by carrying out the following steps:

(i) draw the vector diagram to scale, showing all forces operating;
(ii) label the forces using Bow's notation;
(iii) begin with one force and join up the forces "head-to-tail" so that each commences from the end of the previous one and all arrowheads "chase each other" around the resulting polygon;
(iv) the closing line of the polygon then is the resultant.

Note that the arrowhead on this resultant opposes the direction of all other vector arrows.

Figure 3.11 illustrates how the polygon of forces is applied to a four-force system to determine its resultant.

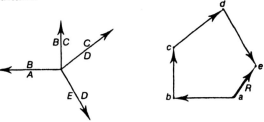

(a) SPACE DIAGRAM   (b) FORCE DIAGRAM

Figure 3.11   Each vector in the force polygon is parallel to the force it represents in the space diagram; the magnitude of each force is proportional to the length of its vector.

COMPONENTS OF A FORCE

If we look again at the parallelogram of forces or the triangle of forces we see that the principle can be very easily reversed. We could start with a *resultant force* and find, using the parallelogram or the triangle, any two component forces we need.

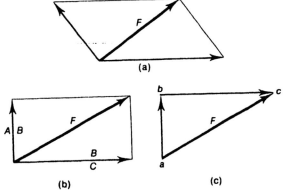

Figure 3.12   (a) Two possible components of the force *F*; (b) and (c) show the vertical and horizontal components, *AB* and *BC*, of the force *F* obtained by using the parallelogram and the triangle of forces.

In engineering practice it is often necessary to discover the *horizontal* and *vertical* components of a force, but any polygon can be drawn to provide forces of required magnitude and direction to *replace* or *oppose* a given force.

The components of a force can be obtained either graphically or analytically. Analytical techniques are often simpler, particularly when horizontal or vertical components are required. This is illustrated in Figure 3.13.

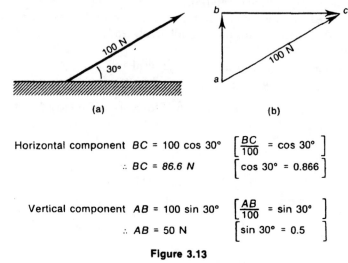

(a)   (b)

Horizontal component BC = 100 cos 30° $\left[\dfrac{BC}{100} = \cos 30° \right]$

∴ BC = 86.6 N $\left[\cos 30° = 0.866 \right]$

Vertical component AB = 100 sin 30° $\left[\dfrac{AB}{100} = \sin 30° \right]$

∴ AB = 50 N $\left[\sin 30° = 0.5 \right]$

Figure 3.13

ANALYTICAL RESOLUTION OF THREE OR MORE CONCURRENT FORCES

It is not always convenient to draw the polygon of forces to scale and so obtain the resultant force graphically. In many instances it is simpler and more exact to calculate the resultant force mathematically or, more specifically, to calculate the *components of the resultant force*. This procedure is very suitable if the horizontal component or the vertical component of a certain resultant force is required, or if the resultant needs to be expressed in terms of its components.

A systematic procedure must always be adopted to avoid confusion:
 (i) first, analyse the problem and sketch the free-body diagram, showing all active forces;
 (ii) select and state a suitable sign convention;
 (iii) resolve each force into its components in the selected directions (often horizontal and vertical components are used);
 (iv) determine the components of the resultant force by adding all components acting in each selected direction (i.e. ΣF_x, ΣF_y or ΣF_H and ΣF_V).
 (v) if required, combine these two components to obtain the resultant of the particular force system being analysed.

As the forces involved will often be weight-forces, choosing positive force vertically downwards, $\downarrow^+$, may simplify the mathematics involved by reducing the number of minus signs in the calculation. The choice is quite arbitrary and some problems will be more simply solved using a different convention.

Consider the three forces AB, BC and CD shown in Figure 3.14. Obviously these are best resolved into horizontal and vertical components which can then be added to obtain components of the resultant.

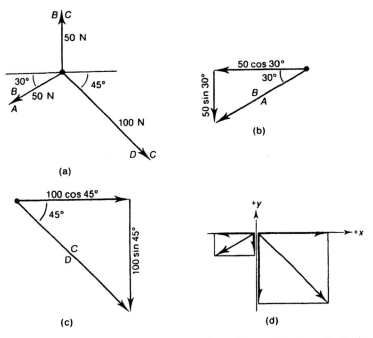

(a)

(b)

(c)

(d)

Figure 3.14　(a) Shows three concurrent coplanar forces; (b) shows the horizontal (F_H) and vertical (F_v) components of the force AB; (c) the F_H and F_v components of the force CD; (d) shows both sets of horizontal and vertical components on x-y axes.

Let vertical down forces be positive, ↓⁺ . Let horizontal forces to the right be positive ⁺→ .

Force	Horizontal component	Vertical component
AB	$-\ 50\cos 30° = -43.3$	$+\ 50\sin 30° = +25$
BC	zero　　$=$　　0	$-\ 50$　　　$= -50$
CD	$+100\cos 45° = +70.7$	$+100\sin 45° = +70.7$
Resultant AD	$\Sigma F_H = +27.4$	$\Sigma F_v = +45.7$

Thus, the horizontal component of the resultant is 27.4 newtons acting to right and the vertical component is 45.7 newtons acting vertically downwards. These can be combined to give the resultant force, if necessary (refer to Figure 3.15).

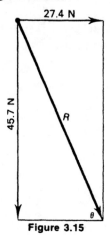

Figure 3.15

From Figure 3.15, using Pythagoras' theorem

$$R^2 = (27.4)^2 + (45.7)^2$$
$$R = \sqrt{(27.4)^2 + (45.7)^2}$$
$$R = 53.3 \text{ N}$$
$$\text{Tan } \theta = \frac{45.7}{27.4}$$
$$= 1.667$$
$$\therefore \theta = 59°$$

Thus the resultant force is 53.3 N acting down to the right at 59° to the horizontal, 59°↘ .

This type of analytical solution can be applied to any system of three or more concurrent forces. However, as Figure 3.16 shows, it is not always desirable to use horizontal and vertical components. In this case it is better to resolve the weight-force W in directions parallel (y) and perpendicular (x) to the given plane, giving components $F_x = -W\cos\theta$ and $F_y = +W\sin\theta$. Obviously the F_y component ($W\sin\theta$) determines the acceleration of the block of ice down the shute.

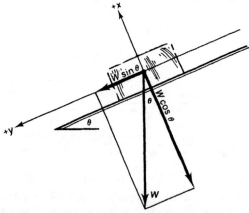

Figure 3.16   A block of ice on a wet metal shute (almost zero friction).

EQUILIBRANT FORCE

The resultant force replaces the given force system and it shows the overall effect of all of the forces acting on the body. It therefore follows that if the resultant is opposed by an equal and opposite force—that is, a force having the same magnitude, same line of action but opposite direction—then the body no longer has any tendency to move. It is in *equilibrium*, and the opposing force is called the *equilibrant*. Having calculated a resultant, therefore, we have also found the equilibrant (same magnitude, same line of action, same point of application, opposite sense). This is shown in Figure 3.17.

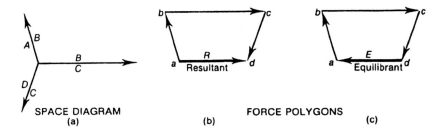

SPACE DIAGRAM
(a)

FORCE POLYGONS

(b)   (c)

Figure 3.17 The distinction between the resultant and the equilibrant forces.

The force polygon in Figure 3.17 (c) is said to be in equilibrium. If any one of the forces were removed, it would no longer be in equilibrium. *Therefore any one of the forces could be the equilibrant for all the others.* Here then is a simple method of remembering whether you have found an equilibrant or a resultant. The equilibrant is indistinguishable from any other force in the polygon as all the vectors "chase the tail" of the next. However, the vector representing the resultant force goes the opposite way to all other forces present. From an engineering point of view, the equilibrant becomes important in designing a foundation, an anchorage, or a support as it is always necessary to know what resistive ability must be provided to maintain a state of equilibrium for a particular structure.

There are forces acting on all forms of matter, often without any obvious result. This book is under the influence of its own weight-force which is pressing down upon your desk, but while it remains balanced by an equal and opposite reactive force exerted by the desk no motion occurs and the force system is said to be in *equilibrium* (see Figure 3.18). Thus, we can say that the sum of the vertical forces is zero, or

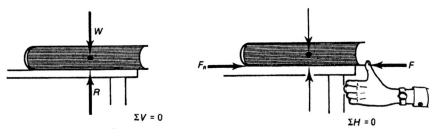

Figure 3.18   **Figure 3.19**

$\Sigma F_V = 0$. Similarly, it is possible to exert a horizontal force on the book without sliding it across the table (press your finger lightly as shown in Figure 3.19). In this case the external applied force, F, is resisted by the reactive force of friction, F_r, and the force system remains in equilibrium (see Figure 3.19). Note, that, as the force from your finger is removed, the reactive frictional force, F_r, becomes zero.

As there are no other forces acting on the book, it is therefore obvious that, for this concurrent force system, the conditions for equilibrium are:

$$\Sigma F_H = 0; \quad \Sigma F_V = 0.$$

Expressed generally, where components of a concurrent force system are resolved into the x and y directions, the system is in *equilibrium* if

$$\Sigma F_x = 0 \quad \text{and} \quad \Sigma F_y = 0.$$

Spring balances, a board, and some string can provide the practical proof of otherwise theoretical equilibrium conditions applying to a range of problems

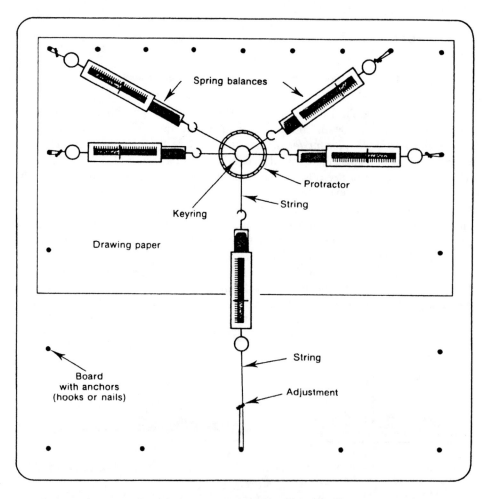

Figure 3.20 A simple practical experiment. The forces are always in equilibrium.

associated with concurrent coplanar forces. The angles between the forces can be read with a protractor behind the string, and the forces are read directly from the spring balances (graduated in newtons). Use this simple arrangement to check, experimentally, the results obtained analytically and graphically from some of the problems given at the end of this chapter.

Important
When resolving a force system analytically, a thorough working knowledge of *basic trigonometry* is essential.

The sine rule
$$\frac{a}{\sin A} = \frac{b}{\sin B} = \frac{c}{\sin C}$$

and the cosine rule
$$c^2 = a^2 + b^2 - 2\,ab\cos C$$

are often useful, and it is strongly recommended that the trigonometrical values for 30°, 45°, 60° angles, and their derivation from first principles, be committed to memory.

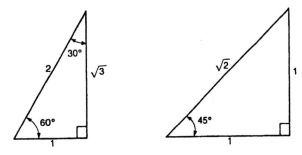

Angle	Sine	Cosine	Tangent
30°	$\dfrac{1}{2}$	$\dfrac{\sqrt{3}}{2}$	$\dfrac{1}{\sqrt{3}}$
45°	$\dfrac{1}{\sqrt{2}}$	$\dfrac{1}{\sqrt{2}}$	$\dfrac{1}{1} = 1$
60°	$\dfrac{\sqrt{3}}{2}$	$\dfrac{1}{2}$	$\dfrac{\sqrt{3}}{1} = \sqrt{3}$

Figure 3.21 Derivation of sine, cosine and tangent for 30°, 45° and 60° angles.

In the early stages of problem solving, it is highly recommended that problems be solved both graphically and analytically. This not only provides a convenient check on the results but, more importantly, helps to develop important concepts and to verify the correctness of the techniques.

After some practice is gained at solving problems by both methods, it will become easier to recognise which particular method will provide the quickest acceptable solution. Graphical solutions are often quicker, *but not always*, and they do not always provide the required accuracy.

METHOD OF SOLVING PROBLEMS

The method of attack is all important in solving an engineering problem. In order to clarify the situation it is necessary first to isolate the problem from extraneous information, then:

(i) write down the data and the required result;

(ii) draw the necessary diagrams and state the principles and the fundamental equations which apply;

(iii) apply these principles and equations.

(iv) make a statement of the conclusions reached.

There are surprisingly few fundamental concepts associated with the science of Mechanics. The knowledge of the method of application of these concepts to the wide variety of situations in which they occur comes easily with practice.

SAMPLE PROBLEM 3/1

Two tugs pull an ocean liner up the harbour. The tension in each hawser is 5 meganewtons and the angle between them is 30°. Find the effective resultant pull on the liner.

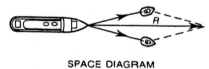

SPACE DIAGRAM

Problem 3/1

Graphical Solution:

To solve this problem graphically, it is necessary first to sketch the problem, showing all the information given. This sketch is known as the *space diagram*. From this diagram, a free-body diagram is drawn, showing only the forces under consideration. In this case it will contain only the active forces, *the tensions in the hawsers*, and not the reactive forces of the liner's resistance to being towed. If this diagram is drawn to a suitable scale, it can become the force diagram which produces the graphical solution.

FORCE DIAGRAM

FREE-BODY DIAGRAM

Problem 3/1

First select a suitable scale (say 10 mm = 1 MN). Draw vectors to this scale, at 30° to one another, representing the forces of the tensions in the hawsers. By drawing lines parallel to each of these forces, complete the parallelogram. Draw the diagonal from the origin. This represents the resultant force in magnitude and direction. Finally state the answer, $R = 9.7$ MN.

Analytical Solution:

Sketch the space diagram as before, with same regard for proportion.

To solve the problem we will need to calculate the length of the diagonal R and the angle it bears to one of the other forces. There are many ways of solving this mathematically, such as by the use of Pythagorean theorem, the sine rule, the cosine rule, and so on.

Only by constant practice in solving many problems, will the simplest method become obvious. Here we elect to use the cosine rule.

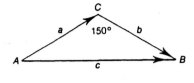

Problem 3/1

$$c^2 = a^2 + b^2 - 2ab \cos 150°$$
$$c^2 = a^2 + b^2 2ab \cos 30°$$
$$= 25 + 25 + 2 \times 5 \times 5 \times 0.866$$
$$\therefore R = \sqrt{93.3} = 9.7 \text{ MN}$$

SAMPLE PROBLEM 3/2

A telegraph pole has four wires radiating from it in the directions shown. If the tension in each wire is 2 kN, what is the resultant force on the pole?

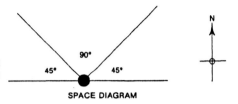

SPACE DIAGRAM

Problem 3/2

Graphical Solution:
Using Bows notation, name all the forces on the space diagram. Starting with any force, draw vectors representing these forces to a suitable scale. (The forces may be drawn in any order, but a logical sequence is simplest.) If each force is started from the end of the previous one, then the resultant is represented in magnitude and direction by the line joining the starting point (*a*) to the terminal point (*e*). This is measured and found to be 2.8 kN at 90° to the force *AB*.

The resultant force on the pole is therefore a pull of 2.8 kN in a northerly direction.

FORCE DIAGRAM

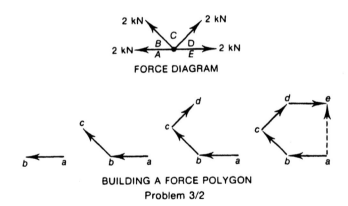

BUILDING A FORCE POLYGON
Problem 3/2

Analytical Solution:
It is convenient to tabulate the forces, and resolve them into components at right angles. The algebraic sum of these components will then give us the components of the resultant.

Force	↑Component	⟶ Component
ab	0	− 2
bc	+ 1.4	− 1.4
cd	+ 1.4	+ 1.4
de	0	+ 2
ae (resultant)	+ 2.8	0

The resultant is the same as before, 2.8 kN north.

SAMPLE PROBLEM 3/3

Water is perhaps the oldest form of transport. Before the advent of steam, barges were towed along rivers and canals by horses walking along the towpath beside the canal.

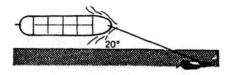

If a horse exerts a pull of 3 kN on the rope, what effective force is pulling the barge along the canal, parallel to the bank? What force is attempting to pull the barge towards the bank?

Problem 3/3

Analysis:
As we wish to discover the effective forces in specific directions, it will be convenient to resolve the 3-kN force into two components at right angles to each other. If we choose one component parallel to the bank, the other will be at right angles to the bank. Finding the magnitude of these components will produce the required result, as neither component will have any influence in the direction of the other.

Graphical Solution:
Draw a vector to represent the 3-kN force to any suitable scale. From one end of this line, draw lines parallel to the bank, and at right angles to the bank. From the other end of the vector, draw lines parallel to each of these, so completing a rectangle, with the original vector as the diagonal. The sides of the rectangle are the required components of the original force and are measured to determine the magnitude.

Using the
parallelogram of forces

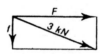

Problem 3/3

The alternative way is to form a triangle of forces as shown, instead of the rectangle—the result is the same as the magnitude of the forces required is scaled from either of these force diagrams.

Using the
triangle of forces

Problem 3/3

Calculation:
Sketch the force diagrams as before, but calculate the length of the vectors using basic trigonometry.

$$F = 3000 \cos 20° = 2819 \text{ N}$$
$$f = 3000 \sin 20° = 1026 \text{ N}$$

Result:
The effective forces are approximately 2.8 kN pull along the canal and 1 kN pull towards the bank. ∎

SAMPLE PROBLEM 3/4

If the tension in one guy rope of the ridge tent shown is 140 N, what is the load on the tent pole? What is the tension in the other guy rope?

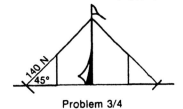

Problem 3/4

Analysis:

Consider the junction of the two guys and the tent pole. Three forces are present. We know the magnitude of one of these, and the direction of all of them. This is enough to sketch a force diagram, which can then be solved graphically, or calculated.

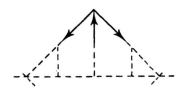

Problem 3/4

Graphical Solution:

First draw the known force to scale, parallel to the direction in which it acts. From each end of this known force, draw lines parallel to the lines of action of the other two forces. They will intersect at a point where the force polygon so built is representative of a force system in equilibrium. If we now add arrows to the polygon we find that the pole is pushing up on the junction to maintain equilibrium. It only remains to scale off the lengths of the two new sides of this triangle, to determine the magnitude of the forces. The fact that the forces are not necessarily coplanar is immaterial.

Calculation:

The mathematical solution of this triangle is equally simple. It is suggested that the student should try several methods for comparison.

Result:

The tent pole has a load of 200 N (compression). The other guy rope has the same tension as the first, 140 N. ∎

REVIEW PROBLEMS

(Use $g = 10$ m/s^2 unless otherwise specified.)

3/5
Determine the magnitude and direction of the resultant of the two concurrent forces shown.

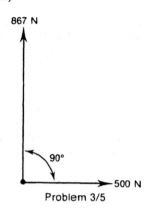

Problem 3/5

3/6
A barge is towed along a canal by means of a rope attached to a tractor. The direction of movement of the barge is kept parallel to the canal wall by off-setting the rudder. The rope makes an angle of 20 degrees with the canal. If the tension in the rope is 20 kN what is the effective towing force?

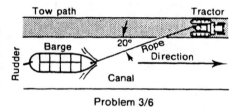

Problem 3/6

3/7
Determine graphically and analytically the horizontal and vertical components of the force acting at O.

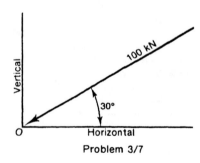

Problem 3/7

3/8
What are the components of the 12 kN force in directions Ox and Oy?

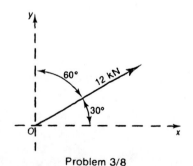

Problem 3/8

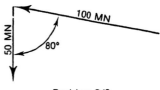

3/9
Replace the system of forces shown in the diagram by a single force which will have the same effect.

Problem 3/9

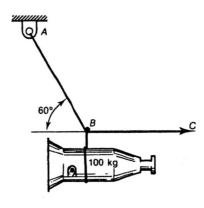

3/10
The figure shows a gearbox which has a mass of 100 kg. During a lifting operation it is supported in the position shown by a rope AB and a rope BC. Determine the tension in each rope.

Problem 3/10

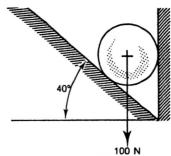

3/11
A steel ball rests in a groove the sides of which are smooth. One side of the groove is vertical, while the other side is at 40° to the horizontal.

If the ball has a mass of 10 kg, find the reaction on each wall of the groove.

Problem 3/11

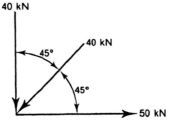

3/12
Find the magnitude and direction of the resultant of the three forces shown in the diagram.

Problem 3/12

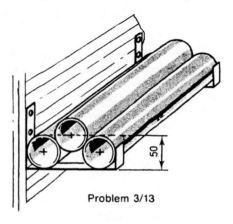

3/13

Three greasy steel pipes of 50 mm outside diameter are stacked in a wall rack as shown. Each pipe has a mass of 25 kg. Discover the forces acting on each bracket.

Problem 3/13

3/14

The car shown in the diagram is held in a bog by forces of 1 kilonewton on each side. Three ropes are attached to the car and forces as shown are applied to the ropes. Will these forces remove the car from the bog? Justify your result.

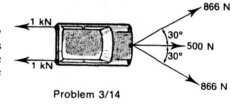

Problem 3/14

3/15

An ocean liner is being towed by three tugboats as shown. The tension in each cable is 5000 MN. (a) Determine graphically the resultant force acting on the bow of the liner. (b) If the tugboats cannot operate safely when the angle between the cables is less than 10°, where should the tugboats be located in order to produce the largest resultant force? (c) What is the magnitude of this force?

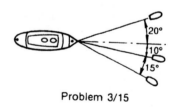

Problem 3/15

3/16

Find the resultant x component and the resultant y component of the force system shown.

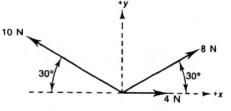

Problem 3/16

3/17
Determine the resultant of the three forces on the welded joint shown in the diagram.

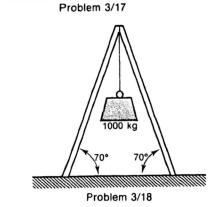

Problem 3/17

3/18
A pair of shear legs is used to support a load of 1000 kg. Determine the force in each of the legs. (The legs and load rope are in the same vertical plane.)

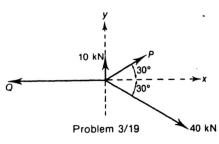

Problem 3/18

3/19
State the conditions of equilibrium of any coplanar system of forces.

   If the force system shown is in equilibrium, find the value of the forces *P* and *Q*.

Problem 3/19

3/20
Determine the forces in each of the members *A* and *B*.

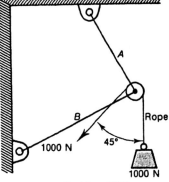

Problem 3/20

3/21

A smooth hook is attached to the end of a rope which is looped around a stormwater pipe of 500 mm diameter and 250 kg mass. Disregarding friction, determine the angle θ and the tension in the rope.

Problem 3/21

3/22

Determine the forces in the jib and tie of the jib crane shown.

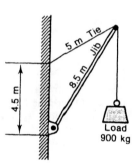

Problem 3/22

3/23

In each of the systems of forces shown, the condition of equilibrium may or may not exist. Determine the force necessary for equilibrium if that condition does not already exist.

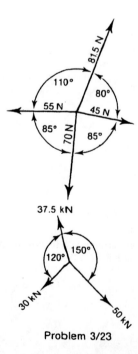

Problem 3/23

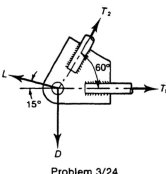

3/24
Two forces, L and D, of magnitude $L = 1000$ N and $D = 1200$ N are applied to the connection shown. Knowing that the connection is in equilibrium, determine the tensions T_1 and T_2.

Problem 3/24

3/25
One of the joints of a pin-jointed frame is shown in the diagram. Describe the forces in members A and B.

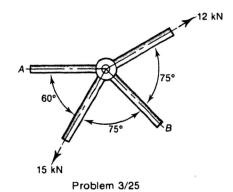

Problem 3/25

3/26
A container and its contents have a mass of 10 tonnes. Determine the shortest chain sling which may be used to lift the container if the tension in the sling is not to exceed 100 kN.

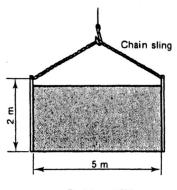

Problem 3/26

3/27

Mürren is reached from Lauterbrünnen by a funicular railway which travels up and down a slope of 45°. The cable passes from one car through guide rollers between the rails, around the winding drum, and back down to the other car. Each car acts as a counterweight for the other, thereby reducing the power needed to wind a loaded car up the mountain. The difference in elevation between the two Swiss villages is approximately 1000 m (Lauterbrünnen 646 m, Mürren 1647 m). Each loaded car has a mass of 5.6 tonnes and the cable connecting the two cars has a mass of 2.8 kg/metre.

   Determine the approximate tension in the cable when there is a loaded car waiting at each village:

(a) near the car at Mürren;
(b) near the car at Lauterbrünnen;
(c) on each side of the winding drum.

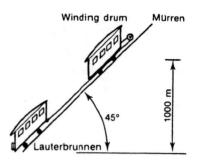

Problem 3/27

3/28

Fred's new 1-tonne car is being unloaded at the docks. A guide rope is attached to the crane hook *A* so that the wharf labourers can swing the car on to the dock. The crane cable is 10° off vertical and the guide rope is 30° to the horizontal. Find the tensions in cable and rope.

Problem 3/28

3/29

A man raises a pole of mass 20 kg by pulling on a rope. Find the tension, *T*, in the rope.

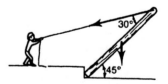

Problem 3/29

4

Moments

$$M = Fd$$
$$\Sigma M = 0$$

The forces we have considered so far have all tended to cause or to resist *translation*, that is, to cause or to resist movement from one place to another. Many forces, however, tend to cause *rotation* as well. The downward pressure on a bicycle pedal is exactly resisted by the upward pressure of the road, and yet the force system is not in equilibrium. The bicycle moves forward, because the downward force on the pedal causes *rotation*.

The harder you push on the pedal, the greater is the tendency to rotate the wheel, and we measure this tendency very simply by multiplying the *force* on the pedal, by the *length* of the crank. The multiplication of these two unrelated dimensions, appears at first glance to give us a meaningless unit, the *newton metre* (N m), but the result is known as the *torque*, the *turning moment*, or the *moment of the force*.

$$\text{Moment} = \text{force} \times \text{distance}$$
$$M = Fd$$

Moment calculations are useful for the following

(1) For the purposes of the *comparison* of two turning effects; the bigger the number, the greater is the tendency to cause rotation. (A turning moment of 10 N m has a greater turning effect than a moment of 8 N m.)

We are all familiar with the torque wrench* used for tightening bolts and nuts to a uniform tension. The scale is graduated directly in newton metres and we do not need to know the magnitude of the force (pull) on the handle, or the length of the handle. Provided the pointer moves to the same place on the scale each time,

---

*The torque wrench in your home workshop is probably graduated in the British Gravitational Units, the ft lb. The conversion factor is 1.355, e.g. 3 ft lb is approximately 4 N m.

the same torque will be applied to each bolt, and we assume that the same tension will have been produced. (Refer to Sample Problem 4/1.)

(2) For the purposes of *discovering* either the magnitude of the force or the length of the moment arm required to produce a desired turning effect.

In many cases, actual rotation does not occur, and we need to discover the force required to produce *equilibrium*. In this situation we know *where* the force is to be applied, and therefore its distance from the assumed axis of rotation. By dividing the turning moment by this distance, we get the force required $\left(\dfrac{\text{N m}}{\text{m}} = \text{N}\right)$.

Similarly, we may have a certain force available, and wish to know where to apply it in order to maintain equilibrium. By dividing the moment by the force, we again revert to a tangible dimension, length $\left(\dfrac{\text{N m}}{\text{N}} = \text{m}\right)$.

This is the distance from the assumed axis of rotation, at which the force would have to be applied.

Moment equilibrium occurs when any force tending to cause rotation is exactly balanced by some other force producing a turning moment equal in magnitude but opposite in direction, that is, for moment equilibrium, the algebraic sum of the turning moments is zero.

$$\Sigma M = 0$$

SIGN CONVENTION

As *torque* is a vector, the sense of the rotation (clockwise or anti-clockwise) is as important as the magnitude of the moment. When adding these vectors, we can use scalar algebra, provided we adopt a suitable sign convention to indicate the sense of the torque. We may call the clockwise direction positive and the anti-clockwise

Figure 4.1

direction negative (or vice versa) provided the convention used in any particular calculation is stated. This is usually done by writing (ↄ⁺) or (ↄ⁺) at the beginning of the calculation.

Being a vector, the moment can be represented graphically, and vector addition techniques can be used to achieve graphical solutions. The vector's *direction* is along the axis about which the moment is taken, and the *sense* of the vector is specified by the *right-hand rule* (the fingers of the right hand are curled in the direction of the tendency to rotate—the thumb then points in the direction of the vector). A small curl is sometimes used to distinguish a moment vector from a force vector. Moment vectors obey all the usual rules of vector combination.

Vector representation and addition of moments for coplanar forces is unnecessary, as the use of scalar algebra is quicker and more accurate, provided a suitable sign convention is used.

SAMPLE PROBLEM 4/1

SPACE DIAGRAM   ↓*Pull*

Calculate the pull required on the handle of the torque wrench to produce a torque of 120 N m. The handle has an effective length of 0.5 m.

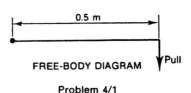

FREE-BODY DIAGRAM   ↓Pull

Problem 4/1

Calculation:
Draw the free-body diagram showing active vectors. Consider the turning moment about the axis of the bolt. Then

$$\text{torque} = \text{force} \times \text{length of moment arm}$$
$$120 \text{ Nm} = \text{pull} \times 0.5 \text{ m}$$
$$\therefore \text{pull} = \frac{120 \text{ Nm}}{0.5 \text{ m}}$$
$$= 240 \text{ N}$$

Result:
A pull of 240 newtons will be required on the handle of the torque wrench.   ■

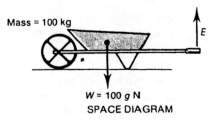

Mass = 100 kg

$W = 100\,g$ N

SPACE DIAGRAM

Problem 4/2

SAMPLE PROBLEM 4/2

Determine the effort required to just raise the handles of the wheelbarrow shown, if the load and the barrow have a total mass of 100 kg concentrated at their common centre of mass as shown.

Analysis:

Draw the free-body diagram showing all vectors. As the nature of this problem does not appear to warrant mathematical precision, we can feel justified in using $g = 10$ m/s^2 to calculate weight-forces.

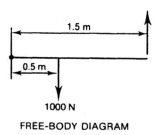

1.5 m

0.5 m

1000 N

FREE-BODY DIAGRAM

Problem 4/2

Consider the turning moments about the axis of the wheel. The combined weight of the barrow and its load will exert a force of approximately 1000 newtons vertically down on a line 0.5 m from this axis, tending to rotate the barrow clockwise about the wheel. This tendency is equalled by the tendency of the *effort* to rotate the barrow about the wheel in an anti-clockwise direction. This will produce a condition of moment equilibrium ($\Sigma M = 0$).

Calculation:

$$\text{clockwise moment} = \text{force} \times \text{distance (from axis)}$$
$$= 1000 \text{ N} \times 0.5 \text{ m}$$
$$= 500 \text{ N m}$$

To support the barrow, the anti-clockwise moment provided by the effort (E) will need to equal the clockwise moment

$$E \times 1.5 \text{ m} = 500 \text{ N m}$$
$$E = \frac{500 \text{ N m}}{1.5 \text{ m}}$$
$$= 333 \text{ N}$$

Result:

A force of approximately 333 newtons would be required to raise the handles of the barrow. (This is the force needed to lift a mass of about 33.3 kg—only 1/3 of the original load.) ■

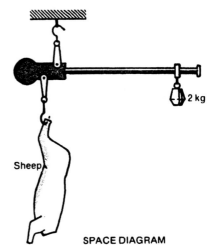

SPACE DIAGRAM

SAMPLE PROBLEM 4/3

The ancient steelyard shown is used to determine the mass of a sheep. If the arm balances when the 2-kg balance weight is 600 mm from the fulcrum, what is the mass of the sheep? Use $g = 10$ m/s^2.

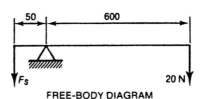

FREE-BODY DIAGRAM

Problem 4/3

Analysis:

Draw the free-body diagram. The 2-kg balance weight will exert a downward force of approximately 20 newtons, tending to rotate the arm of the steelyard clockwise about its fulcrum. This must be equalled by the turning moment produced by the weight-force of the sheep. By taking moments about the fulcrum, we can discover this weight-force, and hence the mass of the sheep.

Calculation:

$$\text{clockwise moments} = \text{force} \times \text{distance}$$
$$= 20 \text{ N} \times 0.6 \text{ m}$$
$$= 12 \text{ Nm}$$
$$\text{anti-clockwise moments} = \text{force} \times \text{distance}$$

but
$$\Sigma M \circlearrowright = \Sigma M \circlearrowleft \quad (\text{or } \Sigma M = 0)$$
$$\therefore F_s \times 0.05 = 12 \text{ Nm}$$
$$F_s = \frac{12 \text{ Nm}}{0.05 \text{ m}} = 240 \text{ N}$$
$$\text{force exerted by sheep} = 240 \text{ N}$$

Result:

$$\text{mass of sheep} = 24 \text{ kg}$$

(The minor error introduced initially, by using the value of 10 m/s^2 for g instead of 9.8 m/s^2, has cancelled itself out exactly, as anyone who would prefer to use 9.8 m/s^2 can prove. We are only comparing masses, and the steelyard (beam balance) would work just as well on the moon.) ∎

PARALLEL FORCES

So far we have considered moments about an axis around which it was obvious that rotation could occur. However, there are many situations which can be resolved easily by considering the turning moments involved, even though no rotation is possible (or desirable). In these problems, we take moments about any convenient axis and, if we choose the axis intelligently, we may be able to ignore one or more of the unknown forces. For example, consider the following situation.

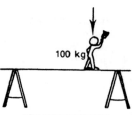

SPACE DIAGRAM

SAMPLE PROBLEM 4/4

A painter is standing on a light plank, supported by two trestles as shown. It is required to determine the load on each trestle.

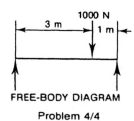

FREE-BODY DIAGRAM

Problem 4/4

Analysis:

If we remove the right trestle, plank and painter will revolve clockwise about the left support, so the right trestle must be exerting an upward force on the plank which prevents this *rotation*. If we take moments about the left support, we can discover the magnitude of this force, i.e. we will discover the load on the right trestle.

By choosing the left support as a moment axis, we can ignore the unknown upward force exerted by it, as this force has no moment about that axis.

Calculation:

Use $g = 10 \, \text{m/s}^2$ for convenience.

$$\text{downward force exerted on plank by painter} = 100 \, \text{kg} \times g$$
$$= 1000 \, \text{N}$$
$$\text{clockwise moment about } L = 1000 \, \text{N} \times 3 \, \text{m}$$
$$= 3000 \, \text{N m}$$
$$\text{anti-clockwise moment about } L = \text{upward force exerted by}$$
$$\text{right trestle} \times 4 \, \text{m}$$

For equilibrium, $\Sigma M = 0$

$$\therefore \text{anti-clockwise moment} = 3000 \, \text{N m}$$
$$\therefore \text{upward force on plank by right trestle} \times 4 \, \text{m} = 3000 \, \text{N m}$$
$$\therefore \text{upward force } (F_R) = \frac{3000 \, \text{N m}}{4 \, \text{m}}$$
$$\therefore F_R = 750 \, \text{N}$$
$$\text{Load on right trestle} = 750 \, \text{N}$$

Similarly, if we take moments about the right trestle, we can discover the load on the left trestle.

$$1000\,\text{N} \times 1\,\text{m} = F_L \times 4\,\text{m}$$
$$\therefore F_L = \frac{1000\,\text{N m}}{4\,\text{m}}$$
$$F_L = 250\,\text{N}$$

Result:

Load on left trestle = 250 N
Load on right trestle = 750 N

The sum of these two forces is 1000 N and is equal to the weight-force exerted by the painter; $\therefore \Sigma F_V = 0$. This is a useful check on the accuracy of the mathematics. ■

SAMPLE PROBLEM 4/5

A tower crane is supporting a load of 2 tonnes, 12 metres from the tower. (This tends to topple the crane—i.e. the boom tends to rotate clockwise about the slewing turntable, or the base in a vertical plane.) Where must the driver position the counterweight to lift the load safely? Assume $g = 10\,\text{m/s}^2$.

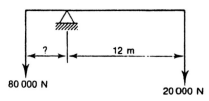

SPACE DIAGRAM

FREE-BODY DIAGRAM

Problem 4/5

Calculation:
First: Analyse the forces involved.
Downward force exerted on boom
by 2-tonne load $= 2000\,\text{kg} \times g$
                $= 20\,000\,\text{N}$

Downward force exerted on boom
by 8-tonne counterweight $= 8000\,\text{kg} \times g$
                $= 80\,000\,\text{N}$

Draw the free-body diagram as shown.

Next: Choose a suitable moment axis.
If we choose the base of the crane as an axis, we will be able to ignore all the forces acting *on* this point as they have no turning moment *about* this point. These will include the self-weight of the crane and the reactive forces produced by it and its loads.

Take moments about the base.

$$\text{clockwise moment} = \text{force} \times \text{distance from base}$$
$$= 20\,000 \times 12$$
$$= 240\,000\,\text{N m}$$
$$\text{anti-clockwise moment} = 80\,000 \times \text{distance from base (N m)}$$

For equilibrium, $\Sigma M = 0$
$\therefore$ anti-clockwise moment = clockwise moment = 240 000 N m
i.e. $80\,000 \times$ distance = 240 000 N m
$$\therefore \text{distance from base} = \frac{240\,000\,\text{N m}}{80\,000\,\text{N}}$$
$$= 3\,\text{m}$$

Result:
The counterweight should be positioned 3 m to the left of the centreline of the tower. ■

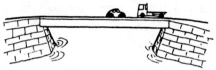

SAMPLE PROBLEM 4/6

A car and a truck are stationary on the narrow bridge as shown. The truck has a gross mass of 7 tonnes and the car 1 tonne. The bridge structure has a mass of 20 tonnes. Determine the load on each of the abutments (i.e. determine the reactions at the supports). Assume $g = 10$ m/s^2.

SPACE DIAGRAM

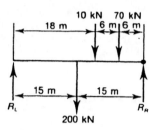

FREE-BODY DIAGRAM

Problem 4/6

Calculation:

First: Analyse the forces involved.

Assume that the mass of each element in the system exerts its weight-force vertically downwards through its centre of mass at the positions shown.

$$\text{force exerted by car} = 1000\,\text{kg} \times g = 10\,000\,\text{N} \ (10\,\text{kN})$$
$$\text{force exerted by truck} = 7000\,\text{kg} \times g = 70\,000\,\text{N} \ (70\,\text{kN})$$
$$\text{force exerted by bridge} = 20\,000\,\text{kg} \times g = 200\,000\,\text{N} \ (200\,\text{kN})$$

Draw the free-body diagram as shown.

Next: Choose a suitable moment axis.

If we choose one of the abutments as an axis, we will be able to ignore the unknown force acting *through* this axis, as it has no moment *about* this axis. Let us arbitrarily choose the *right abutment*. If we were to remove the left abutment, the bridge would rotate anti-clockwise about the right abutment.

Take moments about this axis:

$$
\begin{aligned}
\text{anti-clockwise moments} &= \text{force} \times \text{distance} \\
&\quad (\textit{each}\ \text{force} \times \textit{its}\ \text{distance from the axis}) \\
&= \text{truck} \times 6\,\text{m} + \text{car} \times 12\,\text{m} + \text{bridge} \times 15\,\text{m} \\
&= 70\,\text{kN} \times 6\,\text{m} + 10\,\text{kN} \times 12\,\text{m} + 200\,\text{kN} \times 15\,\text{m} \\
&= 420\,\text{kN m} + 120\,\text{kN m} + 3000\,\text{kN m} \\
&= 3540\,\text{kN m} \\
\text{clockwise moment} &= R_L \times 30\,\text{m}
\end{aligned}
$$

For moment equilibrium, $\Sigma M = 0$ (or clockwise moments = anti-clockwise moments)

$$R_L \times 30 = 3540\,\text{kN m}$$
$$R_L = \frac{3540\,\text{kN m}}{30\,\text{m}}$$
$$= 118\,\text{kN}$$

For force equilibrium, $\Sigma F_V = 0$

Sum of forces exerted vertically down = 280 kN

Sum of forces exerted vertically up must equal the above sum (i.e. also 280 kN)

$$\therefore R_R = 280 - 118$$
$$= 162\,\text{kN}$$

As a check on the accuracy of these calculations we can take moments about the left abutment (we *should* get the same result).

$$\text{clockwise moments} = (200 \times 15) + (10 \times 18) + (70 \times 24)$$
$$= 3000 \quad + 180 \quad + 1680$$
$$= 4860\,\text{kN m}$$
$$\text{As } \Sigma M = 0, R_R = \frac{4860}{30}\,\text{kN}$$
$$= 162\,\text{kN}$$

Result:

$$R_L = 118\,\text{kN} \quad \text{and} \quad R_R = 162\,\text{kN}$$

i.e., for the conditions considered, the *static load* on the left abutment is 118 kN and that on the right abutment is 162 kN. ∎

You will have noticed in the preceding examples that, in isolating the problem, we have ignored some forces in the system, but in each case the force ignored passed through the axis of rotation. It is possible to do this because this force has no tendency to cause rotation about that axis, and therefore has no moment about that axis.

$$\text{Force} \times \text{distance} = F \times 0 = 0$$

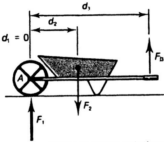

Figure 4.2 The force F_1 has no tendency to cause rotation about the moment axis A (the wheel axle).

You will also have noticed that the distance used, known as the *moment arm*, was the distance measured at right angles to the line of action of the force. The reason for this is demonstrated when we consider the effect of a pull on the steering wheel shown. The maximum moment occurs with the hand at 3 o'clock but, provided the magnitude and direction of the pull remain unchanged, the magnitude of the turning moment progressively diminishes until, in the limit, when the line of action of the force passes through the axis of rotation, the turning effect on the wheel is zero (refer to Figure 4.3).

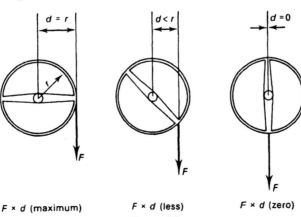

| $F \times d$ (maximum) | $F \times d$ (less) | $F \times d$ (zero) |

Figure 4.3

Occasionally, the length of the moment arm is not known, and in order to calculate the moment of the force, we will do one of two things:
(1) calculate the length of the moment arm, and multiply by the known force; or
(2) calculate components of the force, and multiply by known perpendicular moment arms.

SAMPLE PROBLEM 4/7

Calculate the moment of the force F about the point A

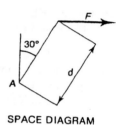

SPACE DIAGRAM

Calculation (1): (Moment arm)

$$\text{moment of force} = \frac{\text{known}}{\text{force}} \times \frac{\text{perpendicular}}{\text{distance (moment arm)}}$$

$$= F \times \frac{\sqrt{3}}{2} d$$

$$= \frac{\sqrt{3}}{2} Fd$$

Calculation (2): (Components)

If we choose suitable components, so that component A is at right angles to the known moment arm, and component B passes through the centre of moments, we can ignore component B, as it has no moment (about the centre of assumed rotation).

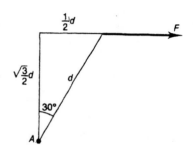

$$\text{moment of force} = \frac{\text{perpendicular component}}{\text{of known force}} \times \frac{\text{known}}{\text{distance}}$$

$$= \frac{\sqrt{3}}{2} F \times d$$

$$= \frac{\sqrt{3}}{2} Fd \quad \text{(same as (1))}$$

Problem 4/7

(The component $\frac{1}{2}F$ that was neglected passes *through A*, and therefore has no moment *about* A.)■

Sample Problem 4/7 illustrates the *principle of moments*, and shows the basis of *Varignon's theorem*, which states that *the algebraic sum of the components of a force equals the moment of the force.*

SAMPLE PROBLEM 4/8

CALCULATION OF ONE COMPONENT

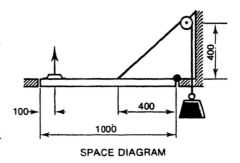

SPACE DIAGRAM

Determine the mass of the counterweight required to raise the 50-kg trapdoor, when a pull of approximately 100 N is applied to the handle. Assume $g = 10\,\text{m/s}^2$.

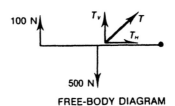

FREE-BODY DIAGRAM

Analysis:

The counterweight will produce a tension in the cable to provide a lifting force for the trapdoor. If we resolve this force into its horizontal and vertical components, we see that only the vertical component T_V is tending to raise the door. The horizontal component T_H is being absorbed by the hinge reaction, and can therefore be ignored in this problem.

Force tending to close the door equals

$$\text{Mass of door} \times g = 50\,\text{kg} \times g = 500\,\text{N}$$

Forces tending to open the door equal to 100 N pull on the handle plus the vertical component of the cable tension.

Take moments about the hinge. Assume $+\!\circlearrowleft$

$$500\,\text{N} \times 0.5\,\text{m} = -250\,\text{N m} \qquad \text{(counter-clockwise)}$$
$$100\,\text{N} \times 0.9\,\text{m} = +90\,\text{N m} \qquad \text{(clockwise)}$$
$$\text{torque still required} = -250\,\text{N m} + 90\,\text{N m} = +160\,\text{N m}$$

This is to be provided by the vertical component of the cable tension.

$$T_V \times 0.4\,\text{m} = 160\,\text{N m}$$
$$T_V = \frac{160}{0.4}\,\text{N}$$
$$= 400\,\text{N}$$

but
$$T_V = T\sin 45°$$
$$\therefore T = 565.6\,\text{N}$$

$$\text{mass of counterweight required} = \frac{565.6}{g} \approx 57\,\text{kg}$$

Result:

The counterweight should have a mass of *approximately* 57 kg. (We have not taken into account friction at the hinge or the pulley, or the energy lost in bending and straightening the cable, so it is pointless trying to give an answer to any greater order of accuracy than shown.) ∎

SAMPLE PROBLEM 4/9

CALCULATION OF BOTH COMPONENTS

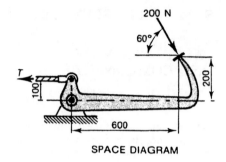

SPACE DIAGRAM

Determine the tension in the tractor brake cable when a force of 200 N is applied to the pedal.

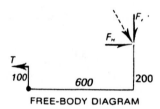

FREE-BODY DIAGRAM

Analysis:

From the measurements available, it would be inconvenient to calculate the length of the true moment árm, so we will resolve the force into suitable components. However, no suitable components pass through the obvious moment axis (the pivot) and therefore we cannot ignore either component, and must include the moment produced by both of them in our calculations. If we resolve the force into its horizontal and vertical components, then the algebraic sum of the moments produced by each component will be equal to the moment produced by the original force (Varignon's theorem).

Take moments about the pivot   Assume $+$

$$\text{horizontal component} = 200 \cos 60° = 100 \text{ N}$$
$$\text{moment produced by horizontal component} = 100 \text{ N} \times 0.2\,\text{m}$$
$$= + 20 \text{ N m} \quad \text{(clockwise)}$$
$$\text{vertical component} = 200 \sin 60° = 173.2 \text{ N}$$
$$\text{moment produced by vertical component} = 173.2 \text{ N} \times 0.6\,\text{m}$$
$$= + 103.9 \text{ N m} \quad \text{(clockwise)}$$

Thus,
$$\Sigma M = + 123.9 \text{ N m} \quad \text{(clockwise)}$$

Under equilibrium conditions, the tension in the brake cable acting 100 mm from the pivot will produce an anti-clockwise moment which will exactly oppose the clockwise moment created by the pressure on the pedal.

$$T \times 0.1 \text{ m} = 123.9 \text{ N m}$$
$$T = 1239 \text{ N}$$
$$= 1.24 \text{ kN}$$

Result:

The tension in the cable is 1.24 kN.

■

SPACE DIAGRAM

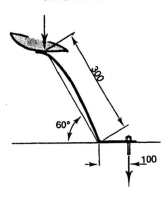

SAMPLE PROBLEM 4/10

CALCULATION OF
THE LENGTH OF
MOMENT ARM

What is the tension in the bolt supporting the tractor seat when it is supporting a 100-kg operator? Assume $g = 10$ m/s^2.

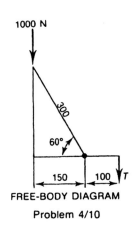

FREE-BODY DIAGRAM

Problem 4/10

Analysis:
Take moments about the lower bend in the spring leaf. The true moment arm then equals $0.3 \cos 60° = 0.15$ m

$$T \times 0.1 \text{ m} = 1000 \text{ N} \times 0.15 \text{ m}$$
$$T = 1000 \times \frac{0.15}{0.1} \text{ N}$$
$$= 1500 \text{ N}$$
$$= 1.5 \text{ kN}$$

Result:
For the position shown, the bolt tension is 1.5 kN. However, as the seat spring deflects, the moment arm will get progressively longer, increasing the tension in the bolt. ∎

SAMPLE PROBLEM 4/11

CALCULATION OF THE LENGTHS OF MOMENT ARMS

The flagpole bracket is attached to the brick wall with a masonry anchor as shown. Determine the minimum tension, *T*, required in the bolt. The pole is of uniform section and has a mass of 5 kg. The flag has a mass of 2 kg. Assume $g = 10$ m/s².

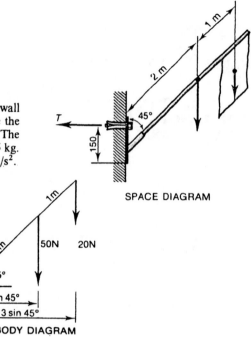

SPACE DIAGRAM

FREE-BODY DIAGRAM

Problem 4/11

Analysis:

The weight-force of the pole and the flag tend to rotate the assembly clockwise about the bottom of the bracket. The bolt provides an equal and opposite tendency to rotate the assembly anti-clockwise about the same axis. To take moments about this axis we will need to

(a) calculate the components of the weight-forces, at right-angles to the known moment-arm (the pole), or

(b) calculate the lengths of the moment-arms, at right-angles to the known forces. We will arbitrarily choose to do the latter.

Moment arms

$$\text{centre of mass of the pole} = 2\sin 45° = 1.4 \text{ m from the wall}$$
$$\text{centre of mass of the flag} = 3\sin 45° = 2.1 \text{ m from the wall}$$

Forces   Assume $g = 10$ m/s²

$$\text{force exerted by the pole} = 5 \text{ kg} \times g = 50 \text{ N}$$
$$\text{force exerted by the flag} = 2 \text{ kg} \times g = 20 \text{ N}$$

Take moments   Assume ↻

$$\text{clockwise moments} = (50 \text{ N} \times 1.4 \text{ m}) + (20 \text{ N} \times 2.1 \text{ m})$$
$$= + 70 \text{ N m} \qquad + 42 \text{ N m}$$
$$= + 112 \text{ N m}$$
$$\text{anti-clockwise moments} = T \times 0.15 \text{ m} \ (= -112 \text{ N m})$$
$$T = -\frac{112}{0.15} \text{ N}$$
$$= -747 \text{ N}$$

Result:

The bolt will carry a minimum tension of ≈ 750 N before being locked up tight. (In practice, there would be more than one bolt installed, and each would be designed to carry far more than this calculated load. A suitable "safety factor" is chosen by the design engineer for each case.) ∎

REVIEW PROBLEMS

(Use $g = 10$ m/s ^{2} unless otherwise specified.)

4/12

The maximum pull that a man can comfortably exert on the handle of a spanner is about 250 N. Determine how long the handle should be if the maximum torque to be exerted by the spanner is 50 N m.

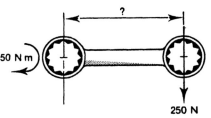

Problem 4/12

4/13

Determine the minimum force F required to move the brake lever. Friction in the connections may be neglected.

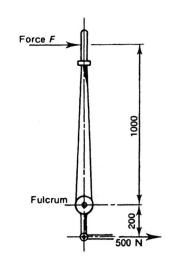

Problem 4/13

4/14

You are removing old fence-posts with a crowbar, which has a mass of 8 kg. The force required to lift the post is 5000 N. What force will you need to apply to the end of the crowbar to raise the post?

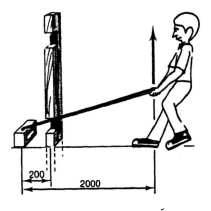

Problem 4/14

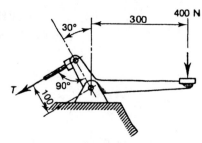

4/15
A 400-N force is required to operate the foot pedal shown. Determine the tension, *T*, in the rod.

Problem 4/15

4/16
You have suffered a flat tyre on your new car, which does not have a jack. The car has a mass of 1 tonne (equally distributed on the four wheels) and your girlfriend has a mass of 50 kg. Where should you place the rock under the lever, so that your girlfriend may sit comfortably while you change the wheel?

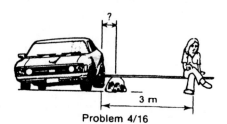

Problem 4/16

4/17
A clamp is used to secure a work piece to the table of a milling machine. If the tightening of the nut induces a force of 1000 N on the bolt, determine the clamping force exerted on the work piece.

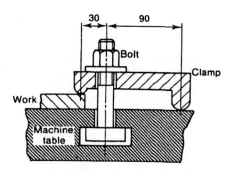

Problem 4/17

4/18
The bar-bell has a total mass of 50 kg and you have raised it with both hands to the position shown. Determine the tension in your biceps.

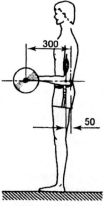

Problem 4/18

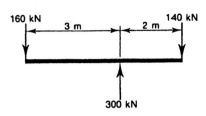

Problem 4/19

4/19
Is the bar, loaded as shown, in a state of equilibrium? Give reasons for your answer in terms of $\Sigma F = 0$, and $\Sigma M = 0$.

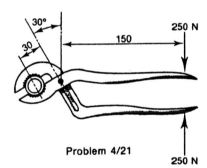

Problem 4/20

4/20
The lever is just balanced on the fulcrum. What is the magnitude of the force *F*?

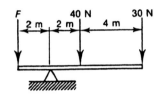

4/21
What pressure is exerted on the pipe when the handles of the multi-grips are squeezed as shown?

Problem 4/21

4/22
(a) Find the pull exerted on the nail by a pull *P* of 200 newtons on the handle.
(b) If the nail's resistance to movement is 1.5 kN, what is the force *P* required on the handle of the hammer, to extract it?
(c) Describe where the block should be placed, to minimise the chances of breaking the handle of the hammer.

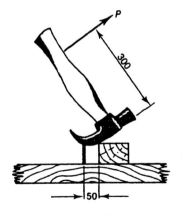

Problem 4/22

4/23

The see-saw is 2 metres long. Tom (15 kg) sits on one end and Dick (18 kg) sits on the other. Where should Harry (12 kg) stand, to keep the see-saw level?

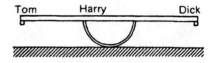

Problem 4/23

4/24

Calculate the tension, T, in the chain of the bicycle when the 50-kg rider puts all his weight on one pedal.

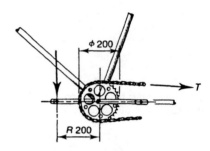

Problem 4/24

4/25

For the pivoted clamp shown in the sketch, determine the clamping force, F, if the force exerted by the bolt when the nut is tightened is 2.4 kN.

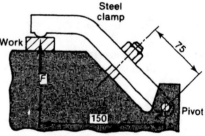

Problem 4/25

4/26

You are trying to push a piece of carpet under the front feet of the refrigerator, so you can slide it to a new place in the kitchen. The refrigerator has a mass of 150 kg. What horizontal force will you need to exert at the top of the refrigerator, to rock it back just far enough to get the carpet under? (Assume the weight of the refrigerator is equally distributed on all four feet).

Problem 4/26

4/27
Determine the extra load on the front wheels of the loader when it is carrying 2 m³ of earth with a mass of 1200 kg/m³.

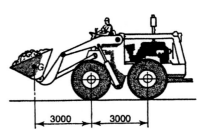

Problem 4/27

4/28
Determine the magnitude of the pull, *P*, when the forces shown in the diagram are applied to the lever.

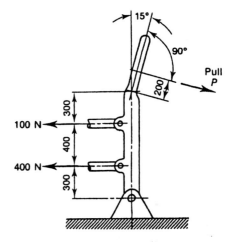

Problem 4/28

4/29
The pin-jointed framework shown in the diagram supports a load of 50 kilonewtons. Determine
(a) the tension in the guy wire; and
(b) the vertical and horizontal components of the force exerted on the pin *P*.

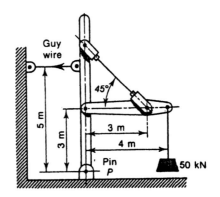

Problem 4/29

4/30
One of a pair of life-boat davits is illustrated in the figure. If all of the vertical thrust is taken at the top pivots, determine the reactions at the supports if the boat has a mass of 1 tonne.

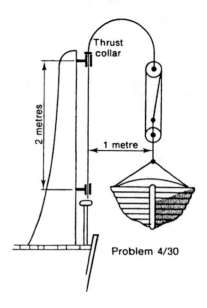

Problem 4/30

4/31
A 1-tonne caravan is attached to a 1.5-tonne car by a ball-and-socket coupling. Determine
(a) the reactions at each of the six wheels when the car and caravan are at rest;
(b) the change in load on each of the car wheels due to the caravan.

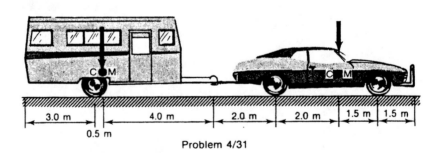

Problem 4/31

4/32
Tom, Dick, and Harry are sitting on the garden seat as shown. The seat has a mass of 40 kg.
(a) What is the load supported by each of the two legs of the seat?
(b) If the base of each leg is 50 mm × 300 mm, what pressure does each exert on the ground?

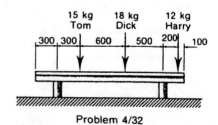

Problem 4/32

5

Non-Concurrent Forces

A 300-tonne crane in the BOS (basic oxygen steelmaking) plant at BHP, Newcastle: the photograph on p. 85 shows the massive crane hooks, the equaliser beam and the supporting cables.

The hooks and equaliser beam have a mass of 54 tonnes: the photograph on p. 86 shows the teeming process — pouring 200 tonnes of molten steel into ingot moulds.

$$\Sigma F = 0$$
$$\Sigma M = 0$$

By now it is apparent that force systems acting upon bodies can be of several different types. For instance, coplanar concurrent forces have been examined in Chapter 3 while some simple systems of parallel forces were analysed in Chapter 4 in relation to moments of forces.

While concurrent forces have their lines of action passing through a common point, the lines of action of parallel forces (of course) do not intersect. Parallel force systems are one example of *coplanar non-current force systems*, in which *the lines of action of the forces present do not intersect in a common point*. Non-concurrent forces can be either parallel, or, more commonly, angled to each other.

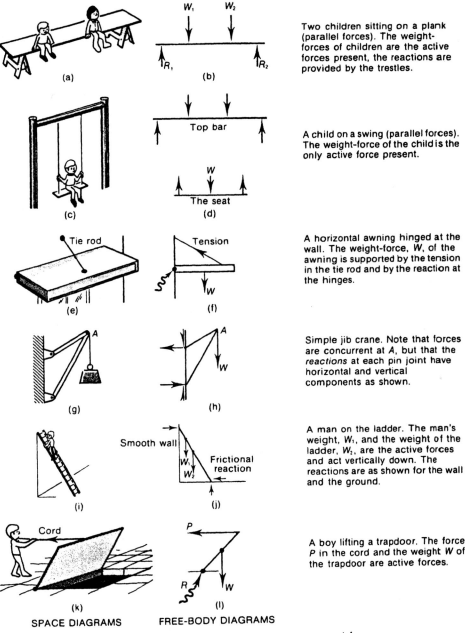

Figure 5.1 Examples of coplanar non-concurrent forces.

RESULTANTS OF NON-CONCURRENT FORCES

In some instances it is necessary to determine the resultant force that would replace a given system of non-concurrent coplanar forces. This can be done either analytically or graphically, the choice of method depending upon the complexity of the problem and the accuracy required.

SAMPLE PROBLEM 5/1

Consider the universal beam loaded as shown in the diagram. The vertical loads of 600 N and 900 N are exerted on the universal beam by the two vertical columns, and the vertical reactions R_1 and R_2 provided by the two walls hold the beam in equilibrium. *Since we are interested in finding the resultant force on the beam, the reactions R_1 and R_2 are neglected in the calculation.* That is, resultant forces are calculated using only the *active vectors* present. Draw the free-body diagram, showing only the active forces present (the two vertical loads).

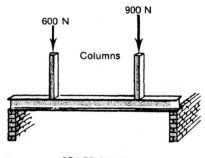

SPACE DIAGRAM

Problem 5/1

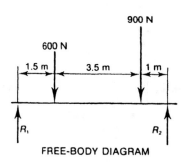

FREE-BODY DIAGRAM

Problem 5/1

ANALYTICAL SOLUTION

Analysis:

The resultant force will have the same effect on the beam as the two given loads (the active forces); that is,

(i) the vertical effect ΣF_V must be the same as the combined vertical effects of the two loads;

(ii) the horizontal effect ΣF_H must be the same as the combined horizontal effects of the two loads;

(iii) the turning effect ΣM about any point must be the same as combined effects of the two loads.

Calculation:

ΣF_V Assume $\downarrow^+$

$$\Sigma F_V = + 600 + 900$$
$$\therefore \Sigma F_V = 1500 \text{ N} \downarrow$$

ΣF_H Assume $\overset{+}{\longrightarrow}$

Since there are no horizontal forces, ΣF_H must be zero, that is, the resultant force is purely vertical.

ΣM Assume $\overset{+}{\circlearrowleft}$ moments about the left-hand support (A)

Let the resultant $(R = 1500 \text{ N from above})$ act x metres from the left-hand support. The moment of the resultant must equal the sum of the moments of the two loads.

$$(600 \times 1.5) + (900 \times 5) = 1500 \ x \ (\text{N m})$$
$$\therefore 900 \ \ + \ \ 4500 \ \ \ = 1500 \ x \ (\text{N m})$$
$$\therefore x = 3.6 \text{ m}$$

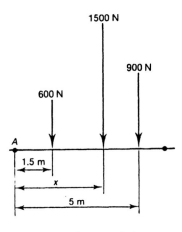

Problem 5/1

Result:
The resultant force on the beam is 1500 newtons acting vertically downwards 3.6 metres from the left-hand supporting wall. ■

SAMPLE PROBLEM 5/2

The timber beam supported as shown has loads applied to it by means of the two props positioned as indicated in the space diagram. Determine the resultant force on the beam. The free-body diagram incorporates only the active vectors.

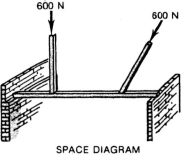

SPACE DIAGRAM

Problem 5/2

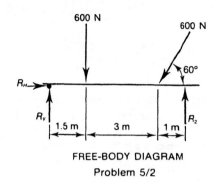

FREE-BODY DIAGRAM
Problem 5/2

ANALYTICAL SOLUTION

Analysis:

Since the resultant force is required, the reactions at the two walls are not considered (however, it should be noted that, because of the wall design, the left-hand support can provide a horizontal reaction while the right-hand wall cannot).

Since one force (load) is angled, it is necessary to find its horizontal and vertical components. Let these be F_H and F_V.

$$F_H = 600 \cos 60° = 300 \text{ N}$$
$$F_V = 600 \sin 60° = 520 \text{ N}$$

Now redraw the force diagram replacing the angled force by its components.

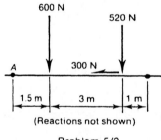

(Reactions not shown)

Problem 5/2

Calculation:

Since the loads have vertical and horizontal components, the resultant force will also have a vertical and horizontal component.

$$\Sigma F_V = 600 + 520$$
$$= 1120 \text{ N } \downarrow^+$$
$$\Sigma F_H = 300 \text{ N } \leftarrow^+$$

Thus, the resultant has components of 1120 N acting vertically down and 300 N acting horizontally to the left.

To find its position use ΣM. Assume $^+\circlearrowright$

The sum of the moments of the components of the resultant about any point must equal the sum of the moments of components of the applied loads about that same point. Take moments about the left-hand support (A).

$$\Sigma M \text{ for loads} = (600 \times 1.5) + (520 \times 4.5) + (300 \times 0)$$
$$= 900 + 2340$$
$$= 3240 \text{ N m clockwise}$$

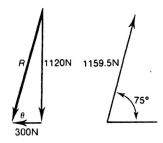

Problem 5/2

Let resultant act x metres from the left-hand support (A).

$$\Sigma M \text{ for resultant} = (1120 \times x) + (300 \times 0)$$
$$= 1120x \text{ N m}$$

Thus, $1120x = 3240 \text{ N m}$

$$\therefore x = \frac{3240}{1120}$$
$$\therefore x = 2.89 \text{ m}$$

Thus, the resultant acts 2.89 metres from the left-hand support (A).

The components of this resultant can now be combined.

$$R = \sqrt{1120^2 + 300^2}$$
$$= 1159.5 \text{ N}$$
$$\operatorname{Tan}\theta = \frac{1120}{300}$$
$$= 3.7333$$
$$\therefore \theta = 75°$$

Result:

The resultant force on this beam has a magnitude of 1159.5 N, it acts 2.89 metres from the left-hand support, and is inclined at 75° to the beam, acting downwards and to the left. ∎

The types of analytical techniques used to solve Sample Problems 5/1 and 5/2 are obviously somewhat involved and become tedious where many angled forces are involved. This is well illustrated in Sample Problem 5/3. However, the method does provide accurate results, and can be summarised in the following steps:

  (i) each active force is analysed into its horizontal and vertical components (or components in selected x and y directions mutually at right angles);

 (ii) the vertical component of the resultant is found from ΣF_V (or the y component from ΣF_y);

(iii) the horizontal component of the resultant is found from ΣF_H (or the x component from ΣF_x);

(iv) the position (point of application) of the resultant is found using ΣM;

 (v) the components obtained in (ii) and (iii) are combined to find the resultant force and its line of action.

SAMPLE PROBLEM 5/3

A bar is loaded as shown. Find the resultant force acting.

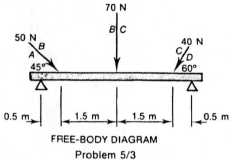

FREE-BODY DIAGRAM
Problem 5/3

Analysis:
In this problem all the relevant information will be tabulated systematically, this being the best procedure to follow for complex problems. The method summarised in the previous five points will be followed in constructing this table.

Calculations:

Force		$F_H = F\cos\theta$		$F_V = F\sin\theta$		$M_0 = F_V \times d = F\sin\theta\, d$	
N	θ^*					about LH support	
AB	50	45°	$50\cos 45° =$	35.4	$50\sin 45° =$	35.4	$35.4 \times 0.5 = +\ 17.7$
BC	70	90°	$=$	0	$=$	70	$70 \times 2 = +140$
CD	40	60°	$-40\cos 60° = -20$		$40\sin 60° =$	34.6	$34.6 \times 3.5 = \quad 121.1$
R			$\Sigma F_H = +15.4$		$\Sigma F_V = +140.0$		$\Sigma M = \quad 278.8$
							$\therefore 140x = \quad 278.8$
							$\therefore x = \quad 1.99\,\text{m}$

*Note that θ is the angle of each force to the beam.

Thus, the resultant has the components
$$R_H = 15.4\ \text{N} \rightarrow$$
$$R_V = 140\ \text{N} \downarrow$$

These can now be combined:

Thus
$$R = \sqrt{140^2 + 15.4^2}$$
$$R = 140.8\ \text{N}$$

and
$$\tan\theta = \frac{140}{15.4}$$
$$= 9.0909$$
$$\therefore \theta = 83.7°$$

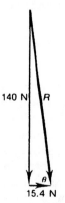

Problem 5/3

Result:
The resultant force on the beam is 140.8 newtons 1.99 metres from the left-hand support acting at 83.7° downwards to the right. ∎

The foregoing analytical method is obviously laborious, particularly when many inclined forces are acting. A graphical solution is much simpler and whereas its "order of accuracy" is not as high as the analytical method, it is much easier to do, and the risk of a major error is small.

Once the method is understood, accuracy well within the limits of normal engineering practice is attainable, provided reasonable drawing procedures are followed.

The graphical method used relates the space diagram and the force polygon with strings (or rays) from a common pole point. The diagram resulting from this relationship is known as a *funicular polygon*. (The word *funicular* is derived from the Latin and French, meaning cord, string or cable, etc.).

THE FUNICULAR POLYGON

When forces are concurrent, it is a simple matter to determine a resultant or an equilibrant graphically by drawing the force polygon to scale (refer to Figure 5.2).

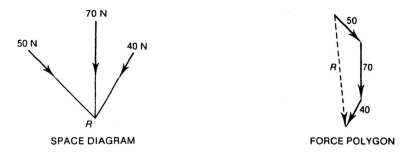

Figure 5.2

When the forces are not concurrent, the graphical methods of determining a resultant (or equilibrant) are only slightly more complex. The force polygon is drawn to scale as before, and the magnitude, direction and sense of the closing vector are measured off (refer to Figure 5.3). It is then only necessary to discover *where this resultant force AD acts.*

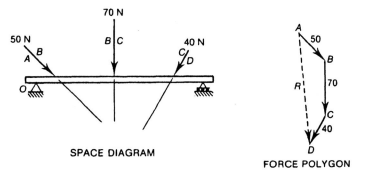

Figure 5.3

Graphically resolve each force into concurrent components at any pole point *O*. The resultant will then have components concurrent at the same point (refer to Figure 5.4 (a)).

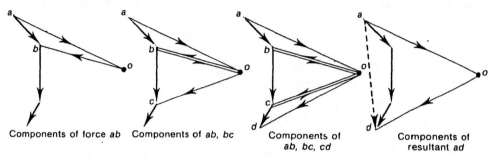

Components of force *ab*   Components of *ab*, *bc*   Components of *ab*, *bc*, *cd*   Components of resultant *ad*

Figure 5.4(a)

Transferring the directions of each of these components to the free-body diagram will provide a point on the line of action of the resultant. As its magnitude and direction are already known from the force polygon, it can now be drawn in to scale (refer to Figure 5.4 (b)).

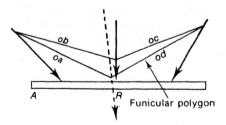

Funicular polygon

Figure 5.4(b)

The final diagram is known as the *funicular polygon*. It is a quick, simple graphical method of solving otherwise tedious problems. Note that the answer obtained here is the same as the answer for Sample Problem 5/3. The resultant, when scaled off, is ≈ 141 N and acts ≈ 2 metres from *A*, the left-hand end.

SAMPLE PROBLEM 5/4

Determine the resultant of the two given forces and locate its line of action.

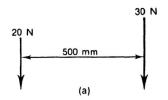

(a)

Procedure:
(1) Draw a free-body diagram to scale.

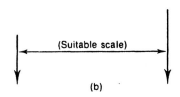

(Suitable scale)

(b)

(2) Name the forces using Bow's notation. (Place a capital letter on either side of each force in free-body diagram.)

(c)

(3) Draw the force polygon to scale—it is a straight line. (*Note:* the resultant is *ac* with a magnitude of 50 N acting vertically downwards.)

(d)

(4) Resolve each force into components concurrent at any point *o*. (*Note: ob* and *bo* are equal and opposite component forces.)

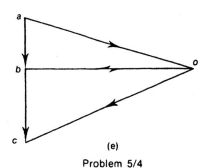

(e)

Problem 5/4

(5) Transfer the directions of each of these components to the free-body diagram by drawing lines parallel to the components of each force.

Make them intersect somewhere on the line of action of their own original force.

Lines ∥ to *ao* and *bo* intersect on *AB*

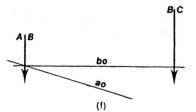

(f)

Lines ∥ to *bo* and *co* intersect on *BC*

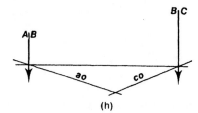

(g)

Lines ∥ to *ao* and *co* intersect on *AC*

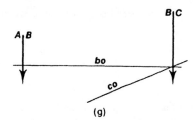

(h)

Problem 5/4

But *AC* is the resultant.

Therefore the resultant passes through the point where *ao* and *co* intersect. It can now be drawn in, and its location scaled from the drawing.

Result:
The resultant is a vertical force of 50 N acting vertically down at a point 300 mm to the right of force *AB*.

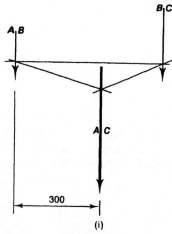

(i)

Problem 5/4

The accuracy of the result depends largely on the accuracy of the drawing methods employed. Thus, if reasonable accuracy is needed, due attention should be paid to accurate drawing techniques. If absolute accuracy is required, then analytical methods must be used, and the solution calculated.

EQUILIBRANTS

The equilibrant force is always equal and opposite to the resultant force. Thus, once the resultant of a non-concurrent force system has been found, the equilibrant (or balancing) force is the force having the same magnitude, the same direction and the same point of application as this resultant, *but the opposite sense*.

CONDITIONS OF EQUILIBRIUM

While it is obvious that the *equilibrant force* always maintains the force system from which it was derived in equilibrium, it is equally obvious that many force systems act on bodies that are *supported* in such a way that equilibrium is maintained by the action of these supports. For example, a house is supported on foundations, a bridge may be supported on piers, and the trapdoor in Figure 5.1 (k) is (partly) supported on a hinge along one of its edges.

From the knowledge gained in previous chapters (particularly Chapters 3 and 4) you can see that the equilibrium of *any force system* is dependent on two things:
(a) that there is no unbalanced force present, $\Sigma F = 0$;
(b) that there is no unbalanced moment present, $\Sigma M = 0$.
These requirements are expressed mathematically as

$$\Sigma F_x = 0$$
$$\Sigma F_y = 0$$
$$\Sigma M = 0$$

where x and y are the principal directions in which components of all forces present, including reaction forces, act.

These formal statements of the conditions required for force equilibrium can be applied to any given force system. For example, $F_x = 0$ and $F_y = 0$ were used in problems involving concurrent forces in Chapter 3. They are easily referred to the three basic combinations of forces as set out below.

The Two-force System

If there are only two forces acting on a body, for equilibrium the forces must be equal and opposite, and have the same line of action (i.e. to be collinear). Only then will $\Sigma F = 0$. (Obviously there is no unbalanced moment.)

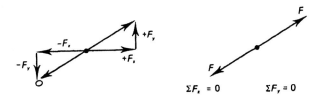

Figure 5.5

The Three-force System

If there are three forces acting on a body in equilibrium, the forces must be either concurrent (Figure 5.6 (a)) or parallel (Figure 5.6 (b)) (and the resultant of any two of the forces must be equal and opposite to the other force).

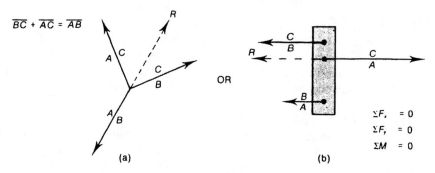

(a)   OR   (b)

Figure 5.6

The Four-force System

If there are four forces acting on a body in equilibrium, the resultant of any two of the forces must be equal, opposite and collinear with the resultant of the other two forces. (This principle can be extended to include any number of forces.)

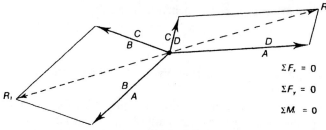

Figure 5.7

REACTIONS AT SUPPORTS

In order to analytically determine the reactions at the supports, the basic equations of equilibrium

$$\Sigma F_x = 0$$
$$\Sigma F_y = 0$$
$$\Sigma M = 0$$

can be applied to a wide range of problems involving supported structures subject to systems of non-concurrent forces. This is true because the reactions at the supports are the forces providing for the *equilibrium* of the body being supported.

A number of different types of supports are commonly used in structures. Some of these supports can provide a reaction force in one direction only, while others can provide a reaction in *any* direction. Common types of supports are shown in Figure 5.8.

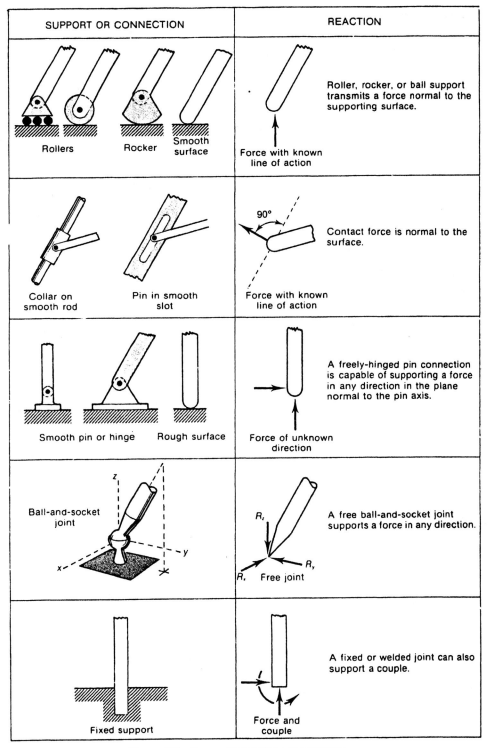

Figure 5.8 Types of reactions at supports.

SAMPLE PROBLEM 5/5

A universal beam is loaded as shown. Determine the reactions at the supports .

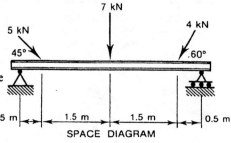

SPACE DIAGRAM

Problem 5/5

ANALYTICAL SOLUTION

Analysis:

The method involves

(i) drawing a free-body diagram showing all the active and reactive forces present. Note that the hinge support can provide a reaction in any direction (that is, a reaction having a horizontal as well as a vertical component) while the roller support can provide only a reaction at 90° to the direction of the roller surfaces (that is, a vertical reaction only in this instance);

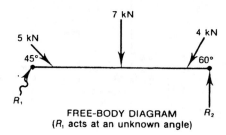

FREE-BODY DIAGRAM
(R_1 acts at an unknown angle)

Problem 5/5

(ii) resolving all angled forces into horizontal and vertical components, including the unknown reaction R, at the hinge support, which is given the components R_H and R_V.

(iii) Using $\Sigma F_V = 0$, $\Sigma F_H = 0$ and $\Sigma M = 0$, to calculate the components of each reaction. These equations must be satisfied since the beam is *in equilibrium* under the influence of the total force system present.

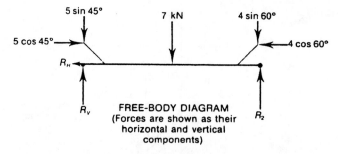

FREE-BODY DIAGRAM
(Forces are shown as their
horizontal and vertical
components)

Problem 5/5

Calculation:

$\Sigma F_H = 0$ Assume $\xrightarrow{+}$

$$- R_H + 5 \cos 45° - 4 \cos 60° = 0$$
$$\therefore R_H = 1.54 \, \text{kN (the fact that the answer}$$

is positive indicates that
our choice of direction
was correct).

$\Sigma M = 0$ Assume $\circlearrowright$ moments about left-hand support. (Note that all horizontal components have zero moments.)

$$+ (R_V \times 0) + (0.5 \times 5 \sin 45°) + (2 \times 7) + (3.5 \times 4 \sin 60°) - 4 R_2 = 0$$
$$R_2 = 6.97 \, \text{kN} \uparrow$$

$\Sigma M = 0$ Assume $\circlearrowright$ moments about right-hand roller support. (Again, note that all horizontal components have zero moments.)

$$(R_2 \times 0) - (4 \sin 60° \times 0.5) - (7 \times 2) - (5 \sin 45° \times 3.5) + 4 R_V = 0$$
$$\therefore R_V = 7.03 \, \text{kN} \uparrow$$

$\Sigma F_V = 0$ Assume $\uparrow^{+}$ (This is a check only.)

$$5 \sin 45° + 7 + 4 \sin 60° - R_V - R_2 = 0$$
$$R_V + R_2 = 14$$
$$7.03 + 6.97 = 14 \qquad \text{(check is correct)}$$

Thus,
$$R_2 = 6.97 \, \text{kN} \uparrow$$

R_1 has the components
$$R_H = 1.54 \, \text{kN} \leftarrow$$
$$R_V = 7.03 \, \text{kN} \uparrow$$

Combining the components of R_1, using a triangle of forces

$$R_1 = \sqrt{7.03^2 + 1.54^2}$$
$$R_1 = 7.2 \, \text{kN}$$

$$\tan \theta = \frac{7.03}{1.54}$$
$$= 4.56$$
$$\therefore \theta = 77.6°$$

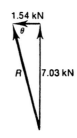

Problem 5/5

Result:
The left-hand reaction is a force of 7.2 kN at 77.6° to the horizontal acting upwards to the left, and the right-hand reaction is a vertically upward force of 6.97 kN.

   Note: The reaction R is often expressed in terms of its two rectangular components, in which case the answer can be written

$$R_1 \begin{cases} R_H = 1.54 \, \text{kN} \\ R_V = 7.03 \, \text{kN} \uparrow \end{cases}$$
$$R_2 \qquad = 6.97 \, \text{kN} \uparrow$$

GRAPHICAL SOLUTION
(1) Draw the free-body diagram to scale showing all active forces and the reactions. Note that while the direction of the arrow at the roller support can be shown, the reaction at the hinge is completely unknown. (refer to Diagram (a)). Label all forces using Bow's notation.
(2) Construct the force polygon using the forces *AB*, *BC*, and *CD* and the direction of the vertical reaction *DE* (refer to Diagram (b)).
(3) From a selected origin *o* draw the rays *oa, ob, oc* and *od* on the force polygon.

(4) Superimpose these rays on to the scaled free-body diagram, starting with the ray *oa*. Since *oa* has to pass through a point on the line of action of the unknown reaction *EA*, it must be drawn through *the centre of the hinge pin* as shown in Diagram (a).

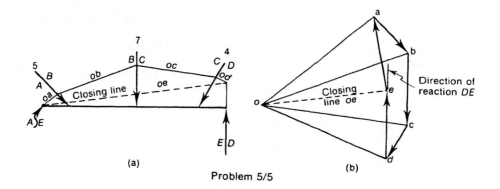

(a)　　　　　　　　(b)

Problem 5/5

(5) Determine the closing line *oe* of the funicular polygon and superimpose it back on to the force polygon. Where *oe* interesects with the direction (line of action) of the vertical reaction *DE* gives the point *e* and the other reaction *EA* can now be drawn in.

Result:
By direct measurement from the force polygon
　　　　　reaction at hinge = 7.2 kN at approximately 78° to the horizontal
　　　　　　　　　　and acting to the left, *i.e.* 78° ＼ ;
　　　reaction at roller = 7 kN vertically upward.
■

SAMPLE PROBLEM 5/6

Determine analytically and graphically the re-actions at the supports of the beam loaded as shown in Sample Problem 5/1.

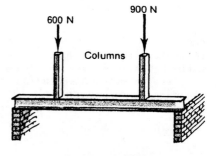

SPACE DIAGRAM

Problem 5/6 (a)

ANALYTICAL SOLUTION
Analysis:
Since there are no horizontal or inclined loads, neither wall has to provide a horizontal reaction component. That is, *both reactions are vertical.*

The analytical solution depends upon the use of the equations $\Sigma F_H = 0$, $\Sigma F_V = 0$ and $\Sigma M = 0$, while the funicular polygon is used to obtain the reactions graphically. (Both solutions, of course, depend upon the fact that the beam is *in equilibrium*.)

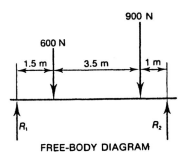

FREE-BODY DIAGRAM

Problem 5/6 (b)

Calculation:

Consider $\Sigma F_H = 0$. Since there are no horizontal forces or components, this equation does not need to be used (even though, of course, it is *satisfied* by the system of forces present).

Consider $\Sigma M = 0$. If moments are taken about the left-hand reaction (R_1), R_1 itself will have a zero moment and is thus eliminated from the calculation.

$\Sigma M = 0$ About R_1 ↺
$$(R_1 \times 0) + (600 \times 1.5) + (900 \times 5) - 6R_2 = 0$$
$$\therefore R_2 = 900 \text{ N}$$

Now, consider moments about the right-hand reaction, R_2

$\Sigma M = 0$ About R_2 ↺
$$(R_2 \times 0) - (900 \times 1) - (600 \times 4.5) + 6R_1 = 0$$
$$\therefore R_1 = 600 \text{ N}$$

Consider $\Sigma F_V = 0$ (This is a useful check on R_1 and R_2.) Assume +↑
$$600 + 900 - R_1 - R_2 = 0$$
$$600 + 900 - 600 - 900 = 0$$
$$\therefore R_1 \text{ and } R_2 \text{ are correct}$$

Result:

The left-hand reaction R_1 is 600 newtons acting vertically upward and the right-hand reaction is 900 newtons also acting vertically upward.

GRAPHICAL SOLUTION

(1) Draw the free-body diagram to scale showing all the active forces and the directions of the reactions. Label all forces using Bow's notation (refer to Diagram (c)).

(2) Construct the force polygon *abc* and, from a selected point *o*, draw in the rays *ao, bo, co* (refer to Diagram (c) below). The force polygon is a straight line and the point *d* must obviously lie somewhere between *e* and *a*.

(3) Superimpose the rays *ao, bo* and *co* on to the free-body diagram (Diagram (c)) beginning with the ray *ao*, which must commence from a point on the line of action of the force *DA*.

(4) The closing line *od* of the resulting funicular polygon is thus apparent; this is transferred on to the force polygon (Diagram (c)) where it coincides with the ray *ob*. This is an uncommon situation.

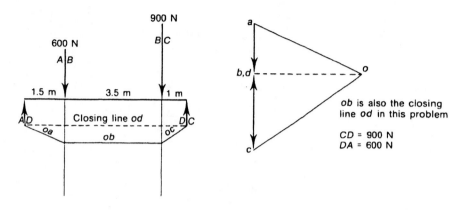

Problem 5/6 (c)

Result:

The reactions can thus be read from the completed force polygon as

$$R_1 = DA = 600 \text{ N}$$
$$R_2 = CD = 900 \text{ N}$$

These are the same answers as were previously obtained analytically.   ■

As shown in Chapter 2, the bathroom scales actually measure force, but are calibrated in mass units for convenience. So are the kitchen scales. Recalibrate a pair of bathroom or kitchen scales to read in force units (newtons) by multiplying each kilogram graduation by g ($g = 9.806\,605$ standard). You could perhaps use $g = 10$ if that seems easier, and this is probably within the order of accuracy of the scales anyway. Add knife-edge supports and a lightweight beam as shown in the diagram. If you now adjust the scales to read zero, you can ignore the weight-force of the beam and the knife-edge supports.

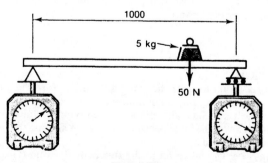

Figure 5.9

By subjecting the beam to various weight-forces (a 1 kg mass exerts about 10 newtons force) the forces exerted on the supports can be read directly from the scales. These forces are the vertical reactions at each support, pushing back up on the beam, and maintaining equilibrium ($\Sigma F_v = 0$).

Uniformly distributed loads (UDL) can be simulated with regularly spaced small masses, or produced accurately with a continuous strip of steel or lead.

Determine analytically, and check experimentally with the equipment described above, the reactions at the supports in the situations as shown in the Figure 5.10.

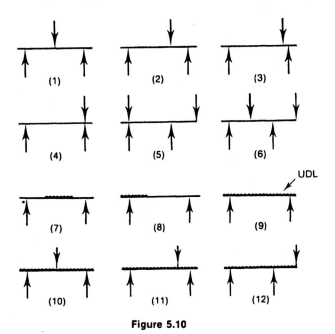

Figure 5.10

The calculated reactions of the Sample Problem 5/6 can readily be verified using this experimental set-up and some loads (weights) of suitable sizes.

SAMPLE PROBLEM 5/7

The universal beam of type 250UB34 located in the ceiling of a building is loaded as shown. The designation 250UB34 means that the universal beam (UB) has a depth of 250 mm and a mass of 34 kg per metre. *Thus the weight-force of this beam is 34 × 6 × 9.8 = 1999.2 N. We will round this out to 2 kN.* This weight-force acts through the centre of mass of the beam as indicated on the diagram.

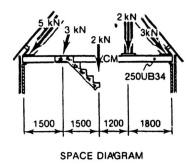

SPACE DIAGRAM

Problem 5/7 (a)

Modifications to the supports of this beam are planned, and it has been suggested that the beam could perhaps be temporarily supported on one prop. Determine graphically the proper location for such a prop, find its attitude, and the probable force it would be required to carry (i.e. completely determine the equilibrant).

Procedure:

(1) Draw a free-body diagram to scale.

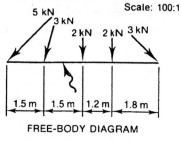

Scale: 100:1

FREE-BODY DIAGRAM

Problem 5/7 (b)

(2) Name the forces using Bow's notation.

Problem 5/7 (c)

(3) Draw a force polygon to scale.

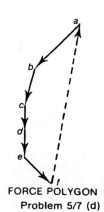

FORCE POLYGON

Problem 5/7 (d)

Note: Resultant is *af*
    Equilibrant is *fa*
    Its magnitude is 13.2 kN
    Its attitude is 82°
    Its line of action is yet to be determined.

(4) Resolve each force in this force polygon into two components, concurrent at any point *o*.

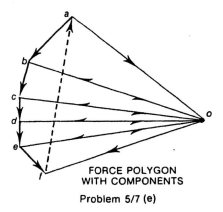

FORCE POLYGON
WITH COMPONENTS

Problem 5/7 (e)

(5) Transfer the attitudes of these components to the free-body diagram by drawing lines parallel to the components of each force. Make them intersect somewhere on the line of action of their original force (*ao* and *bo* intersect on *A B*). It is often convenient to extend the lines of action beyond the free-body diagram for clarity. (Refer to Diagram (f).)

A systematic procedure must be followed. It is important to follow the notation order. It is essential that the lines drawn parallel to each component start and finish on the lines of action of the forces to which they relate. For example, the line parallel to *co* starts on force *BC* and stops on force *CD*.

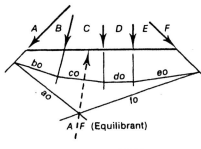

Problem 5/7 (f)

(6) The intersection of *ao* and *fo* locates a point on the line of action of the equilibrant *FA*. As the magnitude has already been discovered from the force polygon, the equilibrant can now be drawn in position on the free-body diagram, and its location scaled off.

Result:
The single prop would need to be placed 2.5 m from the left wall, at an angle of approximately 80° to the beam. The compressive load in the prop would be about 13 kN. ■

SAMPLE PROBLEM 5/8

(Refer to the previous problem.)

Due to difficulty in ensuring the ability of a single prop to restrain the horizontal component of the load from the staircase, which may alter as the prop takes the load, it is decided to use two props (or toms), one of which will be braced to enable it to accept any likely horizontal component. They are to be placed 0.5 metres from each wall. Determine graphically the load in each prop.

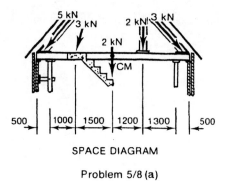

SPACE DIAGRAM

Problem 5/8 (a)

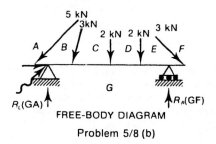

FREE-BODY DIAGRAM

Problem 5/8 (b)

Analysis:

In this case there are two unknown forces. One support cannot accept horizontal forces, therefore the line of action of its force (R_R) must be vertical only. The other support (R_L) is braced and so can accept horizontal forces. This is awkward, as it makes its line of action unknown, and the force polygon cannot be completed without additional information.

A suitable funicular polygon can provide the extra information needed.

Procedure:

The force polygon is drawn as far as the information will permit.

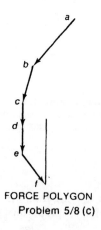

FORCE POLYGON

Problem 5/8 (c)

Concurrent components are added to this incomplete force polygon.

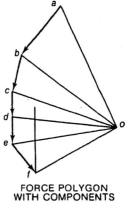

**FORCE POLYGON
WITH COMPONENTS**

Problem 5/8(d)

From this, a funicular polygon is drawn on the free-body diagram, starting at the only known point on the line of action of the awkward force—the *fixed support*. The final line of this polygon, *go*, will close the polygon back to the starting point—the fixed support.

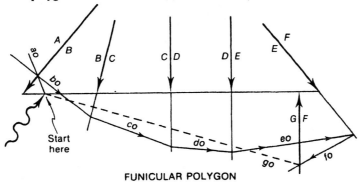

FUNICULAR POLYGON

Problem 5/8 (e)

This line will be parallel to the missing component on the incomplete force polygon. This component can now be drawn, so locating the point *g* on the vertical force *FG*.

The angled force *GA* can now be drawn to complete the force polygon, making it possible to measure the magnitude of both forces, and the attitude of the reaction at the fixed support (R_L).

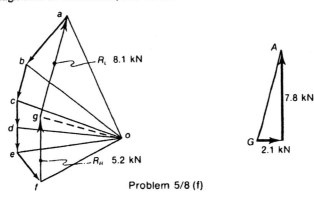

Problem 5/8 (f)

Result:

The right-hand prop, R_R, will carry a vertical force of 5.2 kN. The left-hand prop, R_L, will have a force of 8.1 kN at 75° to the beam. (Resolving this into its horizontal and vertical components shows a vertical force of 7.8 kN and a horizontal force of 2.1 kN to the right.)   ■

SAMPLE PROBLEM 5/9

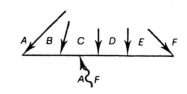

Problem 5/9 (a)

To determine analytically

(i) the position and attitude of the one prop required to support the universal beam discussed in Sample Problem 5/7 and to calculate the load that this prop must carry (i.e. to determine the equilibrant of the given force system);

(ii) the loads in each of the two props referred to in Sample Problem 5/8 (that is, to determine the reaction at the two given supports, only one of which can provide a horizontal component).

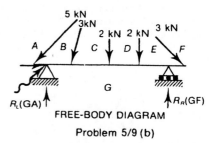

FREE-BODY DIAGRAM
Problem 5/9 (b)

Both calculations can be carried out simultaneously up to a point using a table of calculations based particularly upon the type of approach used in Sample Problems 5/5 and 5/6.

Analysis:
The method involves the following procedure.

(i) The drawing of the two free-body diagrams (as shown in Diagrams (a) and (b)) which must include the unknown forces, in one case the equilibrant *FA*, and in the other case the unknown reactions *FG* and *GA*. The reaction *GA* is also represented by its horizontal and vertical components, labelled R_H and R_V.

(ii) The resolution of all active forces into their horizontal and vertical components after suitable sign conventions have been adopted. In this case, we have assumed $\pm\!\!\rightarrow$ and $+\!\!\uparrow$. (Results are shown in columns 2 and 3 of the Table.)

(iii) The determination of the position of the resultant (or the equilibrant) using ΣM. The moment of the vertical component of the resultant equals the sum of the moments of the vertical components of all the active forces (loads). This is shown in column 4 of the Table.

(iv) In the case of the reaction forces at the supports we know that the resultant horizontal component is to be taken up by the left-hand support. This component already appears in our Table ($\Sigma F_H = -2.1$).

(v) By taking moments about each point of support in turn, the vertical component for each is found, since $\Sigma M = 0$.

Table of Results

Force		$F_H = F\cos\theta$ (kN)	$F_V = F\sin\theta$ (kN)	$M_0 = F_y \times d\,(F\sin\theta \times d)$ (kNm)
kN	θ	$\xrightarrow{+}$	$+\downarrow$	$\curvearrowright$ About o (d in metres)
AB	5	50°	-3.2	$+3.8$	$=0$
BC	3	75°	-0.8	$+2.9$	$2.9 \times 1.5 = +\ 4.4$
CD	2	90°	0	$+2$	$2 \times 3\ \ = +\ 6$
DE	2	90°	0	$+2$	$2 \times 4.2 = +\ 8.4$
EF	3	50°	$+1.9$	$+2.3$	$2.3 \times 6\ \ = +13.8$
FA	(Resultant)		$F_H = -2.1$	$F_V = +13$	$13 \times x = +32.6$
					$\therefore x = \quad 2.5\,\text{m}$
AF	(Equilibrant)		$+2.1$	-13	$2.5\,\text{m}$

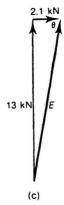

(c)

Problem 5/9

Result (1):
From the Table, the equilibrant has components of 2.1 kN acting to the right and 13 kN acting upward. From a triangle of forces the equilibrant is thus a force of 13.2 kN acting upward to the left at $\approx 81°$ to the beam at a distance of 2.5 metres from the left-hand end.

Calculation:
Calculate vertical components in each of the props (that is, the forces FG and R_V).

$\Sigma M = 0$ Moments about left-hand point of support $\curvearrowright$ (kN m)
$$-(3.8 \times 0.5) + (1 \times 2.9) + (2 \times 2.5) + (2 \times 3.7) - (5R_R) + (5.5 \times 2.3) = 0$$
$$\therefore R_R = 5.2\,\text{N}$$

$\Sigma M = 0$ Moments about right-hand point of support $\curvearrowleft$ (kN m)
$$+(2.3 \times 0.5) - (2 \times 1.3) - (2 \times 2.5) - (2.9 \times 4) + (5R_V) - (3.8 \times 5.5) = 0$$
$$\therefore R_V = 7.8\,\text{kN}$$

Check:
$$\Sigma F_V = 0$$
$$R_V + R_R = 7.8 + 5.2$$
$$\therefore R_V \text{ and } R_R \text{ are correct}$$

Result (2):
Reaction at right-hand prop: 5.2 kN $\uparrow$

Reaction at left-hand prop: $\begin{cases} R_H = 2.1 \ \rightarrow \\ R_V = 7.8 \ \uparrow \end{cases}$

(When combined, reaction provided by left-hand prop is a force of ≈ 8.1 kN whose line of action is inclined at $\approx \angle\ 75°$ to the beam. ∎

REVIEW PROBLEMS

(Use $g = 10$ m/s^2 unless otherwise specified.)

5/10

Can the three forces, the lines of action of which
are shown, be in equilibrium? (Give reasons for
your answer.)

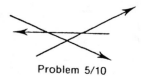

Problem 5/10

5/11

For the parallel force system shown, determine:
 (i) the magnitude;
 (ii) direction; and
(iii) location of the resultant force.

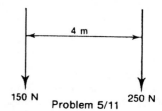

150 N   Problem 5/11   250 N

5/12

Determine graphically the magnitude and location
of the resultant of the three parallel forces shown.

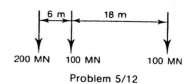

200 MN   100 MN   100 MN

Problem 5/12

5/13

A uniform bar with a mass of 5 kg has vertical
forces applied to it as indicated.
 (i) What additional force must be applied; and
(ii) along what line of action must it be applied to
    hold the bar in equilibrium?

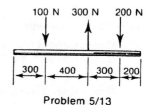

Problem 5/13

5/14

Find the reactions to the beam shown in the
diagram.

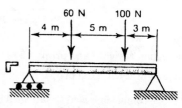

Problem 5/14

5/15
For the beam shown, find the reactions R_L and R_R. The reaction R_L may be expressed either in terms of its horizontal and vertical components, or in terms of its magnitude and direction.

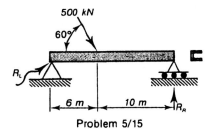

Problem 5/15

5/16
A man raises a pole with a mass of 20 kg by pulling on a rope. Find the tension, T, in the rope and the reaction at A.

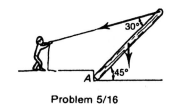

Problem 5/16

5/17
Find the magnitude and direction of the reactive forces at the supports of the beam shown in the diagram.

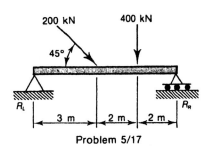

Problem 5/17

5/18
The truss shown supports portion of a drive-in theatre screen. Given that the wind loadings are equal to a 30 kN force on A and a 20 kN force on B as shown, determine the reactions at the supports C and D. Ignore any other forces including the weight-force of the truss itself.

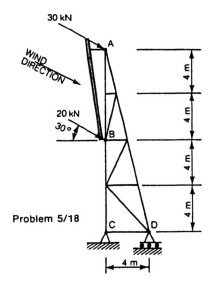

Problem 5/18

5/19

The bell crank shown is in equilibrium. Determine the reaction at the pivot pin, and the force *F*.

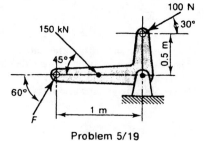

Problem 5/19

5/20

Find the single force which could replace the three given forces without altering the overall effect of the loading on the steel plate shown.

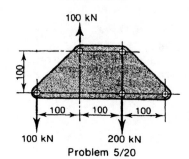

Problem 5/20

5/21

Graphically determine the tension in the cable *B*, and the reaction at the pivot *O*.

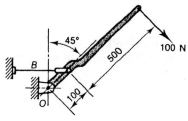

Problem 5/21

5/22

The bar shown in the sketch is not in equilibrium, Determine the magnitude and location of the force that must be added to the given system of forces to achieve equilibrium.

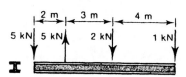

Problem 5/22

5/23

A car with a mass of 1 tonne and a wheelbase of 2.5 m is supported on ramps as shown in the diagram. The rear wheels have the handbrake applied and are chocked for safety. The front wheels are free to roll. Discover the reaction at each wheel, assuming the chocks take no load.

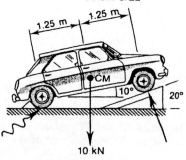

Problem 5/23

5/24
A triangular support frame in a structure is loaded as shown. Graphically and analytically determine the reactions at the supports A and B.

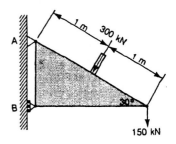

Problem 5/24

5/25
Each of the beams shown below is in equilibrium. However, in each case not all the information is shown. Complete the diagrams by adding the missing information.

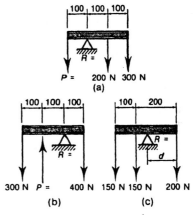

$P =$ 200 N   300 N
(a)

300 N $P =$ 400 N   150 N 150 N   200 N
(b)   (c)

Problem 5/25

5/26
Determine the reactions at B and C for the beam and loading shown. The beam has a mass of 100 kg.

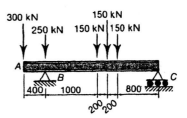

Problem 5/26

5/27
A light crane supports a 3-tonne load as shown in the diagram. Determine the tension, T, in the cable and the reaction at the pivot A.

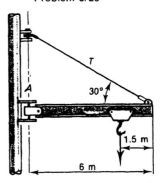

Problem 5/27

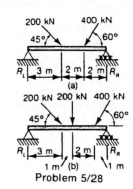

5/28

Graphically determine the magnitude and direction of the forces at the supports of the beams shown in (a) and (b).

Problem 5/28

5/29

The radio tower shown in the diagram is 2 m square and 14 m high and it has a mass of 5 tonnes. It is supported against horizontal wind loads by four guy wires attached 9 m above the central base B. Its effective projected area of 10 m² is subjected to horizontal wind pressure of 600 N/m². When the wind blows from right to left, only guy wire AC is active. Determine the vertical and horizontal reaction components at A and B and the tension T in AC.

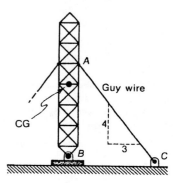

Problem 5/29

5/30

In the pivot drive, the weight of the motor is used to maintain tension in the belt. When the motor is at rest, the tensions T_1 and T_2 may be assumed equal. The mass of the motor is 75 kg and the diameter of the drive pulley is 100 mm. Assuming that the weight of the platform AB is negligible, determine the tension in the belt and the reaction at the pivot C when the motor is at rest.

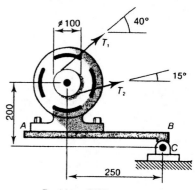

Problem 5/30

5/31

Determine the force F on the pistons of the two hydraulic cylinders A necessary to start raising the load of 1500 kg. Neglect the weights of the bucket and arms compared with the load and take the centre of gravity of the load to be at G. Determine the shear forces in the pivot pins at B.

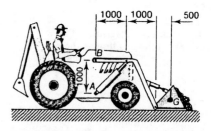

Problem 5/31

6

Couples

A couple can be balanced only by another couple

Forces occur in many different combinations. Some of these combinations arise frequently enough to be given a family name, for example,

collinear forces are forces in one line;
coplanar forces are forces in one plane;
concurrent forces are forces acting at one point.

One particular combination of coplanar forces is of sufficient interest to deserve its own particular name. The combination of two equal and opposite forces with parallel lines of action is known as a *couple*. The plane in which the forces lie is known as the *plane of the couple*, and the perpendicular distance between their lines of action is known as the *arm* of the couple. Common examples of a couple include those shown in Figure 6.1. These are *active* couples. Both forces are active and the action of the couple results in rotation.

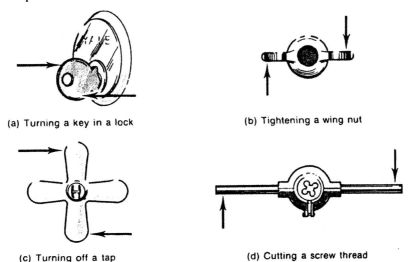

(a) Turning a key in a lock

(b) Tightening a wing nut

(c) Turning off a tap

(d) Cutting a screw thread

Figure 6.1 Active couples.

Many couples are *reactive* and actual rotation does not occur. They result when one force causes a reaction which is not in line with the original active force. Take, for example, the diving board shown in Figure 6.2 (a). For vertical equilibrium, the diver's weight-force must be resisted by an equal and opposite force. The only place at which this reaction can occur is the block. The combined effect of the weight-force of the diver and the vertical reaction at the block is a couple, which acts *on the block* and the block must provide an equal and opposite reactive couple which acts *on the load*. This illustrates the fact that part of the total reaction at the fixed end of any *cantilever* (a rigid beam supported at one end only) is a *moment* or *couple*.

Similarly, a street light exerts a weight-force on the pole, which is resisted exactly by the pole (see Figure 6.2 (b)), and the branch of the tree is held up by the trunk of the tree—two equal and opposite forces, but not in line (see Figure 6.2 (c)). In both of these situations, reactive couples are induced in the supports by the action of the active forces present.

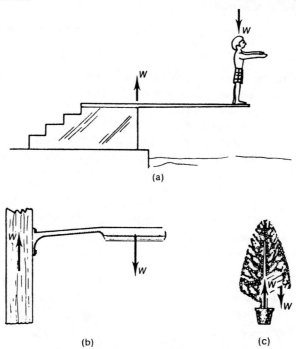

(a)

(b)   (c)

Figure 6.2 Reactive couples.

A couple has the following characteristics.*
(1) *Resultant force.* The resultant force is zero. Therefore, a couple does not tend to produce translation, but only tends to produce *rotation*, i.e.

$$\Sigma F_x = 0 \qquad \Sigma F_y = 0 \qquad \Sigma M \neq 0$$

---

*The following simple definitions may assist you to remember about couples:
(i) *every couple has its moments on a plane;* and
(ii) *the closer the arms of the couple the less the torque.*

(2) *Resultant moment.* The moment of a couple is the product of one of the forces and the perpendicular distance between them, and is the same for all points in the plane. (This is illustrated in Figure 6.3.)

$$M = Fd$$

(3) *Equilibrium.* A couple can only be balanced by another couple of equal and opposite moment, as ΣM will then be zero.

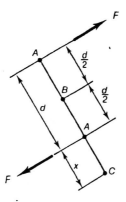

A) $\Sigma M = F \times d$ (other F has no moment about A, as it passes through A)

B) $\Sigma M = F \times \dfrac{d}{2} + F \times \dfrac{d}{2} = F \times d$

C) $\Sigma M = F \times (d+x) - (F \times x) = F \times d$

Figure 6.3   The moment of a couple.

A couple is a *turning* moment, but equivalent turning moments can involve very different forces, both in magnitude and direction. In the case of the street lamp shown in Figure 6.4, the lamp exerts a weight-force f over the distance D, producing a moment of $M = fD$. This *active moment* is resisted by an equal and opposite *reactive moment* produced by the pull of the mounting bolt F and the equal and opposite push of the bottom of the bracket acting over a distance d. Thus, the equivalent reactive moment provided at the mounting is $M = Fd$. As the situation is in equilibrium, these two moments are in *moment equilibrium* and $Fd = fD$. The magnitude of the forces F in the mounting will, however, be very much bigger in magnitude than the weight-force f since the distance d is so much less than the length of the lamp bracket D.

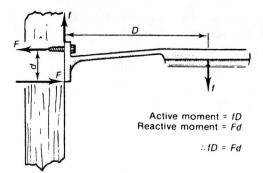

Active moment = fD
Reactive moment = Fd

$\therefore fD = Fd$

Figure 6.4 Active and reactive equivalent moments.

The concept of *equivalent couples* can readily be demonstrated using two pieces of flat board, four spring balances graduated in newtons, some cup hooks and nails and a few lengths of fine wire (refer to Figure 6.5).

The top board is pegged loosely to the baseboard with removable pins. Equal and opposite parallel forces are applied to a pair of hooks A with the two spring balances, thus providing a couple which tends to rotate the board. The magnitude of this couple is calculated from the formula $M = fd$ (where f is the reading on the spring balances and d is the distance between the hooks). A couple of equal magnitude (but opposite sense) is applied to the hooks B using the other two spring balances to provide suitable forces. When these two couples are exactly equivalent, the removable pins may be withdrawn quite easily, and the top board *will not move*.

The experiment can be repeated using cup hooks any distance apart positioned anywhere on the movable board.

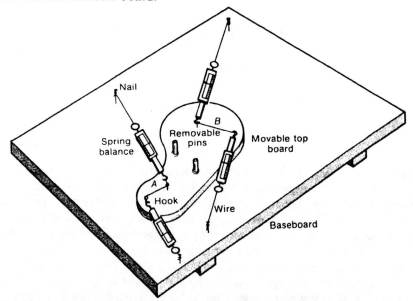

Figure 6.5 Experimental proof of equivalent couples.

RESOLVING A FORCE INTO A FORCE AND A COUPLE

The street-lamp shown in Figure 6.6 (a) exerts a total weight-force f_1 on its mounting bracket.

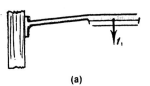

(a)

The bracket exerts an equal force f_2 on the pole (the bracket in fact transfers the total weight-force f_1, to the pole). (See Figure 6.6 (b).)

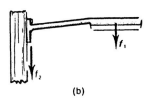

(b)

The pole reacts with a force f_3 on the bracket, this reactive force f_3 being of equal magnitude but opposite in direction to the original weight-force f_1. (See Figure 6.6 (c)). Thus, f_2 now becomes the weight-force on the pole while f_1 and f_3 form a couple of magnitude fD.

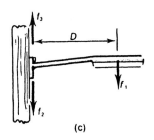

(c)

The original force f_1 (the weight-force of the street lamp) has now been replaced by a force of the same magnitude, acting down the pole (f_2) and the couple, fD, which acts clockwise on the pole. This combination of vectors is known as a *force-couple system* and is written symbolically as shown in Figure 6.6 (d). It is sometimes most useful to replace a force with an equivalent force-couple system when solving problems.

(d)

Figure 6.6

SAMPLE PROBLEM 6/1

The metal plate of mass 10 kg is supported on the two smooth pins as shown. Assuming that these pins are rigid (i.e. do not bend) calculate the resisting couple which they must provide in order to support the plate. Determine also, the shear force in each pin. Assume $g = 10$ m/s^2.

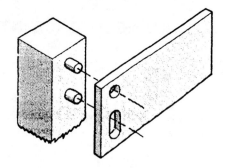

SPACE DIAGRAM

Problem 6/1

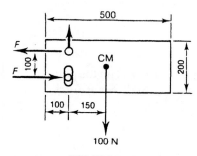

FORCE DIAGRAM

Problem 6/1

Analysis and Calculations:
The weight-force of the plate, acting through its centre of mass, tends to rotate the plate clockwise. If we take moments about the top pin we can discover the magnitude of this moment.

$$M = fd$$
$$= 100\,N \times 0.15\,m$$
$$= 15\,Nm$$

This turning moment must be resisted by an equal and opposite moment provided by the pins, which is, therefore, also 15 Nm.

The horizontal shear force F in the bottom pin can now be found.

$$M = fd$$
$$15\,Nm = F \times 0.1\,m$$
$$F = 150\,N$$

The horizontal shear force in the top pin must also be 150 N ($\Sigma F_H = 0$) but the top pin is also carrying the vertical load of 100 N as well ($\Sigma F_V = 0$) and so the total load is the vector sum of these two forces.

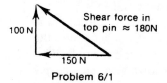

Problem 6/1

Result:
The force-couple provided by the pins is composed of a simple horizontal shear force in the bottom pin of 150 N and a resultant shear force in the top pin of approximately 180 N being the vector sum of the vertical and the horizontal shear. ∎

SAMPLE PROBLEM 6/2

A swing-wing jet fighter pivots its wings in a manner similar to the simplified version shown in the diagram. If the maximum effective force which occurs on the wing is equivalent to 5 MN at the position shown, calculate the force-couple system acting at the pivot (i.e. determine the shear force in the pivot pin and the load on the thrust bearing).

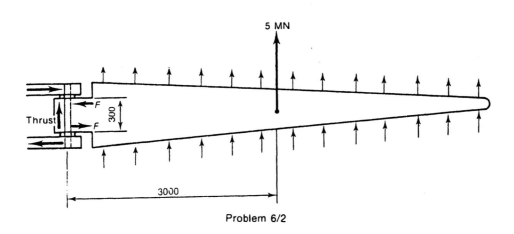

Problem 6/2

Analysis and Calculations:

The forces on the wing tend to rotate it anti-clockwise. The magnitude of this turning moment is

$$M = fd$$
$$= 5 \text{ MN} \times 3 \text{ m}$$
$$= 15 \text{ MN m}$$

The couple acting at the pivot pin must also be 15 MN m ($\Sigma M = 0$) and so the shear forces in the pin can be calculated

$$M = Fd$$
$$15 \text{ MN m} = F \times 0.3 \text{ m}$$
$$F = 50 \text{ MN}$$

The load on the thrust bearing is equal to the total load on the wing, i.e. 5 MN ($\Sigma F_V = 0$).

Result:

The force-couple system at the pivot pin constitutes a force of 5 MN and a couple of 15 MN m, which puts the pivot pin in shear with a force of 50 MN, that is,

■

RESOLVING A FORCE-COUPLE INTO A SINGLE FORCE

The procedure of replacing a force with an equivalent force-couple system is easily reversed.

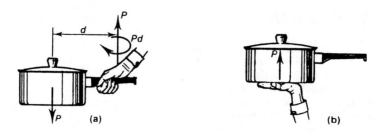

Figure 6.7

Lifting a saucepan of weight-force P by the handle, as shown in Figure 6.7 (a), requires the force P (to provide the vertical lift) and the couple Pd (to prevent it from rotating and thus spilling). Both of these vectors are provided by your hand and arm as shown.

The force-couple system supplied by one hand can easily be replaced with the force P only, provided by the other hand under the saucepan (unless it is hot). This is shown in Figure 6.7 (b).

SAMPLE PROBLEM 6/3

The concrete ledge is cantilevered beyond the supporting wall as shown. The force-couple system produced at the wall face A consists of a vertical force of 300 N and a moment of 135 N m clockwise. Determine the position of the centre of mass, CM, of the ledge.

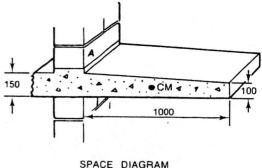

SPACE DIAGRAM
Problem 6/3

Analysis:
The force-couple system at the wall is obviously produced by the weight-force of the ledge acting through its centre of mass, CM. Therefore, the problem is to replace the force-couple system at A with a single force acting through the centre of mass, CM.

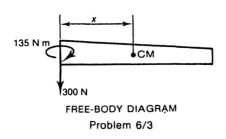

135 N m

x

CM

300 N

FREE-BODY DIAGRAM

Problem 6/3

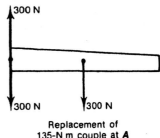

300 N

300 N 300 N

Replacement of
135-N m couple at **A**
by two extra 300-N forces

Problem 6/3

Calculation:
The force acting through CM is obviously 300 N. Let CM be x mm from A.
Moment of 300 N force acting through CM about $A = Fd = 300 \times x$
But this is equal to the moment at A

$$300x = 135 \text{ N m}$$
$$x = \frac{135}{300} \text{ m}$$
$$= 0.45 \text{ m}$$
$$= 450 \text{ mm}$$

Result:
The centre of mass is 450 mm from the wall face A. ■

The ordinary door (Figure 6.8 (a)) is probably one of the best examples of a reactive couple. The reactions at the supports of the door are easily analysed as they obviously occur at definite places, the hinges. Safe design of these supports ensures that either hinge is more than strong enough to carry the total weight-force of the door, or the physical design will be such that the vertical load can only occur on one particular support which has been designed to carry it, for example, the gate pivots shown in Figure 6.8 (b).

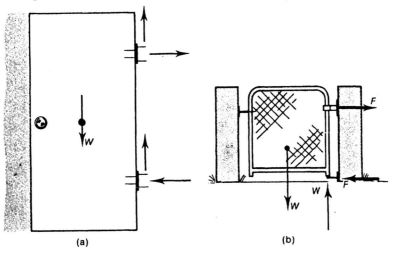

(a) (b)

Figure 6.8 Reactions at hinges and pivots.

Even though the door shown in Figure 6.8 (a) is subjected to several forces of considerable magnitude, it is not rising, falling, moving sideways, or rotating. This is because all the basic conditions for equilibrium are being met. These are:
(a) the sum of all the horizontal forces is zero, or $\Sigma F_H = 0$;
(b) the sum of all the vertical forces is zero, or $\Sigma F_V = 0$;
(c) the sum of the moments about any point in the plane is zero, or $\Sigma M = 0$.

SAMPLE PROBLEM 6/4

A door with a mass of approximately 10 kg is hung on two hinges which are 2 metres apart. Calculate the forces on the hinges. Approximate g to 10 m/s^2.

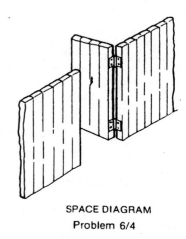

SPACE DIAGRAM
Problem 6/4

Analysis:
Draw the free-body diagram, showing all the active and reactive forces. To simplify the analysis we will assume that the entire vertical reaction is supported by only one hinge. As this is an arbitrary choice, either hinge will do, and the results obtained will apply similarly to both hinges.

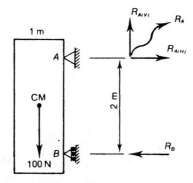

FREE-BODY DIAGRAM

Problem 6/4

The mass of the door will produce a force acting through its centre of mass. This vertically down weight-force will be resisted by an equal and opposite vertical force up. We will assume that this reaction acts through the hinge A. The weight-force acting down and the vertical reaction at A

are equal and opposite forces and constitute a *couple*. This couple can be balanced by an equal and opposite couple and the only place that this equal and opposite couple can occur is at the hinges A and B. As one hinge can only support a horizontal reaction, the horizontal reactions $R_{A(V)}$ and R_B, which provide the reaction couple, are easily calculated.

Calculation:

$$\text{Mass of door} = 10 \text{ kg}$$
$$\text{Weight-force of door} = 10 \text{ kg} \times g$$
$$\approx 100 \text{ N}$$

Vertical component of reaction at A, $R_{A(V)} = 100$ N. $(\Sigma F_V = 0)$

Moment of couple produced by these two forces:
Take moments about hinge A

$$M = Fd$$
$$= 100 \text{ N} \times 0.5 \text{ m}$$
$$= 50 \text{ N m anti-clockwise} \quad .$$

For equilibrium, the moment of the reactive couple must be 50 N m clockwise. As this couple occurs at the hinges, it will again be convenient to take moments about A:

$$M = Fd$$
$$50 \text{ N m} = R_B \times 2 \text{ m clockwise}$$
$$R_B = 25 \text{ N to the left (compression)}$$

Thus,
$$R_{A(H)} = 25 \text{ N to the right (tension)}$$

Result:
The (reaction) forces on the door hinges, expressed in terms of horizontal and vertical components are as follows:

Reaction at top hinge $\left\{ \begin{array}{l} \text{vertical component} = 100 \text{ N} \uparrow \\ \text{horizontal component} = 25 \text{ N} \rightarrow \end{array} \right.$

Reaction at bottom hinge, horizontal component = 25 N ←

The two components of the total reaction at the top hinge could be combined (triangle of forces) into the single angled reaction, if required. ■

SAMPLE PROBLEM 6/5

The diver has a mass of 50 kg and is poised at the end of the light fibreglass diving board of negligible mass. Assuming that he exerts a weight-force of 500 newtons on the end of the board, analyse the forces involved, in the following manner:

(i) resolve the weight-force of the diver into a force and a couple at the roller support B;

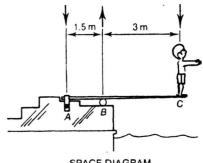

SPACE DIAGRAM

Problem 6/5

(ii) find the force in the anchor strap *A*. (The active couple must be resisted by an equal and opposite couple);

(iii) determine the total load on the roller *B* when it is in the position shown;

(iv) determine the bending moment in the board at the point *B*.

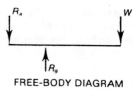

FREE-BODY DIAGRAM

Problem 6/5

Analysis and Calculations:

Draw the free-body diagram (shown above). Since the board has negligible mass, its weight-force can be ignored.

(i) Because the diver is pushing down with a force of 500 N, the roller *B* must be pushing up with a force of 500 N. (In fact, the total upward reaction at the roller is much more than 500 N—this becomes apparent later.) These two forces of 500 N (one up, one down) constitute a couple of 500 N × 3 m = 1500 N m.

At *B*, therefore, there is a force-couple system of 500 N up and 1500 N m clockwise. This is symbolically written as

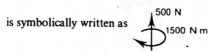

(ii) The clockwise couple induced by the downward force of the diver and the upward force of the roller (1500 N m) is resisted by an equal but anti-clockwise couple induced by the downward force at the anchor strap, and another upward force at the roller. As the moment arm is known (1.5 m) the magnitude of these forces can be determined from $M = Fd$

$$1500 \text{ N m} = F \times 1.5 \text{ m}$$
$$F = 1000 \text{ N}$$

Thus, the vertical force in the anchor strap *A* is 1000 N (tension).

(iii) If the resisting couple has a downward force of 1000 N at *A*, then it must have an upward force at *B* of 1000 N. However, the weight-force of the diver has already induced an upward force of 500 N at *B* so the *total force at B is the sum of these*, i.e. 1000 N + 500 N = 1500 N.

(iv) The bending moment is a couple and is equal to the couple trying to bend the board. At the roller *B* this has already been shown to be 1500 N m clockwise.

Result:

(i) The force-couple system at the roller which replaces the effect of the weight-force of the diver is

500 N
1500 N m

(ii) The force in the anchor strap *A* is 1000 N (tension).
(iii) The total vertical reaction at the roller *B* is 1500 N vertically upwards.
(iv) The moment at the roller is 1500 N m clockwise. ■

(*Note:* The total reaction at the roller and the tension in the strap could readily have been found by taking moments, in turn, about *A* (the strap) and *B* (the roller).)

REVIEW PROBLEMS
(Use *g* = 10 m/s² unless otherwise specified.)

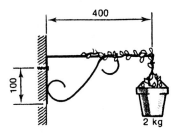

6/6
Determine the forces in the mounting screw for
the pot-plant bracket shown in the diagram.

Problem 6/6

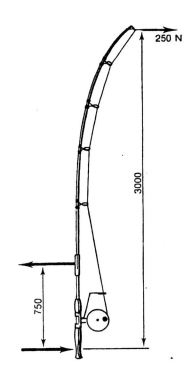

6/7
Fred's "gone fishin'" and hooked a whopper on his
3 m rod. Fred's strength and imagination estimate
the fish to be pulling on the line with a 250 N force.
Fred's left hand is placed on the bottom of the rod
and right hand 750 mm from the bottom of the
rod. What is the force in his left arm while playing
the fish? What is the force in his right arm?

Problem 6/7

6/8
A screw thread is cut with a hand stock and die.
What is the torque or moment being applied if
each hand exerts a force of 100 N and the effective
length of each handle to the axis of the screw is
900 mm?

Problem 6/8

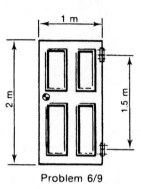

Problem 6/9

6/9
The cottage door has a mass of approximately 20 kg. Determine the loads on the hinges.

6/10
A steel I-beam is cantilevered as shown. If the couple at the support (the bending moment) is not to exceed 100 kN m, calculate the maximum load which can be supported at the end of the beam. (Neglect the mass of the beam).

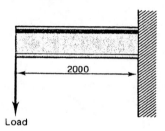

Load

Problem 6/10

6/11
The street lamp has a mass of 12 kg and the bracket which supports it has a mass of 6 kg.
(a) Describe the force-couple at the pole.
(b) Determine the minimum tension in the top mounting bolt of the bracket.

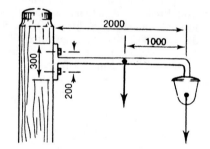

6/12
A wet blanket with a mass of approximately 20 kg is hanging on the outermost line of the rotary clothes line as shown.
(a) Resolve the weight-force produced by this blanket into a force-couple system at the base of the column.
(b) What is the bending moment (i) at ground level? (ii) at peg level?

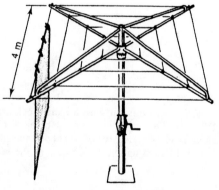

Problem 6/12

6/13

The long-handled shovel is used to lift 10 kg of sand.

(a) Determine the force in each hand for the position shown. What is the moment of the couple exerted by the hands?

(b) If the left hand is moved to a position halfway along the shovel (distance 0.7 m) what is the force in each hand? What is the couple?

(c) Describe the position of the hands for easier · shovelling.

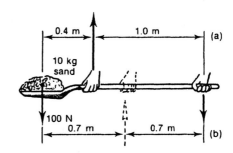

Problem 6/13

6/14

A 50-N pull is required to operate a certain bottle-opener. Describe the force-couple system acting on the bottle top (at the point A).

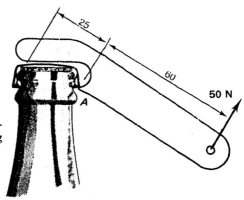

Problem 6/14

6/15

A workbench seat is held in the position shown by a vertical bar and supports a 75-kg man.

(a) Determine the reaction at *A*.

(b) If the inside diameter of the collar is slightly larger than the bar, the collar will bear only at points *B* and *C*. Determine the magnitude of the horizontal and vertical forces developed at *B* and *C*.

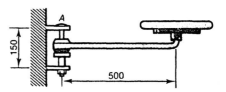

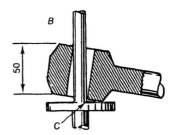

Problem 6/15

6/16
A fish pulls on the line with a force of 60 N. Describe the forces the fisherman will exert while playing the fish.

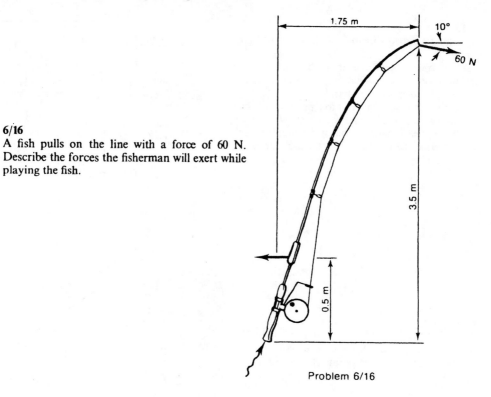

Problem 6/16

6/17
Replace the three forces acting on the gear by an equivalent force-couple system at O.

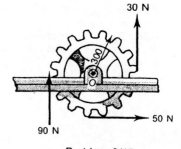

Problem 6/17

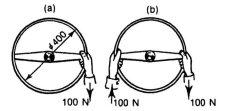

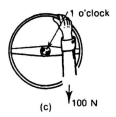

Problem 6/18

6/18

Determine the torque in the steering column of the motor car for the conditions shown. Find the other force completing the couple for (a) and (c).

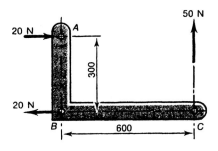

Problem 6/19

6/19

Replace the three forces shown in the sketch by a single force acting at a point along *BC*. The replacing force must have the same overall effect on the member as the combined effect of the three forces shown. Find the location of the replacing force and its magnitude in its new position.

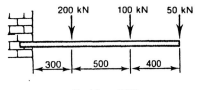

Problem 6/20

6/20

A cantilever beam is loaded as shown. The beam is fixed at the left end and free at the right end. Determine the reactions at the fixed end.

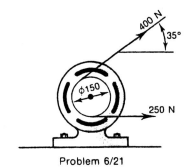

Problem 6/21

6/21

The pulley attached to the electric motor revolves with constant velocity. The belt tensions are found to be 400 N and 250 N on the taut and slack sides of the pulley. Determine
(a) the shear force on the pulley shaft;
(b) the resultant moment produced by the motor armature.

6/22
A steel I-beam is supported and loaded as shown.
Determine the load on the support, and calculate
the magnitude of the couple in the beam
(a) at the support;
(b) half-way between load and support.
Neglect the mass of the beam.

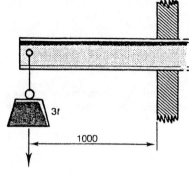

Problem 6/22

6/23
A telephone pole is used to support the ends of
two wires. The tension in the wire to the left is
800 N and, at the point of support, the wire forms
an angle of 10° with the horizontal. Determine the
largest and smallest allowable tension T if the
magnitude of the couple at A may not exceed
600 N m.

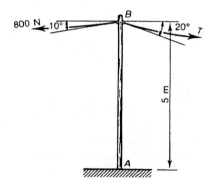

Problem 6/23

6/24
One of a pair of gravity davits is shown in the
diagram. Determine the loads on its rollers A, B,
C, and D, when the lifeboat is slung outboard as
shown. The lifeboat has a mass of 1000 kg and
each davit has a mass of 500 kg with its centre of
mass (CM) as shown.

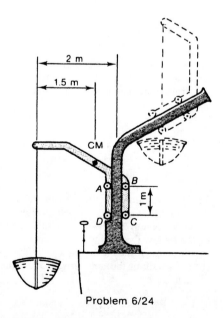

Problem 6/24

6/25

The concrete deck is cantilevered beyond the supporting wall as shown. Determine the force-couple system at the wall face *A* for a typical module (100 mm) of the slab. (Reinforced concrete has a mass of approximately 2400 kg/m^3 giving it a weight density of 24 kN/m^3.)

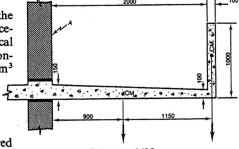

Problem 6/25

6/26

The diagram shows a monorail hoist cantilevered through the wall of a factory, and used to lift light loads with a chain-block. The rolled steel section chosen for the monorail is a 200UB30 (that is, a universal beam, with a nominal depth of 200 mm and a mass of 30 kg per metre). To avoid buckling, this beam should not be subjected to a bending moment greater than 15 kN m. Loads greater than 3 tonnes *will* cause local bending of the bottom flange of the beam, at the trolley wheels.

(a) Suggest suitable safe working load limits for the positions *A*, *B* and *C*.

(b) Will the 50 kg mass of the trolley and chain-block affect these significantly?

(c) What effect will the dead-weight of the beam have on the maximum loads?

Keep in mind that a safety factor of about 4 to 1 is used in this type of design (that is, the beam may bend appreciably (fail) at about 60 kN m) and rationalise your answers accordingly. Your results are to be painted on the side of the beam as a guide to the operators.

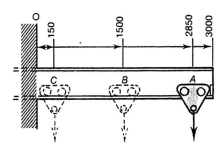

Problem 6/26

6/27

Despite the fact that towing trailers, boats, and caravans forms a regular part of our way of life, Australian and American car designers always seem to put the number plate right behind the towbar, thereby destroying the advantage of the quick-release ball coupling, and requiring the driver to remove part of the towbar each time he uncouples his trailer. This thoughtless design puts a big responsibility on the driver to reassemble his towbar safely each time.

For the recommended 300-N static load shown, calculate the minimum tension in each of the bolts, *A* and *B*, if the other bolt has worked loose. Discuss the likely shock loading conditions when towing with a loose bolt. (*Note*: Calculation of the varying loads occurring in correctly tensioned bolts is beyond the scope of this course, in which we only consider rigid bodies.)

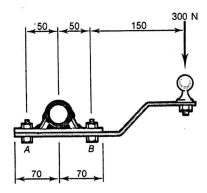

Problem 6/27

7

Frames and Structures

Structural design has been an integral part of life from the moment man first rolled a rock across the mouth of a cave. Perhaps the first structure ever made was when one rock was placed on top of another because one rock alone would not do the job satisfactorily. Our caveman probably chose his rocks on the basis of much the same criteria that we use today when we choose materials for particular purposes. The first consideration in the selection of any piece of material is always: Will it do the job satisfactorily?

Structural design has undergone tremendous changes since caveman days, and trial and error methods have no place in modern engineering design. Engineers can now predict with considerable accuracy the behaviour of most of the materials that they use. Thus, engineers are able to choose the most suitable types of materials for specific applications rather than relying on availability alone. The technique that enables engineers to build the massive structures and machines of today (and the smaller ones used for space travel) relies on the fact that if a suitable material is not available, it can probably be made to the required specifications and in the required sizes as well. The ability to calculate the forces to which each section of a structure will be subjected enables the engineer to produce a *homogeneous design*, where each member is *compatible* with its neighbours—that is, no individual member of the structure or machine is excessively strong or excessively weak (except for a predicted point of failure, such as an electrical fuse).

The engineering structures to be studied in this chapter are limited to simple assemblies of steel bars, most bar assemblies being known as *frames*. When the bars forming a frame are *pin-jointed* together, the frame is usually known as a *truss*.

One of the 330-kilovolt steel transmission towers in the New South Wales electricity grid system. These towers are designed to withstand transverse loads imposed by wind speeds of up to 170 km/h.

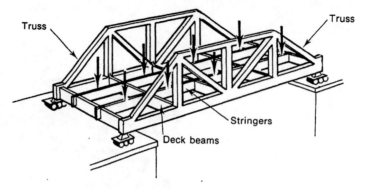

Figure 7.1 These two trusses and associated deck beams form a typical bridge framework.

A *pin-joint* is a connection where no consideration has been given in the design for the members to be able to accept anything but axial forces; that is, pin joints are assumed to be frictionless in operation and thus cannot provide a resisting moment. Pin joints are therefore assumed to be able to resist shear forces only. It is convenient to assume also that frames consist of *rigid bars* which cannot bend, stretch, compress, twist or buckle. Also, it is usual to ignore the mass of these bars in the initial calculations. *Frictionless pin joints and rigid weightless bars are used only as a starting point* in the design of a frame or truss and, once the load in any given bar or member has been determined, it is usually a simple matter to calculate its desirable sizes and probable deformation under load.

Within every building, sometimes well concealed by the exterior cladding, is a frame of timber, steel or concrete, or some compatible mixture of these materials. Each individual part of the frame is called a member and the members fall into two types, *compression members* called *struts* and *tension members* called *ties*.

A member is in *compression* if it is preventing the parts of the frame to which it is connected from coming closer together. A member is in *tension* if it is preventing the parts of the frame to which it is connected from pulling further apart.

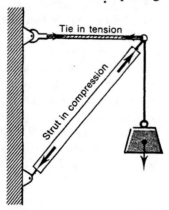

Figure 7.2 Strut and tie.

A *tie* could be made of rope or wire, provided that it is strong enough, as the lack of stiffness is not a disadvantage when the materials are in tension. However, a compression member, or *strut*, needs stiffness as well as strength, and rope and wire have insufficient stiffness to take compressive loads. The stiffness of a strut is not only dependent on the type of material used but also upon its length and cross-sectional shape and area. For example, a small mass can probably be pushed using a relatively short piece of wire, but not with a longer piece of wire or with a piece of rope. Similarly, a pipe section is better than a solid bar having the same cross-sectional area of material, because it is much *stiffer* for the same cross-sectional area.

Whereas ties are often made from thin round bar, a strut carrying the same stress (or load per unit area) would appear to be bigger, as it would be made of pipe, angle, channel, or some even more expensive section such as *rolled hollow square section* (RHS), a *universal column* (UC) or a *universal beam* (UB), all of which possess greater stiffness than the solid section.

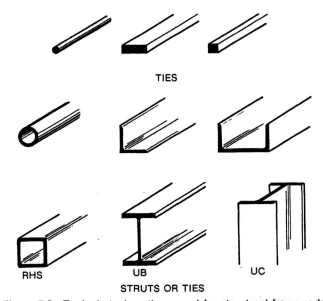

Figure 7.3 Typical steel sections used for structural frameworks.

TYPES OF BAR ASSEMBLIES

There are three basic types of bar assemblies which are readily distinguished from each other; they are known as *unstable* assemblies, *stable* assemblies, and assemblies containing *redundant members*.

(1) *Unstable assemblies*: If an external load can cause the assembly to distort without resistance, the assembly is termed a *mechanism* (see Figure 7.4 (a)).

(2) *Stable assemblies*: If an assembly is suitably braced so that it can accept a certain external load without (appreciable) distortion, and has no more members than are

necessary, then it is stable (see Figure 7.4 (b)). It is also said to be *statically determinate* because the force in each member can be readily calculated using the principles of basic statics (equilibrium of forces) already covered in earlier chapters. This chapter is confined to the study of this type of structure.

(3) *Redundant assemblies*: A redundant structure is also stable but has more members than are apparently necessary. It is therefore *statically indeterminate* because any deformation in one member requires other members to deform in various ways, depending on their *compatibility* (see Figure 7.4 (c)).

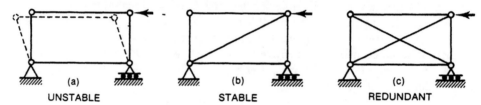

UNSTABLE   STABLE   REDUNDANT

Figure 7.4 Three basic types of bar assemblies

The redundant structure is often employed in practice for a variety of reasons. For example, two ties are usually cheaper than one strut since ties are made from simple, cheap sections. They also usually weigh considerably less, and they certainly have less wind resistance—both were significant factors in the design of the beautiful filigree lace-work of the Eiffel Tower which has hundreds of "redundant" frames, as does the Sydney Harbour Bridge.

Figure 7.5 Aerial view of the Sydney Harbour Bridge showing redundant frames.

There is a popular misconception that a truss is stable if the number of members (m) and the number of joints (j) satisfy the equation m = 2j − 3. This is not necessarily so. It is not possible to determine, merely by counting the bars and joints, whether or not a structure is stable. The truss in Figure 7.6 satisfies the equation m = 2j − 3 but it is obvious that the structure is redundant in one panel, and unstable in another. In the elementary structures with which we deal it is usually easy to see if a structure is stable, unstable or redundant using first principles and without recourse to a formula*.

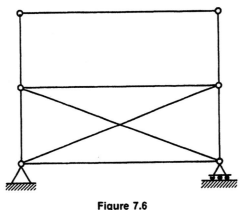

Figure 7.6

An excellent practical demonstration of the stability of a truss can be given using a simple model made from strips of three-ply about 10 mm wide and 200 mm long. A hole is drilled in each end of these members to take a nail which serves as a pin joint. A stable truss is then assembled from several members (strips), and the group is asked to decide whether a certain member of this truss is in tension, compression, or whether it is redundant. One end of this member is then flicked off its pin joint and about half the group discover that they had guessed incorrectly about the type of force present in the member. *This kind of demonstration proves the need to closely*

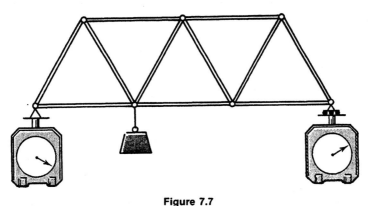

Figure 7.7

---

*The authors' policy is for the students to gain understanding without having to use formulas.

analyse trusses and other structures in order to determine the types of forces present in members. This is also an ideal use for those modified kitchen scales previously described in Chapter 5 which can be used to measure reactions at the supports of trusses carrying vertical loads.

TYPES OF FORCES TO CONSIDER IN FRAME ANALYSIS

Forces to be considered must include all external forces present, including the dead-weight of the structure itself, all forces acting on the structure, and all forces present in the members of the structure which are reactions to the dead weight and the active loads present.

External Forces

Gravitational force. The structural engineer has to make many predictions when designing a frame or structure. For example, while he knows the force exerted by gravity on any mass his structure is designed to support, he can, during preliminary work, predict from experience the probable mass and hence dead weight of the structure itself. The relationship between dead weight and supported loads or forces can vary considerably. Small cranes, for instance, often weigh much less than the loads they carry, but most buildings have a dead weight far greater than the live load that they are expected to support.

Figure 7.8 The mass of this crane is insignificant compared to the load it carries.

During the design process the structural engineer *begins* by assuming that members are weightless. However, after the forces present in any member are calculated, the size of this member can then be determined; this size may then have to be modified in proportion to the final loads involved, including the dead weight of the structure itself. As each member of a frame or structure usually has the same section throughout its entire length, it behaves as a uniformly distributed load (UDL) and the weight-force can be considered as a force acting through its centre of mass (or centre of gravity).

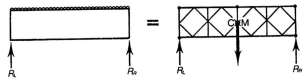

Figure 7.9 As the Bailey bridge shown is totally symmetrical, the reactions R_L and R_R are easily calculated.

Furthermore, if a structure is symmetrical, the mass of the whole can often be assumed to act through the centre of mass of the total structure. It is often most useful to do this when calculating the *reactions* present at the supports of the structure.

Active Loads

These are all other external loads on the structure. The vehicles on the bridge, the load on the crane hook and its counterweight, the weight-force of the roofing material on a building, and the push or pull of the wind loads are all active external loads. The active loads combine with the dead weight of the structure to produce the total load which is to be finally *resisted* at the supports of the structure.

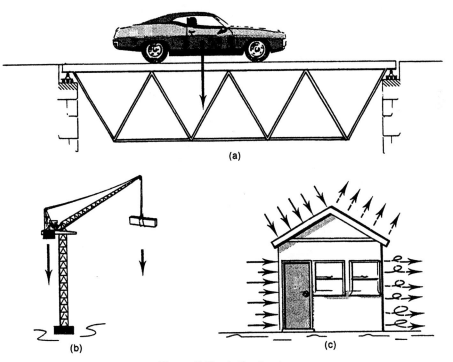

Figure 7.10 Active loads.

Reactive Forces

The load on the supports produces a reactive force in the supports themselves; these reactive forces (reactions) push back on the structure and maintain its equilibrium. Reactive forces are considered to be external forces when calculating the stresses in the members of the structure.

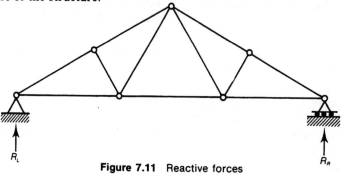

Figure 7.11 Reactive forces

Internal Forces

The design philosophy for a pin-jointed frame assumes that no member is continuous through a joint, and therefore cannot transmit any moment at the pin. *The only force capable of being transmitted through a pin-joint is a direct push or pull along the axis of the member.* In practice, of course, many members *are* continuous through joints and very few connections are truly pin-jointed. However, this assumption simplifies the analysis of forces in each member, as these forces can only be direct *axial forces* of compression or tension. Consequently no design calculations need be done for the more complex conditions of bending or buckling. Welded, riveted and bolted joints are often considered as pin joints in the design, and the fact that they are not adds to the rigidity and stiffness of the final structure. The engineer's knowledge and experience help him to predict when he can substitute welded or riveted joints for pin joints and when he really must have pin joints. The Sydney Harbour Bridge, for instance, is

Figure 7.12 One of the four large pin joints supporting the Sydney Harbour Bridge.

sitting on four large pin joints at the base of the pylons. It is doubtful if they move very much—but they were essential for the construction technique employed, and their design allows for temperature variations in the arch, without placing tremendous bending stresses on the base of the arch.

To avoid having to allow for internal forces due to temperature changes, chemical changes, or deflections and deformations due to imposed loads, *it is often assumed in the design phase that one support is pin-jointed and the other is on rollers.* Again this helps to simplify the design calculations.

Figure 7.13   The steel railway bridge shown here is supported on pin-jointed rollers and the roadway above is supported on ball-jointed rockers. (Circular Quay, Sydney.)

In practice, *both* supports are often rollers or rockers; for example many bridges are supported on roller supports at each end but this does not mean that the bridge is likely to roll down the road! Also, the columns of buildings may act as rocker supports for beams placed upon them. Much of the many kilometres of a continuously welded railway line in an Egyptian desert is on special roller supports, and the story is that, as the sun comes up in the morning, the railway lines only just beat the train into Cairo Station . . . .

It will be obvious from a glance at the standard trusses shown in Figure 7.14 that the basic unit of a stable structure is the *triangle*. To design a truss economically, we must know the forces in each member. Due to the pin joints, these forces can only act directly along the axis of each member (i.e. they are *axial* forces). This allows for

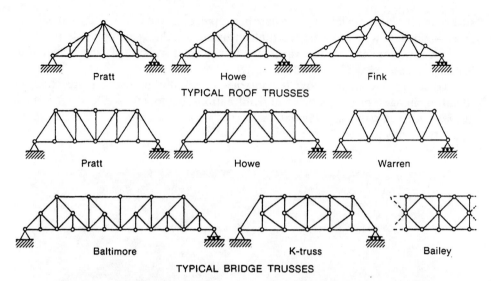

Pratt   Howe   Fink

TYPICAL ROOF TRUSSES

Pratt   Howe   Warren

Baltimore   K-truss   Bailey

TYPICAL BRIDGE TRUSSES

Figure 7.14 These basic trusses are light frameworks named either after a designer, or the place where they were first used, or their shape. In each case the pin-jointed triangle forms the basis of design, and a drawing of the layout of the frame becomes a space diagram of the forces within the structure.

simple graphical solutions of the forces present at each pin joint, as the angles at which all the internal forces act are therefore known—these angles are, of course, the same as the angles of the members themselves.

SAMPLE PROBLEM 7/1

Consider one frame of the trestle supporting one end of the painter's plank as shown. Determine
(a) the load on each wheel;
(b) the compressive force in each angled pipe, and
(c) the tension in the chain.

Problem 7/1 (a)

Analysis:
If the dead weight of the trestle is neglected, then by inspection the reaction at each wheel must be 250 N.

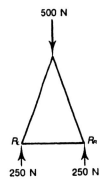

500 N

R_L R_R

250 N 250 N

FREE-BODY DIAGRAM FOR FRAME
Problem 7/1(b)

Now a complete free-body diagram can be drawn for the frame as shown. Force polygons can then be drawn for each of the joints and the forces present in each member can be scaled from these force polygons. Alternatively, these forces can be calculated since the angles between the members are known.

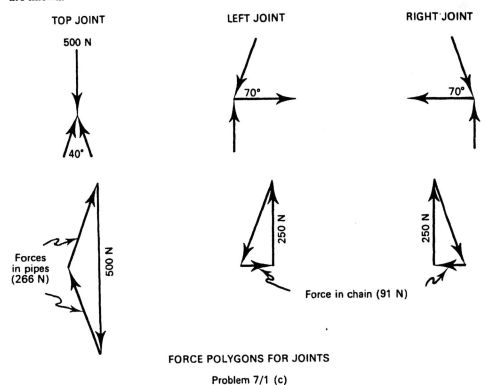

TOP JOINT   LEFT JOINT   RIGHT JOINT

500 N

40°   70°   70°

Forces in pipes (266 N)   500 N   250 N   Force in chain (91 N)   250 N

FORCE POLYGONS FOR JOINTS

Problem 7/1 (c)

Results:
For the conditions shown, the load on each wheel is 250 N; the compressive force in each leg of the trestle is 266 N; and the tension in the chain is 91 N. ∎

Once the magnitude of each force within a truss has been determined, it is a simple matter to discover whether that force is one of tension or compression. In a member that is in tension, the external forces tend to stretch the member and pull on the joint.

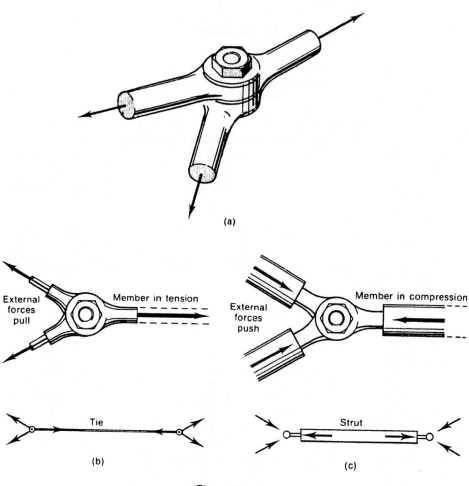

(a)

(b)

(c)

Figure 7.15

The internal (reactive) forces oppose the external forces (as in Figure 7.15 (a)). *Thus a bar or member is in tension if the internal force is directed away from the pin joint as in Figure 7.15 (b).*

Similarly, in a compression member the external forces tend to compress that member, while the internal reactive forces oppose the external forces (as in Figure 7.15 (c)). *Thus a bar or member is in compression if the internal force is directed towards the pin joint* (i.e. pushes on the joint).

Referring back to the results obtained for Sample Problem 7/1, it is thus obvious from the three force diagrams that the angled pipe members are in compression while the chain is in tension (Figure 7.16).

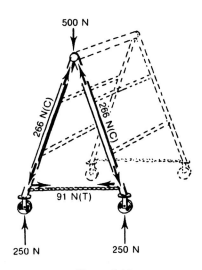

Figure 7.16

Method of Joints

The method used to discover the forces in the members of this simple symmetrical frame was probably obvious. By repeating this process for each joint the forces in the members of much more complex frames can be discovered just as readily. This is known as the *method of joints* and can be applied to all joints in any stable or statically determinate truss. Each joint is analysed in turn, each joint thus involving the simple analysis of a few coplanar, concurrent forces that are in equilibrium. *The analysis is started at any joint where there is at least one known force, and not more than two unknown forces.* The force polygon is drawn for that joint, and the unknown forces scaled or calculated from the resulting force polygon. The information so obtained enables the next joint to be analysed in turn, provided there are now only two unknown forces at that joint. Thus, the force present in each member of the truss is discovered using a joint-by-joint analysis.

SAMPLE PROBLEM 7/2

The simplest and most common form of truss is probably the Warren truss. In the Warren truss shown all the triangles are equilateral, all joints are pin joints, and the dead weight of the members will be ignored in the following calculations.

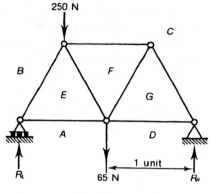

Problem 7/2 (a)

Problem:

To discover the forces in each member of the Warren truss for the loading shown.

Analysis:

Draw a space diagram of the truss, and label using Bow's notation (see above). Now determine the reactions at the supports using any suitable method.

For example, taking moments about R_R:

$$R_L \times 2 = (250 \text{ N} \times 1.5) + (65 \text{ N} \times 1)$$
$$R_L = \frac{375 + 65}{2}$$
$$R_L = 220 \text{ N}$$

$\Sigma F_V = 0$, thus $\qquad R_R = 95 \text{ N}$

Next, analyse any joint with one or two unknown forces. Suppose we consider the joint *ABE* first.

It has one known force, *AB* (the reaction R_L) and two forces of unknown magnitude, *AE* and *BE*. However, the lines of action of the forces *AE* and *BE* are known, since they act directly along the members *AE* and *BE* of the truss.

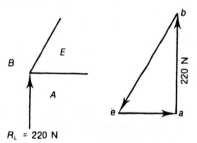

Problem 7/2 (b)

A force diagram for the joint *ABE* is started with the known force *AB* drawn to scale. Lines are then drawn through each end of *AB* parallel to the members *AE* and *BE*. These lines intersect at *E* and thus the magnitudes of the forces in the members *AE* and *BE* can be scaled off or calculated (127 N, 254 N). Arrows can now be put on the force triangle, head-to-tail, starting with the known force. These arrows will indicate the effect of each force on the joint under consideration. If these arrowheads are now transferred to the space diagram close to the joint *ABE*, they will indicate whether each member is in tension or in compression. This information should then be transferred by means of arrowheads to the joints at the other ends of members *AE* and *BE*, thus providing some extra information for the joints *BCFE* and *AEFGD*.

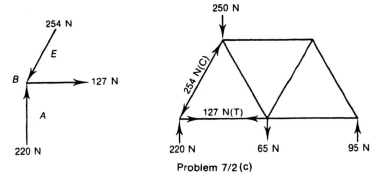

Problem 7/2 (c)

Now another joint can be analysed in the same way. Consider the joint *BCFE*. It now has two known forces *EB* and *BC* and the two unknown forces in the members *CF* and *FE* acting on it.

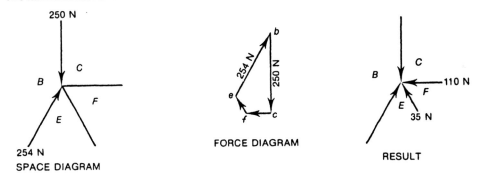

Problem 7/2 (d)

The force polygon of the joint *BCFE* is now drawn to scale as before, and the magnitude of the forces in the members *CF* and *FE* is scaled or calculated from the resulting force polygon (110 N, 35 N). The arrowheads are added head-to-tail to discover the effect of each force on the joint *BCFE*. These arrows are then transferred to the space diagram around the joint *BCFE* and also to the other ends of *CF* and *FE* on the diagram of the truss.

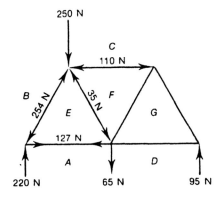

Problem 7/2 (e)

The remaining joints are analysed one-by-one in the same way, and the resulting force diagrams are shown in Table 7.1.

Table 7.1

JOINT	SPACE DIAGRAM	FORCE DIAGRAM	RESULT

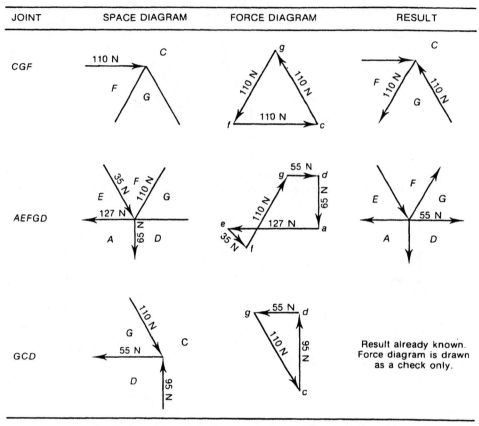

RESULT OF COMPLETE TRUSS ANALYSIS

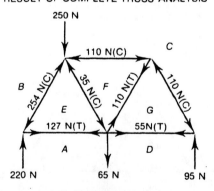

Problem 7/2 (f)

Analysis of the last joint *GCD* does not appear to be necessary as the forces in the members of this joint have already been discovered. However, with the *method of joints* any error in the analysis will, unfortunately, affect all following results and it is possible to get a progressive error in your calculations. *Therefore, it is advisable to analyse the final joint as a self-checking operation as this provides a check on the accuracy of all the preceding work.* ∎

A systematic working procedure must be adopted, and it is most useful to note neatly on the original space diagram the load in each member as it is discovered, together with an indication of whether the member is in tension or compression. Placing of arrows near each pin joint is a convenient way to indicate this. An arrow pointing towards the pin means a push, or compression, while an arrow pointing away from the pin means a pull, or tension, on that joint.

Adding (T) or (C) to the value of the force as it is written on the space diagram is another simple way of designating the type of load in each member. *However, conventions involving the use of + and − are not recommended, as they are too easily included in calculations (in which they have no part).*

Method of Sections

When it is required to discover the force present in only one or two members of a frame, it is often more convenient to use another technique which avoids the laborious step-by-step progression from one joint to the next that is inherent in the method of joints.

This method is known as the *method of sections* and involves passing a section plane through the truss which cuts two or three members, one of which is the member in question. By choosing a suitable moment axis which ignores the forces in the other members cut by this section plane, the force in any member can be readily calculated.

SAMPLE PROBLEM 7/3

Consider the previous Warren truss under the same conditions of loading. It is now required to discover the forces in the members *CF* and *EA* only.

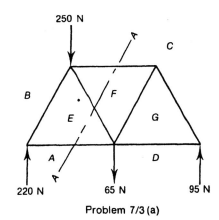

Problem 7/3 (a)

Analysis:
A cutting plane *A-A* is passed through the members *CF*, *FE* and *EA*, separating the truss into two sections. *The equilibrium of one side is now considered.*

Consider the left section only. Under the action of the external loading, this section of the truss would *rotate* if it were not restrained by the forces provided by the right-hand portions of the members *CF*, *FE* and *EA*. Let these forces be *p*, *n*, and *m* as shown, the sense of these three forces being arbitrarily selected by inspection.

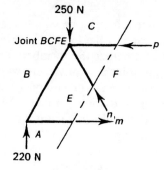

Problem 7/3 (b)

Calculation:

To discover the force *m* in the member, take moments about the joint *BCFE*.

$\Sigma M = 0$ Assume $\curvearrowright$. Consider the frame members to have a length of 2 units.

$$(220 \times 1) - (m \times \sqrt{3}) = 0$$
$$\therefore m = \frac{220}{\sqrt{3}}$$
$$\therefore m = 127 \text{ N}$$

To maintain equilibrium it is thus seen that the force *m* in the member *AE* must act *away* from the joint *ABE*, that is, *the member AE is in tension.*

Now consider the force *p* in the member *CF*. To discover the force *p* in the member *CF* take moments about the joint *AEFGD*, *but consider only the left-hand section as shown.*

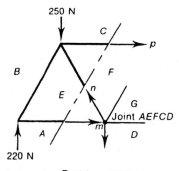

Problem 7/3 (c)

Moments about joint *AEFGD*: $\Sigma M = 0$ Assume $\curvearrowright$

$$(-250 \times 1) + (220 \times 2) + \sqrt{3}p = 0$$
$$p = \frac{-190}{\sqrt{3}}$$
$$p = -109.7 \text{ N}$$

The negative result means that the sense of the force p in CF was incorrectly assumed. The arrow should have pointed in the opposite direction and thus the member *CF* is in compression.

Result:
Partial truss analysis for members *CF* and *EA*.

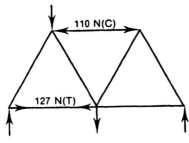

Problem 7/3 (d)

If the force in the member *EF* is required, there are a number of suitable methods which can be used. One is to apply the equilibrium equation $\Sigma F_V = 0$ *to the left section of the truss only.*
$\Sigma F_V = 0$ Assume $+\downarrow$

$$250\ \text{N} - 220\ \text{N} + x = 0$$
$$\therefore x = -30\ \text{N}$$

Therefore a vertical component equal to 30 N vertically upwards is needed for equilibrium. This can only be the vertical component of the force *n* in the member *EF*. Resolution of this vertical component into two forces, one horizontal and the other parallel to the member *EF*, provides the force *n* in *EF*.

$$\frac{30}{n} = \cos 30°$$

$$\therefore n = \frac{30}{\cos 30°}$$

$$\therefore n = 34.6\ \text{N}$$

Problem 7/3 (e)

Thus, the force in the member *EF* is $\approx$ 35 N compression (compare to previous graphical results).

Another method involves taking moments about some other point in the plane of the truss, say, the joint *ABE*.
Moments about joint *ABE*: $\Sigma M = 0$ Assume $\curvearrowleft$

$$(250 \times 1) + (\sqrt{3} \times 109.7) - (\sqrt{3} \times n) = 0$$
$$\sqrt{3} \times n = 250 - \sqrt{3} \times 109.7$$
$$\therefore n = \frac{60}{\sqrt{3}}$$
$$\therefore n = 34.6\ \text{N}$$

Thus, the force in the member *EF* once again is found to be $\approx$ 35 N compression.

250 N

p = 110 N

B

E

n = ?

m = 127 N

A

220 N

Problem 7/3 (f)

Sometimes the simplest method is to apply the *method of joints* to either end of the unknown member. In this case it would need to be the joint *BCFE*, as the other joint, *AEFGD*, has too many unknown forces. ∎

SAMPLE PROBLEM 7/4

A large Christmas decoration is suspended above the road between two tall buildings on the framework shown in the diagram. Determine the load in each member.

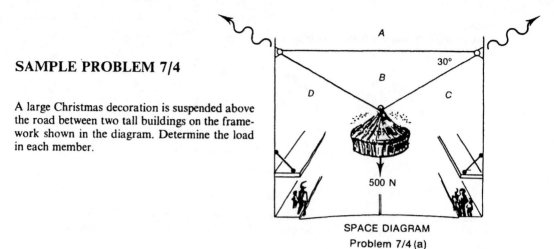

SPACE DIAGRAM
Problem 7/4 (a)

Analysis:
First the force polygon for the joint *BCD* is drawn, and the forces in the members *cb* and *bd* scaled or calculated. (Each is found to be 500 N.)

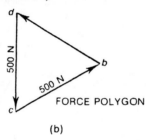

FORCE POLYGON

(b)

Consider the joint *ABC*. The force polygon is begun with the known force *BC* (500 N) and the known direction of the force *AB* (horizontal) but the force *AC* is of unknown magnitude or direction and it is therefore impossible to determine where to position this force in order to close the polygon accurately.

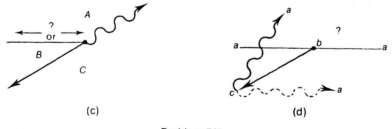

(c)   (d)

Problem 7/4

Result:
This structure is therefore "statically indeterminate" and the member *AB* is "redundant".* The member *AB* is also quite unnecessary as the buildings would be capable of providing the 500 N reaction.   ∎

---

*The term *redundant* is used here in the engineering sense. Redundant members are not usually unnecessary (see Figure 7.17).

Figure 7.17

The tank stand shown in Figure 7.17 has panels with apparently redundant members, and yet careful design has enabled it to still be statically determinate.

None of the diagonal cross-bracing members can be loaded in compression as the rods can only pull on the central connecting ring. Any loads tending to produce compression transfer their effect to the other diagonal of the same panel, in which they induce a tensile stress, which it is designed to accept.

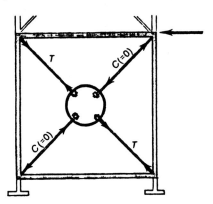

Figure 7.18 A simple method of preventing ties from being placed in compression.

As stiffness does not need to be considered in these ties, they are made of round bars, a much cheaper rolled section than would otherwise be needed if one brace had to resist both tension and compression. This principle is often employed in practice, and not always with joints which actually slide. The redundant tension member is often used to prevent another tension member from being placed in compression by wind loads or other variables.

REVIEW PROBLEMS

(Use $g = 10 \text{ m/s}^2$ unless otherwise specified.)

7/5
The corner of a roof truss is shown connected by a steel plate. The rafter has a compressive load of 2500 N. Determine the loads in the collar tie and in the stud.

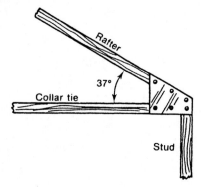

Problem 7/5

7/6
The structure shown in the sketch is in the form of an equilateral triangle. For the loading shown determine:
(i) the reactions R_L and R_R;
(ii) the forces in members AB and BC.

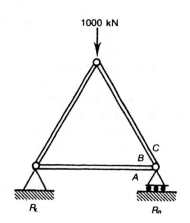

Problem 7/6

7/7
A simple derrick is shown, which supports a load of 1 tonne. Draw free-body diagrams of the load pin C, tie AC, strut CB, and king post AD, showing all forces. Calculate these forces.

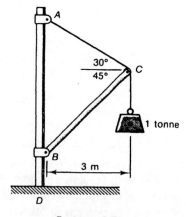

Problem 7/7

7/8
A pin-jointed frame shown in the diagram supports a pulley which is used as illustrated to lift a load. Find;
(a) the magnitude of the force in each member; and
(b) state whether they are in tension or compression.

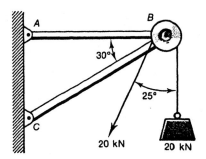

Problem 7/8

7/9
The crane has a mass of 8 tonnes with centre of gravity at G and is held in place by the pin connection at A and the vertical cable BC. Determine the force supported by the pin at A, and the tension in the cable BC.

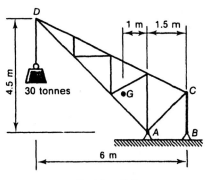

Problem 7/9

7/10
The structure shown in the diagram is in the form of equilateral triangles. For the loading shown determine:
(i) (a) The reaction R_L.
   (b) The reaction R_R.
(ii) (a) The force in member AB.
   (b) The force in member BC.

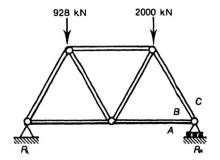

Problem 7/10

7/11

Simple 3 m × 4 m frameworks of mass 5 tonnes are supported vertically as shown. All connections consist of smooth pins, rollers, or short links. In each case, determine whether the reactions are statically determinate and wherever possible, calculate these reactions.

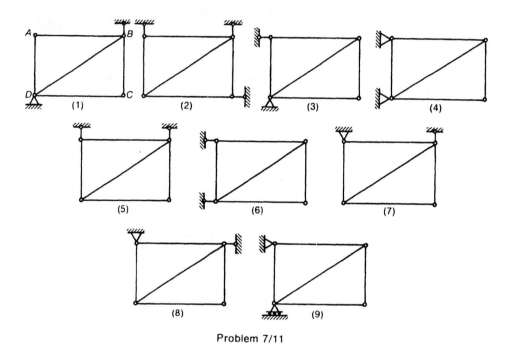

Problem 7/11

7/12

Two bars, *AB* and *BC*, are pivoted at *B*. Both bars lie in a vertical plane and are of negligible weight. The end *C* is pin-jointed to a fixed bearing and end *A* is pinned to a roller which is free to move along a smooth horizontal surface. If a load of 100 kg is hung from *B*, find for the position shown:

(a) the horizontal force *P* required at *A* to prevent the roller moving outwards;

(b) the force in the link *AB*; and

(c) the vertical reaction on the roller.

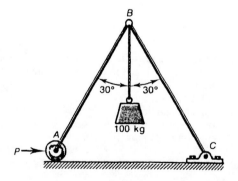

Problem 7/12

7/13

The model truss shown is under test. Determine the force in the members *AD*, *AC* and *CD* when a mass of 100 kg is hung from joint *D*.

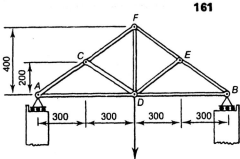

Problem 7/13

7/14

The simple derrick shown in the diagram must lift a 1000-kg vertical load by means of a rope suspended at point *A*.
(a) What is the magnitude of the reaction at *B*?
(b) Determine the forces in the members *AC*, *AB*, and *BC*.

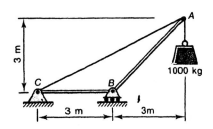

Problem 7/14

7/15

A set of traffic lights is cantilevered over the roadway on a light tubular steel framework. The framework has a mass of 50 kg and the signal lamp unit has a mass of 25 kg. Determine the maximum shear force in the bolts at *A* and *B* and the tensile force in the bolts at *C* and *D*.

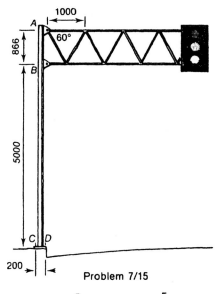

Problem 7/15

7/16

Calculate the forces in all members of the truss shown.

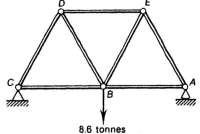

Problem 7/16

7/17
The geometry of a framework, loaded as shown, is given in the diagram. Find graphically the reaction at each support and the nature and magnitude of the force in each bar.

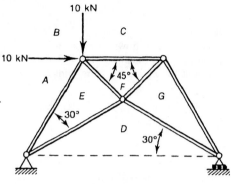

Problem 7/17

7/18
A simple pin-jointed frame is shown in the diagram. Determine
(a) the shear force in the pins at A, B, and D;
(b) the horizontal and vertical components of the reactive force at A;
(c) the horizontal and vertical components of the reactive force at D.

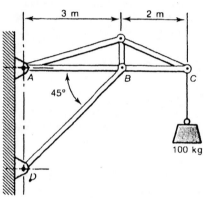

Problem 7/18

7/19
A small truss is supported in different ways as shown. All connections consist of smooth pins, rollers, or short links. For each structure, wherever possible, calculate the reactions, assuming that the magnitude of the force is 10 kN.

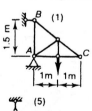

Problem 7/19

7/20
A loaded frame structure is shown in the diagram. Graphically determine the approximate values for the reactions at the supports, and the forces in the members *AB*, *BC* and *CD*.

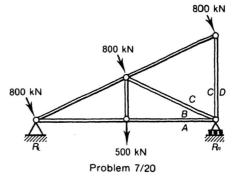

Problem 7/20

7/21
The truss shown is made up of three equilateral triangles loaded as indicated. It is supported by a pin joint at the wall on the right-hand side and by a cable inclined at 30° to the horizontal on the left. Determine
(a) the tension in the cable;
(b) the reaction at the wall;
(c) the nature and magnitude of the force in each bar.

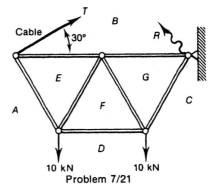

Problem 7/21

7/22
The crane supports a 500-kg load. Find the reactions for each of the three types of supports shown.

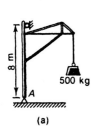

(a)

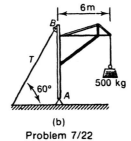

(b)

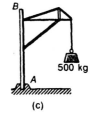

(c)

Problem 7/22

7/23

Find the reactions in the supports and determine the forces in members *FH*, *GH*, and *GI* of the roof truss shown in the diagram.

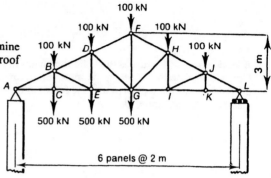

Problem 7/23

7/24

The truss shown in the diagram is subjected to the roof loads and the hoist load shown. Determine the *reactions* R_L and R_R using the funicular polygon.

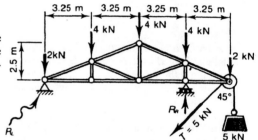

Problem 7/24

7/25

Determine the forces in members *DE* and *HJ* of the truss shown.

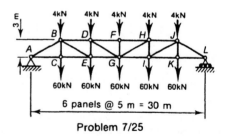

Problem 7/25

7/26

Calculate the forces in members *BC* and *GF* of the cantilever truss shown in the diagram.

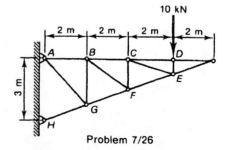

Problem 7/26

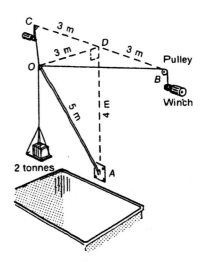

7/27

A ship's jib crane is shown diagrammatically. Determine the approximate compressive force in the boom *OA* and the tension in the cables *OB* and *OC* when they have positioned the boom as shown. The mass of the boom is negligible compared with that of the load and may be ignored in the calculations.

Problem 7/27

7/28

A power line is suspended from the side of a building by the projecting framework shown. The tension in the power line is 6 kN at the insulator. Determine

(a) the tension in the insulator *OD*;
(b) the tension in the member *OA*;
(c) the compressive load in members *OB* and *OC*.

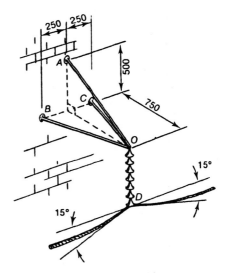

Problem 7/28

8

Frictional Forces

$$\mu = \frac{F}{N} = \tan \phi$$

You are a passenger in a motor car that is being driven at constant velocity along a perfectly straight flat and horizontal section of a concrete road. At a certain instant, the engine of the car is turned off and the car is allowed to *coast* in neutral. The car then slowly comes to a standstill. It is clear that some force must have been acting, which was opposing the initial forward motion of the car, since, from Newton's second law, ($F = ma$), a force is obviously necessary to cause the *deceleration* of the car to rest. This retarding force can only be the sum of the *frictional forces* acting to slow the car down—such things as the air resistance, the friction in the wheel bearings and in the transmission and, most importantly, the frictional forces present between the tyres and the roadway. When the car was travelling at constant velocity, the *power* of the engine was providing a propulsive force that was equal and opposite to the frictional drag acting on the car. Thus, the sum of the forces in the direction of motion was zero, and constant horizontal velocity was maintained.

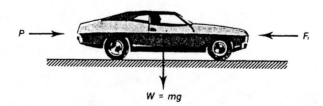

Figure 8.1 A car travelling along a horizontal stretch of road. At constant velocity, the total propulsive force P supplied by the motor is equal to the total frictional force F_r (air resistance plus rolling resistances) opposing motion. Thus the car is in equilibrium since there is no acceleration and therefore no unbalanced force acting.

In the above instance, frictional forces appear to be a tremendous disadvantage; however, if no friction existed between the wheels and the road, the car could not be started when stationary and could not be stopped when it was moving. Also, without friction we could not walk or run. In order to get some idea of this situation try running across a flat stretch of ice sometime, it may prove to be quite an experience!

Another simple demonstration of the existence of frictional forces acting on a body in equilibrium is illustrated in Figure 8.2.

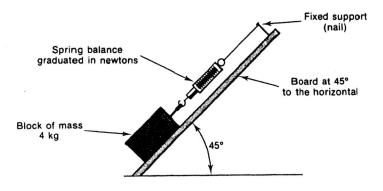

Figure 8.2

A block of known mass, say, 4 kg, is placed on a flat plane inclined at 45° to the horizontal. A spring balance graduated in newtons is attached to one end of the block and its other end is fastened to a nail driven into the plane as shown, so that the spring balance is parallel to the plane. Since the block rests on this plane, the component of its own weight-force acting parallel to the plane tends to cause the body to slide down the plane. Consider a free-body diagram of the *block* (Figure 8.3 (a)).

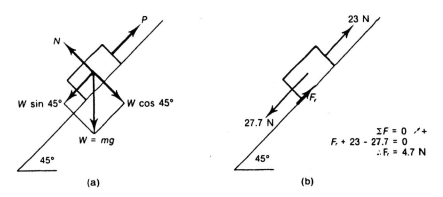

Figure 8.3 Analysis of the forces acting on the block which is sitting on the inclined plane: (a) shows a partial free-body diagram of the block; and (b) shows that since $\Sigma F = 0$ for forces parallel to the plane, the friction force F_r must act as shown and have a value of 4.7 N.

The component of the weight-force acting down the plane is

$$F = W \sin 45° \nearrow$$
$$= mg \sin 45°$$
$$= 4 \times 9.8 \times 0.7071$$
$$\approx 27.7 \text{ N.}$$

where: $m = 4$ kg; and
$g = 9.8$ m/s^2

At first glance it appears that, if the block is not to slide, the force P provided by the spring balance must also be equal to 27.7 N.

Now read the force P from the spring balance. It is less than 27.7 N. Suppose that it is 23 N. *This means that there must be another force of 4.7 N opposing sliding motion down the plane.* This can only be the frictional force F_r acting between the block and the plane as shown in Figure 8.3 (b). *Clearly, this frictional force F_r must be generated by the mutual action of the block and the plane, one upon the other.*

A further simple experiment serves to illustrate most of the facts that need to be understood about frictional forces. A block of metal of known mass is placed on a horizontal surface such as a bench top. A cord and spring balance are attached to the centre of one end of the block as shown in Figure 8.4 so that the block can be pulled along if a horizontal force F is applied to the cord.

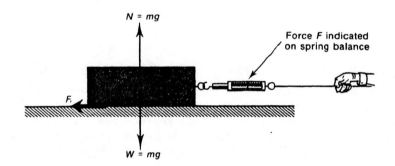

Figure 8.4   A block of mass m resting on a rough horizontal surface. When no force is exerted by the hand on the wire, F is zero and F_r is zero. However, as the pulling force F increases, the frictional force F_r also increases up to a maximum value. When this is exceeded, sliding motion begins.

When the block is at rest, the downward force it exerts on the flat surface is equal to its weight-force, $W = mg$. The reaction of the table on the block is the normal force, N, which is equal in size but opposite in direction to W (Newton's third law).

A small pulling force, F, which is too small to cause movement, is applied to the string and its magnitude is registered on the spring balance. The force F tends to move the block but, since motion does not occur, the force F must be exactly opposed by the frictional force F_r generated between the block and the plane ($\Sigma F_H = 0$, thus $F = F_r$ and no horizontal motion occurs). The pulling force F is now gradually increased until the block is just on the point of sliding. *At this point, the frictional force F_r has reached its maximum value which is known as the force of limiting friction,* and the ratio of F/N is calculated. The experiment is now repeated using blocks of identical material having different masses and different surface areas and, in each instance, the ratio of F/N is calculated at the point of limiting friction. Upon a comparison of the results it will be found that all the ratios of F/N are equal. In other words, irrespective of the mass of the block or its surface area, the ratio of

$$\frac{\text{force just capable of causing motion }(F)}{\text{normal force between the surfaces }(N)} = \text{a constant}$$

However, at the point of limiting friction, $F = F_r$.

Thus,
$$\frac{F}{N} = \frac{F_r}{N} = \text{a constant}$$

where: F = force just causing sliding;
F_r = force of limiting friction; and
N = normal force between surfaces.

This constant is called *the coefficient of static friction* and is given the symbol μ_s.

Therefore,
$$\frac{F_r}{N} = \mu_s$$

or
$$F_r = \mu_s N$$

Once the force of limiting friction is overcome and the block begins to slide freely, the force F can be progressively reduced until a point is reached where, if F is further reduced, sliding of the block ceases. At this point the ratio F/N is again calculated for many different-sized blocks and these ratios are again found to equal a constant. This constant, μ_k, is known as *the coefficient of sliding friction*, and is usually considerably smaller than the coefficient of static friction, μ_s, for the same materials in contact, that is

$$\frac{\text{force just capable of sustaining sliding motion }(F)}{\text{normal force between the surfaces }(N)} = \text{a constant}$$

that is,
$$\frac{F}{N} = \frac{F_r}{N} = k$$

where: F = force just capable of sustaining motion
F_r = force of kinetic or sliding friction
N = normal reaction between surfaces
μ_k = coefficient of kinetic or sliding friction

or,
$$F_r = \mu_k N$$

Important Points About Frictional Forces

(1) *The frictional force always acts to oppose motion* and is always *tangential* to the surfaces in contact.

(2) *The frictional force is, within wide limits, proportional to the normal force existing between the surfaces in contact,* that is

$$F_r = \mu_s N \quad \text{(static)}$$
$$F_r = \mu_k N \quad \text{(dynamic)}$$

However, if the area of contact is very small, the resultant *stress* produced between the surfaces may be sufficient to cause one surface to *seize*, or *weld* itself to the other.

(3) *The frictional force is independent of the area of contact between the surfaces.*

However, this is also true only within certain limits and frictional forces will increase as the *true area* of contact increases. This explains why wide tyres, which have a very large area of true contact with the road, increase the frictional grip of the car on the road (i.e. increase traction).

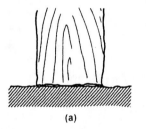   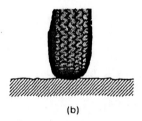

(a)   (b)

Figure 8.5 Surfaces in contact: (a) shows a highly magnified view of two relatively smooth surfaces in contact (note the relatively small area of contact between the surfaces); (b) shows the much larger area of true contact possible between a rubber tyre and the roadway.

(4) *The frictional force, F_r, depends upon the nature and the relative roughnesses of the surfaces in contact.*

(5) *Forces of kinetic (sliding or rolling) friction are always less than forces of limiting friction.*

Thus, a greater force is required to initiate motion against a frictional resistance than is required to sustain the same motion.

It is important to realise that the above five points apply only when *dry surfaces* are in contact. In bearings, slides, etc., where lubricants are present, the most important factor to consider is the internal frictional drag of the lubricant itself.

Table 8.1. Coefficients of Friction For Dry Surfaces

Contacting surfaces	Static friction	Sliding friction
Bronze on bronze	0.2	0.1
Rubber on wood	0.4	0.3
Hardwood on metal	0.6	0.4
Rubber tyres on pavement*	0.9	0.8 $\left(\begin{array}{l}\text{wheels}\\\text{locked}\end{array}\right)$
Asbestos brake lining on cast iron	0.4	0.3

*Note: the coefficient of *rolling friction* between rubber tyres and pavement is much less than either of these values, being about 0.02.

SAMPLE PROBLEM 8/1

A piece of flat steel is clamped in the jaws of an engineer's vice with a clamping force of 600 N. If the coefficient of friction (μ_s) is 0.3, what minimum pull P on the steel would just cause it to move?

SPACE DIAGRAM
Problem 8/1(a)

Analysis: Draw a free-body diagram showing all forces acting on the metal strip.

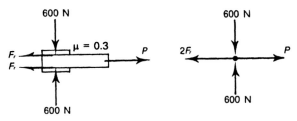

FREE-BODY DIAGRAMS
FORCES ACTING ON THE STEEL STRIP

Problem 8/1 (b)

If the steel is just on the point of moving, the force of limiting friction F_r is generated between each jaw and the steel strip. Since movement is not actually occurring

$$\Sigma F_H = 0, \qquad \therefore \quad P = 2F_r$$

Calculation:

$$\mu_s = \frac{Fr}{N}$$

$$\therefore F_r = \mu_s N$$
$$= 0.3 \times 600$$
$$\therefore F_r = 180 \text{ N}$$
$$P = 2F_r = 360 \text{ N}$$

where: $\mu_s = 0.3$
$N = 600$ N

Result:
The minimum pull that must be applied to cause movement is 360 N. ■

SAMPLE PROBLEM 8/2

A force of 196 N applied through the wheels to the road is just sufficient to keep a 1-tonne motor car moving at constant velocity on a straight level road. Determine the total coefficient of rolling friction present.

196 N

SPACE DIAGRAM
Problem 8/2 (a)

Analysis:
Draw a free-body diagram showing all forces acting on the car.

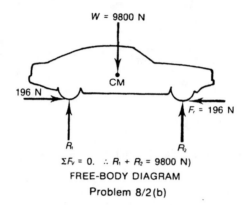

$$\Sigma F_v = 0. \quad \therefore R_1 + R_2 = 9800 \text{ N})$$

FREE-BODY DIAGRAM
Problem 8/2(b)

Since the car is moving with constant horizontal velocity, the force of 196 N is being used to overcome frictional drag, i.e. the total frictional force resisting motion is equal to 196 N.

Calculation:
Normal force between car and road = weight-force of car

$$= mg$$
$$= 9800 \text{ N}$$

where: $m = 1$ tonne
$\quad = 1000$ kg
$\quad g = 9.8$ m/s^2

Now,

$$\mu k = \frac{F_r}{N}$$

$$= \frac{196}{9800}$$

$$\therefore \mu k = 0.02$$

where: $F_r = 196$ N
$\quad N = 9800$ N

■

SAMPLE PROBLEM 8/3

A diesel locomotive has its mass of 250 tonnes distributed over its eight driving wheels. The coefficient of friction between the wheels and the rails is 0.25. What is the greatest pulling force (drawbar pull) that the engine can exert before the wheels begin to slip, if the locomotive is moving at constant velocity?

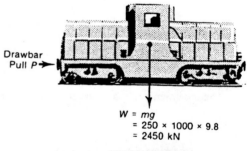

Drawbar
Pull P→

$W = mg$
$\quad = 250 \times 1000 \times 9.8$
$\quad = 2450$ kN

SPACE DIAGRAM
Problem 8/3 (a)

Analysis:
Draw a free-body diagram showing all forces acting on the locomotive.

Since the engine is moving at constant velocity, there is no acceleration and no unbalanced force in the direction of motion. Thus, the maximum frictional force generated equals the greatest pulling force of the locomotive, i.e. $F_r = P$.

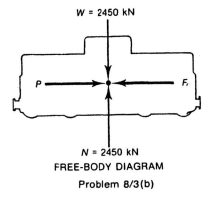

W = 2450 kN

P $\longrightarrow$ $\longleftarrow$ F_r

N = 2450 kN

FREE-BODY DIAGRAM

Problem 8/3(b)

Calculation:

$$F_r = \mu_s N$$
$$\therefore F_r = 0.25 \times 2450 \text{ kN}$$
$$= 612.5 \text{ kN}$$

where: $\mu_s = 0.25$
$$N = 250 \times 1000 \times 9.8 \text{ N}$$
$$= 2450 \text{ kN}$$

Result:
The maximum drawbar pull that can be exerted by the locomotive is 612.5 kN. ∎

SAMPLE PROBLEM 8/4

A ladder of length 8 metres leans up against a smooth vertical wall as shown, its other end resting on hard, rough ground. The ladder has a mass of 20 kg and the coefficient of static friction between the ground and the ladder is 0.4. The ladder is inclined at 60° to the horizontal ground. How far can a man of mass 70 kg climb up the ladder before the ladder begins to slip?

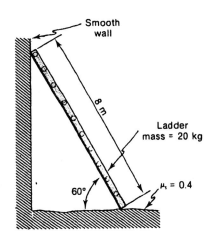

Smooth wall

8 m

Ladder mass = 20 kg

60°

$\mu_s = 0.4$

Problem 8/4 (a)

Analysis:
Since the vertical wall is *smooth*, friction can be ignored, and the wall thus provides a horizontal reaction only. The ground, however, provides a total reaction R which has both a vertical and a horizontal component. The horizontal component is obviously the force of friction resisting motion along the ground. When the ladder begins to slip, this frictional force reaches its maximum (and is thus limiting).

Let the man be x metres up the ladder when limiting friction occurs.

Calculation:
Draw a free-body diagram showing all forces acting on the ladder when friction is limiting.

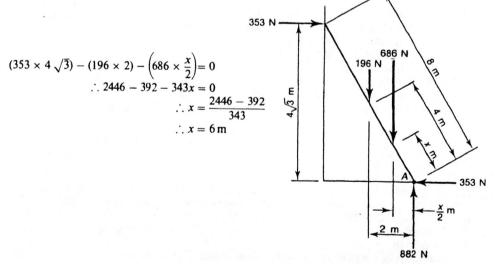

Now,

$$W_1 = \text{weight of ladder} = 20 \times 9.8$$
$$= 196 \text{ N}$$
$$W_2 = \text{weight of man} = 70 \times 9.8$$
$$= 686 \text{ N}$$

Problem 8/4 (b)

Consider $\Sigma F_V = 0$ Assume $\uparrow_+$

$$\therefore R_V - 196 - 686 = 0$$
$$\therefore R_V = 882 \text{ N}$$

Now,
$$F_r = \mu_s N \qquad \text{where } N = R_V$$
$$= 0.4 \times 882 \qquad \qquad = 882 \text{ N}$$
$$\therefore F_r \approx 353 \text{ N}$$

Consider $\Sigma F_H = 0$ Assume $\xrightarrow{+}$.

$$\therefore R_H - F_r = 0$$
$$\therefore R_H - 353 = 0$$
$$\therefore R_H = 353 \text{ N}$$

Now take moments about A, the bottom of the ladder.
For equilibrium, $\Sigma M_A = 0$ Assume $\curvearrowright_+$ Note that all distances depend upon the geometry of a $60°-30°-90°$ triangle.

$$(353 \times 4\sqrt{3}) - (196 \times 2) - \left(686 \times \frac{x}{2}\right) = 0$$
$$\therefore 2446 - 392 - 343x = 0$$
$$\therefore x = \frac{2446 - 392}{343}$$
$$\therefore x = 6 \text{ m}$$

Problem 8/4 (c)

Result:
The man could climb 6 metres up the ladder before it would begin to slip.

FRICTIONAL FORCES ON INCLINED PLANES

If a block of known mass is placed on a horizontal plane and the plane is gradually tilted until the block is just on the point of sliding, then the force acting to oppose motion between the block and the plane at this instant must be the force of limiting friction.

Let the angle of the plane to the horizontal at this instant be θ (Figure 8.6 (a)).

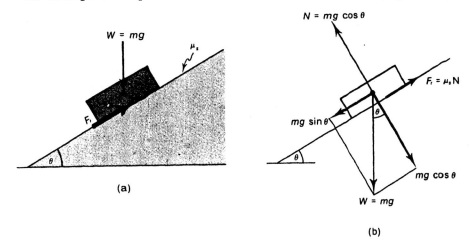

(a)

(b)

Figure 8.6   When the object is on the point of sliding, $\mu_s = \tan \theta$.

It is obvious that, in this situation, the normal force between the block and plane is *not* equal to the weight-force of the block. Instead it is equal to that component of the weight-force which acts perpendicular to the plane. This component has the value $mg \cos \theta$ (Figure 8.6 (b)).

It is also obvious that the force tending to pull the block down the plane is the component of the weight-force of the block which acts parallel to the plane, this component being equal to $mg \sin \theta$. The force of limiting friction Fr, must be equal to this component of the weight-force (since $\Sigma F = 0$ ✗ and no motion results).

Since
$$\mu_s = \frac{F_r}{N}$$
where: $F_r = mg \sin \theta$
$N = mg \cos \theta$.

$$\mu_s = \frac{mg \sin \theta}{mg \cos \theta}$$
$$\therefore \mu_s = \tan \theta$$

In other words, *when a block sitting on an inclined plane is on the point of sliding, the tangent of the angle of the inclined plane measured to the horizontal is equal to the coefficient of static friction.* This angle is known as the *angle of repose.*

Once sliding begins, the angle of the plane can be slowly decreased until a stage is reached where the sliding motion is just on the point of ceasing. At this point, the coefficient of kinetic (sliding) friction is equal to the tangent of the angle of the plane (measured to the horizontal). Since this angle is obviously less than before, $\mu_k < \mu_s$.

It is obvious from the above that no new concepts are required to analyse problems involving friction on an inclined plane and thus no new formulas need to be used.

SAMPLE PROBLEM 8/5

A brick of mass 4 kg rests on the middle of a builder's plank lying on horizontal ground. One end of the plank is raised and, when the angle of the plank is 25° to the ground, the brick just begins to move. After the brick is sliding freely, the end of the plank is gradually lowered until, when the plank is at 16° to the ground, it is found that the brick is just on the point of ceasing to slide. Calculate from first principles μ_s and μ_k for the brick and board.

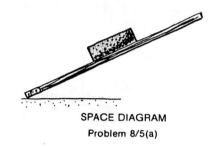

SPACE DIAGRAM
Problem 8/5(a)

(i) *Analysis for μ_s:*
When the block is on the point of sliding down the plane, the component of the weight-force of the brick acting down the plane must equal the force of limiting friction acting up the plane.

Calculation:
Draw a free-body diagram of the block on the plane.

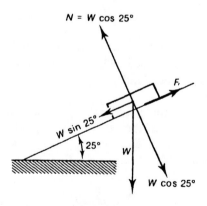

Problem 8/5(b)

$\Sigma F = 0$ Assume $+\nearrow$

$$\therefore F_r - W \sin 25° = 0$$
$$\therefore F_r = mg \sin 25°$$

But,
$$\mu_s = \frac{F_r}{N}$$
$$= \frac{mg \sin 25°}{mg \cos 25°}$$
$$= \tan 25°$$
$$\therefore \mu_s \approx 0.47$$

(ii) *Analysis for μ_k:*
When the sliding motion is about to cease, the component of the weight-force acting down the
plane must equal the force of friction opposing motion.

Calculation:
Draw a free-body diagram of the block on the plane.

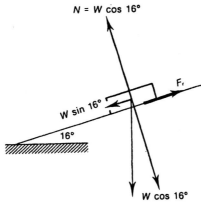

$N = W \cos 16°$

F_r

$W \sin 16°$

$16°$

$W \cos 16°$

Problem 8/5(c)

$\Sigma F = 0$ Assume ↗

$$\therefore F_r - W \sin 16° = 0$$
$$\therefore F_r = mg \sin 16°$$

But,
$$\mu_k = \frac{F_r}{N}$$
$$= \frac{mg \sin 16°}{mg \cos 16°}$$
$$\therefore \mu_k = \tan 16$$
$$\therefore \mu_k \approx 0.29$$

Result:
The coefficient of static friction, μ_s, present is 0.47 and that of sliding friction, μ_k, is 0.29. ■

SAMPLE PROBLEM 8/6

A sand sled of mass 10 kg is used on a sand dune
having a slope of 15°. If the coefficient of friction
present is 0.35, what force must the boy exert on
the tow rope to drag the sand sled up the dune with
constant velocity?

Mass = 10 kg

P

15°

SPACE DIAGRAM

Problem 8/6 (a)

Analysis:
Draw a free-body diagram showing all forces acting on the sled.

Since the sled is moving with constant velocity, the sum of all forces parallel to the surface of the sand dune must be zero. Thus, the pull, P, exerted by the boy on his tow rope must be equal to the sum of the component of the weight-force acting down the plane and the frictional force, F_r, which also acts down the plane (i.e. opposes motion).

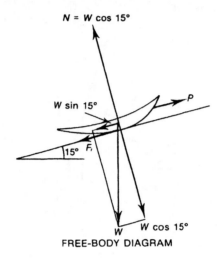

FREE-BODY DIAGRAM

Problem 8/6 (b)

Calculation:
$\Sigma F = 0$ Assume ↗

$$\therefore P - W \sin 15° - F_r = 0$$

But, $F_r = \mu N$, where N = component of weight-force W perpendicular to the surface of the dune,

$$\therefore F_r = 0.35 \times mg \cos 15° \qquad \text{where } \mu = 0.35$$
$$= 0.35 \times 10 \times 9.8 \times 0.966$$
$$\therefore F_r = 33.1 \text{ N}$$

$$P - W \sin 15° - 33.1 = 0 \qquad \text{where } W = mg$$
$$P = 25.4 + 33.1 \qquad\qquad = 10 \times 9.8 \text{ N}$$
$$= 58.5 \text{ N}$$

Result:
The force exerted by the boy on the tow rope was $\approx$ 59 N.   ■

ANGLE OF FRICTION

The methods employed so far in this chapter depend upon resolution of the forces operating into components parallel and perpendicular to the plane along which the frictional force acts. However, a simpler method of calculation, involving the *angle of friction*, is possible for many problems.

Consider the block of mass m units which is on the point of moving along the horizontal plane as shown in Figure 8.7 (a). The block is under the influence of four forces, the weight-force, W, the normal reaction of the plane, N (equal in magnitude to W), the force tending to cause motion, P, and the force of limiting friction, F_r.

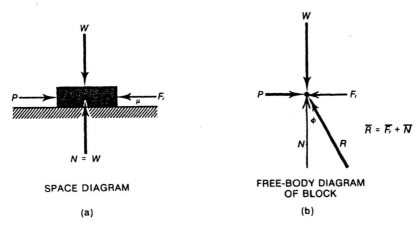

SPACE DIAGRAM

(a)

FREE-BODY DIAGRAM
OF BLOCK

(b)

Figure 8.7 A block on a horizontal plane: (a) shows that when the block is on the point of moving, the horizontal force P equals the frictional force F_r and $N = W$; (b) shows that F_r and N can be replaced by the single reaction R which is inclined at the angle ϕ to the normal reaction N, when tan $\phi = \mu$.

It is possible now to replace the normal reaction, N, and the force of limiting friction, F_r, with the force R, *where R can be considered to be the total reaction between the block and the plane* (Figure 8.7 (b)).

This resultant force, R, acts at the angle ϕ to the vertical force, N, and the value of ϕ can be found from

$$\tan\phi = \frac{F_r}{N}$$

But, $\dfrac{F_r}{N} = \mu_s$ for limiting friction, $\therefore \tan\phi = \mu_s$.

The angle that the total reaction, R, makes with the normal reaction, N, is thus $\phi = \tan^{-1}\mu_s$, *the total reaction, R, always being inclined away from the direction of motion.*

This information can be used to solve a wide range of problems involving limiting friction. For example, consider the block shown previously in Figure 8.4. We can now regard it as being subjected to three forces instead of the four as used previously and, at the point of limiting friction, these three forces are in equilibrium (refer to Figure 8.8 (a)).

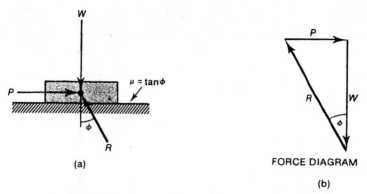

Figure 8.8

Thus, a simple triangle of forces diagram as shown in Figure 8.8 (b) enables the necessary pull, P, to just cause movement to be obtained graphically (or analytically, since $P/W = \tan\phi = \mu$).

SAMPLE PROBLEM 8/7

The force P is being applied by means of the rope to a crate of mass 20 kg in order to slide it along a flat horizontal concrete floor. If the coefficient of static friction is 0.5, determine:

(i) the tension in the rope if the rope is held parallel to the floor;

(ii) the *least value* of the tension that will cause motion, and *the angle of the rope* to the floor when this tension is present.

Analysis and Calculation:

(i) *When the rope is horizontal, i.e. parallel to the floor.*

Draw a free-body diagram of the crate, showing all forces present.

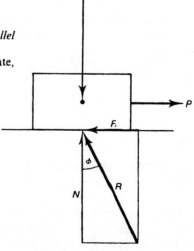

SPACE DIAGRAM

Problem 8/7 (a)

FREE-BODY DIAGRAM

Problem 8/7 (b)

Now replace the force of limiting friction, F_r, and the normal reaction, N, with the total reaction, R, inclined at ϕ to the force N. The angle ϕ is obviously the angle of friction which, in this instance, is equal to $\tan^{-1} 0.5$ (i.e. $26.6°$). The three forces P, W and R are acting on the block and are in equilibrium.

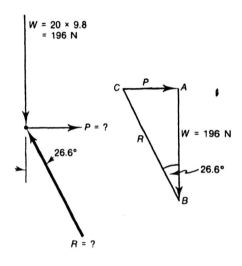

Problem 8/7 (c)

Thus the triangle of forces ABC can be drawn where the vector AC is the necessary force P. This can now be scaled off.

Alternatively, P can be calculated from

$$\frac{P}{W} = \tan 26.6°$$
$$P = W \tan 26.6° \qquad \text{where } W = 20 \times 9.8$$
$$= 196 \times 0.5 \qquad\qquad\quad = 196 \text{ N}$$
$$P = 98 \text{ N}$$

(*Note*. It would have been simpler, in this instance, to calculate P by calculating ΣF parallel to the plane, since, from $\Sigma F = 0$, $P = F_r$, and $F_r = \mu_s N = 0.5 \times 196 = 98$ N.)

(ii) *When the rope is inclined.*
Let the pull P on the rope be inclined at some angle α to the horizontal. Thus, three forces act on the block as shown, the weight-force, W, the angled pull, P, and the total reaction, R.
Now, draw a partial triangle of forces using the magnitude and direction of W (the vector AB) and the direction only of the total reaction, R.

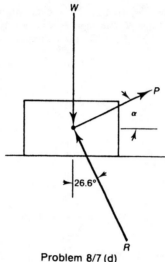

Problem 8/7 (d)

It is obvious that the least value of the force P is when the vector CA is drawn at $90°$ to BC, the total reaction. Thus, the magnitude of the least pull, P, can be scaled off the complete triangle of forces. Alternatively, from the geometry of the triangle of forces ABC

$$\frac{P}{W} = \sin 26.6°$$
$$P = W\sin 26.6° \text{ N}$$
$$= 196\sin 26.6° \text{ N}$$
$$P = 87.8 \text{ N}$$

and the angle $\alpha = 26.6° = \phi$, the angle of friction.

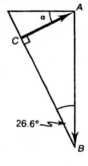

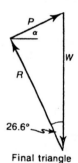

Final triangle
of forces for
W, R and P

Problem 8/7 (e)

Results:
When the rope is horizontal and the block is about to move, the tension is 98 N, but the least tension necessary to cause motion is 87.8 N and this must be applied when the rope is inclined upward at 26.6° to the horizontal. ∎

SAMPLE PROBLEM 8/8

The crate in Sample Problem 8.7 is now to be loaded into a truck by sliding it up a plank as shown. What least pull P applied to the rope will just cause motion up the plank and at what angle to the plank must the rope be in order to use this least pull? The coefficient of static friction between crate and plank is 0.5 and the plank is inclined at 15° to the ground.

SPACE DIAGRAM

Problem 8/8 (a)

Analysis:
Draw a free-body diagram showing all of the forces acting on the crate.

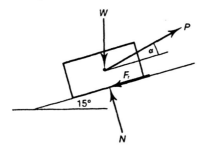

FREE-BODY DIAGRAM
Problem 8/8 (b)

Now replace the normal reaction between the plank and the crate, N, and the force of limiting friction, F_r, with the total reaction, R, which must be inclined at the angle ϕ to the normal reaction as shown. The angle ϕ for limiting friction is the *angle of friction* and *tan ϕ = 0.5.*
Now only three forces act on the crate: the weight-force, W, the pull, P, and the total reaction, R. These are in equilibrium when motion up the plane is impending.

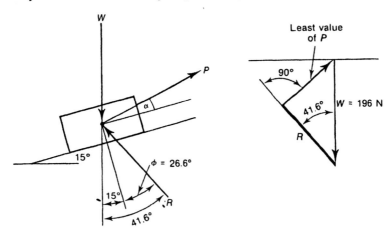

Problem 8/8 (c)

Since the magnitude and direction of W is known, and the direction of R is also known, a partial triangle of forces can be drawn. *As in Sample Problem 8/6, the least value of P is when P is drawn perpendicular to the total reaction, R.* Thus, the magnitude of P can be scaled off from the triangle of forces. Alternatively, P can be calculated from the geometry of the triangle of forces

$$\frac{P}{W} = \sin(15° + \phi)$$

But,
$$\phi = \tan^{-1} 0.5 \quad \text{(since } \phi = \text{angle of friction)}$$
$$\therefore \phi = 26.6°$$
$$\therefore \frac{P}{W} = \sin(15° + 26.6°)$$
$$\therefore P = W \sin 41.6° \qquad \text{where } W = 196 \text{ N}$$
$$\therefore P = 130.1 \text{ N}$$

From the geometry of the problem, it is readily seen that P acts at the angle $(\alpha + 15°)$ to the horizontal. But, again from the final triangle of forces for W, P and R

$$\alpha + 15° = 15° + \phi$$
$$\therefore \alpha = \phi = 26.6°$$

Result:
The least pull, P, necessary to drag the crate up the plank is ≈ 130 N and it must act upwards at an angle of 26.6° to the inclined plank. ∎

Note: Part (ii) of Sample Problem 8/7 and the Sample Problem 8/8 illustrated that the least force necessary to move an object along a plane is always inclined at the angle ϕ to that plane, where ϕ is the angle of friction (i.e. $\tan \phi = \mu_s$).

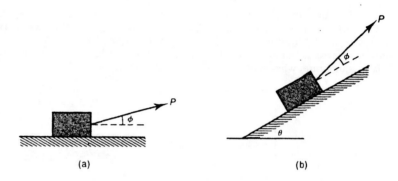

(a)   (b)

Figure 8.9 The least value of P that just causes motion is always inclined at the angle of friction ϕ to the direction of motion. This is true regardless of the frictional coefficient or the masses involved.

FRICTIONAL FORCES AND WEDGES

While the working action of a wedge is essentially a problem of dynamics and usually involves impulsive forces such as sledgehammer blows, forces acting to hold a wedge in equilibrium can be solved by the application of the laws of statics. In such problems it is usual to ignore the weight-forces of the wedges themselves since they are generally small in relation to the other forces involved.

SAMPLE PROBLEM 8/9

While in the process of replacing certain foundations of a house it was found necessary to use a vertical prop and two folding wedges to take the load of part of a floor bearer and its associated floor structure.

If the load taken by the prop was 1 tonne and the wedge surfaces are inclined at 10° to the horizontal, calculate the minimum coefficient of friction necessary between the wedge surfaces in order to prevent the wedges slipping under the applied load.

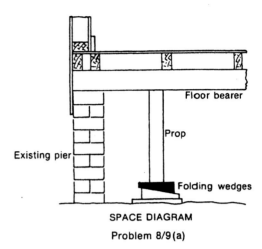

SPACE DIAGRAM

Problem 8/9 (a)

Analysis:

Consider the forces acting on the *top wedge*; it is in equilibrium under the influences of three forces:

(i) the vertically down force exerted by the load, which is equal to $1000 \times 9.8 = 9800$ N.

(ii) the frictional force F_r acting along the inclined surface of the wedge so as to oppose the tendency of this wedge to slide downwards and to the right.

(iii) the normal reaction N acting at 90° to the inclined surface of the wedge.

If the block does not move these three forces must be in equilibrium.

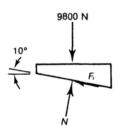

FORCES ON THE TOP WEDGE

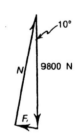

FORCE TRIANGLE
FOR TOP WEDGE

Problem 8/9 (b)

Calculation:

Draw the triangle of forces for these forces acting on the block. From the geometry of this triangle.

$$\frac{F_r}{9800} = \sin 10°$$
$$\therefore F_r = 9800 \sin 10°$$
$$\frac{N}{9800} = \cos 10°$$
$$\therefore N = 9800 \cos 10°$$

But
$$F_r = \mu_s N,$$
$$9800 \sin 10° = \mu_s \times 9800 \cos 10°$$
$$\therefore \mu_s = \frac{9800 \sin 10°}{9800 \cos 10°}$$
$$\mu_s = \tan 10° \approx 0.2$$

where μ_s = coefficient of static friction

Result:

The minimum coefficient necessary is 0.2. Two important assumptions were made in this solution:
  (i) that there was no tendency for the bottom of the prop to slide along the top surface of the wedge and thus no frictional force exists on this surface,
  (ii) there was no crushing of the wedges during the lifting operation. ■

SAMPLE PROBLEM 8/10

A heavy block of mass 500 kg is to be lifted vertically by the system of two wedges as shown. If the coefficient of static friction for all surfaces is 0.3, determine the force P which will cause the system to be on the point of movement.

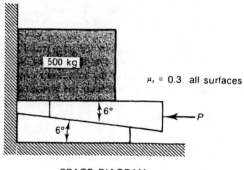

SPACE DIAGRAM

Problem 8/10 (a)

Analysis:

Consider the forces acting on the 500-kg block.

  (i) Since the block is on the point of moving, the angle of friction ϕ_s is equal to $\tan^{-1} 0.3$, which is 16.7°.

 (ii) Consider the forces on the vertical side of the block. They will consist of the friction force F_1 and the normal force N_1, *and they can be replaced by the total reaction R_1 acting at angle $\phi_s = 16.7°$ to N_1.* Note that the friction force F_1 acts down so as to oppose the tendency of the block to rise vertically.

(iii) Consider the forces acting on the bottom surface of the block. They are the friction force F_2 and the normal reaction N_2, and they can be replaced by the total reaction R_2 acting at $\phi_s = 16.7°$ to N_2. Note that the friction force F_2 is directed to the left since *as the top wedge moves to the left the relative movement of the block in relation to this wedge is to the right,* thus F_2 acts to the left so as to oppose motion.

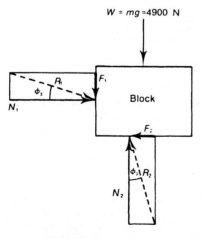

FORCES ON THE BLOCK

Problem 8/10 (b)

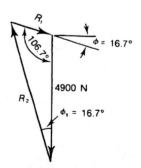

TRIANGLE OF FORCES
FOR THE BLOCK

Problem. 8/10 (c)

Now, consider the forces acting on the top wedge. The angle of friction remains $\phi_s = 16.7°$.
(i) On the top surface the total reaction R_3 replaces the frictional force F_3 and the normal reaction N_3. *This total reaction R_3 must be equal in magnitude but opposite in sense to the total reaction R_2 acting on the block.*
(ii) On the inclined surface, by similar reasoning, the total reaction R_4 replaces the frictional force F_4 and the normal reaction N_4, *or at $(16.7 + 6)° = 22.7°$ to the vertical.*

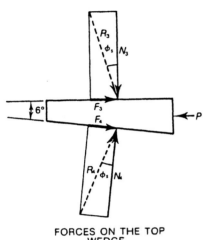

FORCES ON THE TOP
WEDGE

Problem 8/10 (d)

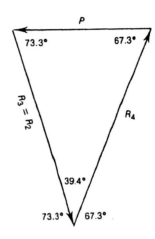

TRIANGLE OF FORCES
FOR THE WEDGE

Problem 8/10(e)

Calculation:
For the block: since W, R_1 and R_2 are in equilibrium, they must satisfy the triangle of forces shown in the diagram.
From the geometry of the triangle and the sine rule

$$\frac{4900}{\sin 56.6°} = \frac{R_2}{\sin 106.7°}$$
$$\therefore R_2 = 5667 \text{ N}$$

For the wedge: P, R_3 and R_4 also satisfy a triangle of forces. Note that $R_2 = R_3 = 5667$ N from the previous calculation.
From the geometry of the triangle and the sine rule:

$$\frac{P}{\sin 39.4°} = \frac{5667}{\sin 67.3°}$$
$$\therefore P = 3899 \text{ N}$$

Result:
The minimum force P necessary to put the wedge system on the point of movement is approximately 3.9 kN. (Note that, once the two triangles of forces were drawn, an answer could have been obtained by graphically scaling R_2 and then P). ■

REVIEW PROBLEMS

(Use $g = 9.8$ m/s^2 unless otherwise specified.)

8/11

A block of stone of mass 100 kg is resting on a horizontal concrete path. Determine the horizontal force P necessary to just cause the block to slide if the coefficient of static friction is 0.5.

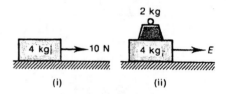

Problem 8/11

8/12

An effort of 10 N is required to cause the 4-kg block to be on the point of sliding.
 (i) What is the coefficient of static friction present between the block and the plane?
(ii) Determine the effort E to just move the block if a 2-kg mass is placed on top of the 4-kg mass as shown in the diagram (b).

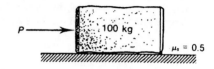

Problem 8/12

8/13

A diesel locomotive is stationary on the track. Given that the mass of the locomotive is 45 tonnes, find the greatest drawbar pull that the locomotive can exert if the coefficient of friction between the wheels and rails is 0.2.

Problem 8/13

8/14

A book of mass 1.5 kg is held vertically and prevented from falling by being grasped by the fingers of one hand. What minimum clamping force must be exerted by the fingers if the coefficient of friction is 0.25?

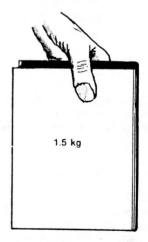

Problem 8/14

8/15

A car of mass 1.2 tonnes is left stationary as shown on a concrete ramp of slope 1 in 3. The bonnet is bumped as the owner passes across in front of the car which then slides a short distance down the ramp. Determine the coefficient of static friction present between the wheels and the concrete.

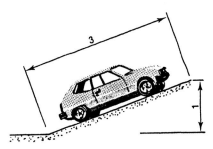

Problem 8/15

8/16

A clamp exerts a normal force of 500 N on three pieces of wood A, B and C, held together as shown in the diagram. The coefficient of friction between the wooden pieces is 0.2 and between the wood and the clamp surfaces is 0.35. What is the maximum tensile force P that can be applied before sliding occurs?

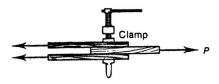

Problem 8/16

8/17

A small box of mass 4 kg is moved with constant velocity across a horizontal surface by a force of 25 N. Determine the coefficient of friction present and the frictional force operating, given that the 25-N force is applied
 (i) horizontally on steel;
(ii) at 30° to the horizontal, on concrete.

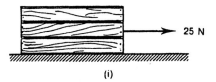

(i)

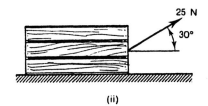

(ii)

Problem 8/17

8/18

A boat of mass 2 tonnes rests on a slip which is inclined at 20° to the horizontal, the coefficient of static friction between the slip-rails and the boat being 0.3.
    Determine the least force, F, needed in the winch cable to move the boat
 (i) up the slip, and
(ii) down the slip with constant velocity, given that the cable is parallel to the rails.

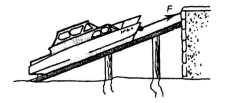

Problem 8/18

8/19
A ladder of mass 20 kg rests against a smooth vertical wall and on a rough concrete path as shown. Draw a free-body diagram of the ladder in this situation and determine the reactions at the wall and the path.

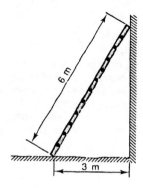

Problem 8/19

8/20
The ladder in Problem 8/19 is now repositioned by moving its end further away from the vertical wall. What is the minimum angle θ possible between the concrete path and the ladder before the ladder slips under its own weight? The coefficient of static friction between ladder and path is 0.6.

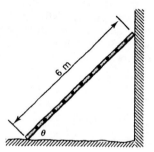

Problem 8/20

8/21
A boy of mass 45 kg places a 7-metre long ladder against a house with the bottom of the ladder 2.5 metres away from the wall. Assume a coefficient of friction at the bottom to be 0.50 and 0.2 between the top of the ladder and the wall. Is the ladder safe against slipping when the boy is three-quarters of the way to the top? Suppose the ladder is at 45° angle; is it safe against slipping? Solve this problem graphically.

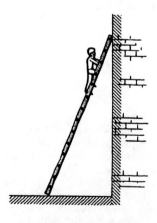

Problem 8/21

8/22

A brake pad operated by the lever system shown is pressed against a wheel with a force of 60 N. If the coefficient of friction between the brake pad and the wheel is 0.6,

(i) determine the tangential frictional force tending to stop the wheel, and

(ii) determine the force P necessary to produce this braking force.

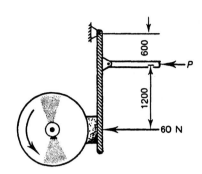

Problem 8/22

8/23

A box of mass 10 kg rests on a plane inclined at 30° to the horizontal as shown. If the coefficient of static friction between box and plane is 0.3, determine the magnitude of the horizontal force P that prevents the box sliding down the plane. [*Hint*: Solve graphically.]

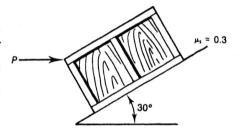

Problem 8/23

8/24

A machine of mass 500 kg is being dragged up a ramp by a block and tackle as shown. The coefficient of friction between the machine and ramp is 0.4 and the ramp is inclined at 10° to the horizontal. Determine, when motion is about to occur up the plane,

(i) the angle α between rope and ramp necessary to produce the least possible tension in the rope;

(ii) the rope tension at this instant.

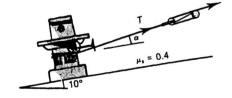

Problem 8/24

8/25

In a certain mechanism a hydraulic ram is used to apply a force P to the end of a wedge-shaped slider as shown. This slider then moves a block of mass 50 kg horizontally across a surface as shown. Determine the value of the force P necessary to put the system on the point of motion, given the coefficient of static friction is 0.23 for all surfaces in contact. (Take $\tan^{-1} 0.23 = 13°$.)

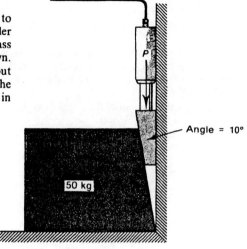

Problem 8/25

8/26

A metal bracket is free to slide on a pipe as shown.
 (i) If the mass of the bracket is 3 kg will it slide down the pipe under its own weight if the coefficient of static friction is 0.45?
(ii) What maximum length of bearing will just prevent the bracket sliding under its own weight?

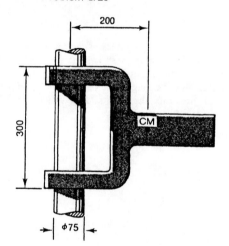

Problem 8/26

8/27

Determine the value of the force P which will just cause the 50-kg block to move. The coefficient of static friction between the block and the plane is 0.25. [*Hint*: Motion may occur by either *sliding*, or by *tipping* about the corner A.]

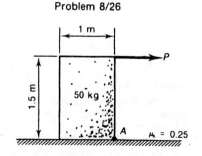

Problem 8/27

8/28

A tractor is used to pull out stumps. The pulling chain is horizontal and 0.6 metres above the ground. The tractor has a mass of 1200 kg of which 400 kg is on the front axle and 800 kg on the rear axle, when it is not pulling. The two axles are 2.2 metres apart. If the chain is pulled with a 6-kN force, then what is the vertical load on each axle? What coefficient of friction is needed between the rear driving tyres and the ground if no slipping is to occur?

Problem 8/28

8/29

A man of mass 81.6 kg walks up a plank *BE* of negligible weight, which rests with the end *B* on a ramp sloping at 30° to the horizontal. It rests on a frictionless roller at *D*, 3 m along the plank from *B*. What must be the minimum value of the coefficient of static friction, if the plank does not slide *down* the slope when the man steps on to the plank at *B*?

Supposing that the man can reach a position *x* from the wall *DC* before the plank will slide *up* the slope, and the coefficient of friction is still the same, *show by a sketch* the forces acting on the plank, and determine the distance *x*. Is it ever possible for the man to reach *D*?

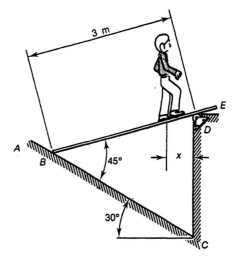

Problem 8/29

The Apollo 15 moonshot. The Saturn V rocket was 110 metres high and generated
33 meganewtons of thrust on lift off.

9

Linear Motion

$$v \text{ (average)} = \frac{u + v}{2}$$

$$v = u + at$$
$$v^2 = u^2 + 2\,as$$
$$s = ut + \tfrac{1}{2}\,at^2$$

Dynamics involves the study of the motion of bodies and of the forces that either cause or alter motion. In most dynamics' problems the size of the body is immaterial and it is usual to regard the body as a dimensionless particle with its mass concentrated at a certain point, called the *centre of mass* or *centre of gravity*.

The study of dynamics is usually divided into two main sections:

(a) *kinematics*, which considers the motion of a body quite apart from the forces either causing the motion or which may be produced as a result of the motion;

(b) *kinetics*, which is concerned with the forces that either cause or result from motion. Thus, whereas in statics we are concerned with balanced force systems (i.e. those in equilibrium), in dynamics we are concerned with the *effect of unbalanced force systems* on bodies. The relationship $F = ma$, derived from Newton's second law, is fundamental to the study of kinetics and is fully considered in Chapter 10, "Forces and Motion".

The motion of a body can be one of three basic types listed below.

(i) *Rectilinear*, or motion in a *straight line*.

(ii) *Curvilinear*, or motion in a *curved path*. The most common example of this type of motion is movement in a circular path; however, the flight of a shell from a heavy gun or the trajectory of a rocket are also common examples of curvilinear motion.

(iii) *Rotation* about a fixed axis is the third basic type of motion and is really a special case of plane curvilinear motion. This is the motion that a pilot experiences when placed in a centrifuge. In this type of motion, we are usually concerned with the study of the forces that constrain the object in its circular path. The pilot in the centrifuge, for example, is restrained by his seat and harness and can be subjected to many times the normal *g* force acting on a human body. His ability to cope with such abnormal forces can be assessed by recording his heartbeat, respiration, and other vital body functions. This type of research is a vital part of all aerospace projects where pilots are used.

When studying dynamics it is simpler to consider rectilinear (straight-line) motion first, since all the formulas that apply here have their direct equivalents in problems of circular motion. To this end, Chapters 9, 10, and 11 are concerned only with problems involving rectilinear motion, or with projectile problems where motions in two directions mutually at 90° to each other can be considered separately.

There are four fundamental concepts that must be fully understood before the solutions to problems of kinematics can be determined. These concepts are *displacement, time, velocity,* and *acceleration*.

DISPLACEMENT (*s*)

If a particle or an object moves from one position to another it is said to have been *displaced*. Both the distance of displacement and its direction must be specified, thus displacement is a vector quantity. Displacement can be either linear or angular, linear displacement being stated in metres and angular displacement in either revolutions or radians (θ). Since displacement is the distance that a body moved in a particular direction, the usual methods of vector addition (parallelogram or triangle of vectors) must be used if a body undergoes a series of displacements.

It is important to note that displacement is often very different from the total distance travelled.

SAMPLE PROBLEM 9/1

If a ship sails due east for 500 kilometres and then heads due north for another 1200 kilometres, find its displacement from its starting position.

Analysis:
From the data, the vector triangle *OEN* can be constructed in which the vectors *OE* and *EN* represent the two distances travelled. The displacement, *s*, of the ship from its starting position is thus equal to the vector *ON*.

Calculation:
From Pythagoras' theorem

$$ON^2 = OE^2 + EN^2$$
$$s^2 = 500^2 + 1200^2$$
$$s = 1300 \text{ km}$$
$$\text{Tan } \theta = \frac{1200}{500}$$
$$= 2.4$$
$$\theta \approx 67°$$

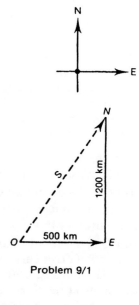

Problem 9/1

Result:
The displacement of the ship from its original position is 1300 km E, 67°N.

SAMPLE PROBLEM 9/2

A ball is fired vertically upwards and reaches a maximum height of 25 metres. What is its displacement when it has fallen 10 metres from its maximum elevation?

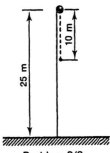

Problem 9/2

Calculation:
Displacement = distance measured from original position
$$= + 25\,m - 10\,m$$
$$= + 15\,m$$

Result:
The displacement of the ball at this instant is 15 metres above its starting position. (It is important to note that the displacement in this problem is quite different from the total distance travelled which is 35 metres.) ∎

SAMPLE PROBLEM 9/3

A Mirage jet fighter leaves Richmond Air Base and flies due east for 300 km. It then turns and flies due north for 400 km, after which the pilot sets a direct course back to base. What was the displacement of the Mirage jet fighter
 (i) after travelling the first 700 km, and
(ii) after landing back at the base?

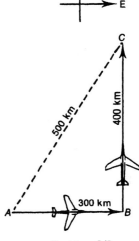

Problem 9/3

Analysis:
Draw the vector diagram *ABC* where *A* is the air base, *AB* is the easterly displacement, and *BC* the northerly displacement. Since displacement is the position relative to the starting position of the motion, after travelling the distance of 700 km the displacement of the Mirage from base is represented by the vector *AC*.

Calculation:
 (i) Resultant displacement = vector *AC*
$$AC = 300^2 + 400^2$$
$$\therefore AC = 500 \text{ km}$$
$$\theta = \tan^{-1} 1.333$$
$$\therefore \theta \approx 58°$$
(ii) After landing at base the displacement is zero.

Result:
The displacement after travelling the first 700 km was 500 km in a direction E 58° N. After arriving back at base the displacement was zero. ∎

VELOCITY (*v*)

The *velocity* of a body is its rate of change of position. Therefore velocity is expressed as displacement per unit time and is a vector quantity. The commonly used time unit is the *second**, and thus linear velocities are expressed as metres per second (m/s), while angular velocities are expressed as radians per second (rad/s) or as revolutions per second (rev/s). (Note that while revolutions per second or revolutions per minute are not SI units they are extremely useful in engineering and will be used where appropriate.)

Speed is the magnitude of a velocity without regard to its direction; it is therefore a scalar quantity only.

When the velocity of an object is being measured, it must be related to some other object (or reference frame). For example, if a car is travelling at 20 m/s in an easterly direction, then this velocity is relative to the earth's surface (or roadway). The driver in the car is also travelling at 20 m/s relative to the earth's surface, but his velocity relative to the car itself is zero.

SAMPLE PROBLEM 9/4

Refer to Sample Problem 9/3. If the Mirage jet flies at constant speed and takes 45 minutes to fly the 700 km to point *C*, what is
(i) its average speed;
(ii) its average velocity, relative to its base?

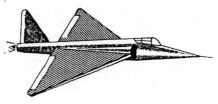

Problem 9/4

Calculation:

$$\text{Average speed} = \frac{\text{distance travelled}}{\text{time taken}}$$
$$= \frac{700}{0.75}$$
$$= 933.3 \text{ km/h}$$

where: distance travelled $= 700$ km
time taken $= 45$ min
- $= 0.75$ h

$$\text{Average velocity} = \frac{\text{displacement}}{\text{time taken}}$$
$$= \frac{500}{0.75}$$
$$= 666.7 \text{ km/h in a direction E } 58° \text{ N.}$$

where: displacement $= 500$ km
time taken $= 0.75$ h

Result:
The average speed was ≈ 933 km/h and the average velocity was ≈ 667 km/h.
(Note that kilometres per hour may be used to express velocities but that this is not a preferred SI unit.) ∎

---

*The *second* is defined as the duration of 9 192 631 770 periods of the radiation corresponding to the transition between the two hyperfine levels of the ground state of the caesium-133 atom.

ACCELERATION (*a*)

If the velocity of a body is changing with time, then an *acceleration* is occuring. Thus, acceleration is the rate of change of velocity with respect to time. The units of uniform linear acceleration are metres per second *per second*, written as m/s².

Angular acceleration will be considered in detail in Chapter 12, the units being radians per second per second, written rad/s², or revolutions per second per second, written rev/s².

A *negative* acceleration is commonly called a *deceleration* or a *retardation*, a minus sign being used in front of the numerical value. For instance, a deceleration of 4 m/s² is written as an acceleration of − 4 m/s². Similarly, a body fired vertically upwards is subjected to a deceleration of 9.8 m/s², which is written as the acceleration of − 9.8 m/s² (or − *g*).

SAMPLE PROBLEM 9/5

A car on a dragstrip accelerates uniformly from a standing start to 180 km/h in 12 seconds. What was its acceleration in m/s²?

Calculation:

$$180 \text{ km/h} = \frac{180 \times 1000}{60 \times 60} \text{ m/s}$$
$$= 50 \text{ m/s}$$
$$\text{acceleration} = \frac{\text{final velocity} - \text{initial velocity}}{\text{time}}$$
$$\therefore a = \frac{50 - 0}{12}$$
$$a = 4.17 \text{ m/s}^2.$$

Result:
The acceleration of the car was 4.17 m/s². ∎

ACCELERATION DUE TO GRAVITY (*g*)

The *acceleration due to gravity*, *g*, (or the *acceleration of free fall*) is the uniform acceleration of bodies induced by the gravitational attraction of the earth. Its value can be found from Newton's law of gravitation which states that two bodies of masses *M* and *m* are mutually attracted towards each other by equal and opposite forces *F*. The force *F* can be found from the relationship $F = G(Mm/r^2)$, where *r* is the distance between the two bodies and *G* is the universal constant termed the *constant of gravitation*. In the case of the object on the earth's surface, the force *F* exerted by the earth on the object then becomes the weight *W* of the object, and *r* the radius of the earth.

Substituting in the equation $F = G \dfrac{Mm}{r^2}$,

we obtain
$$W = G \frac{Mm}{r^2}$$

or
$$W = m \frac{GM}{r^2}$$

where: W = the weight-force of the body

m = the mass of the body

$\dfrac{GM}{r^2}$ = the acceleration due to the gravity at the earth's surface.

Thus, the acceleration due to gravity (g) is a constant that is independent of the mass of the body concerned, and the equation for weight-force becomes $W = mg$.

On or near the earth's surface the only gravitational force of any magnitude is the weight-force of the object concerned. To illustrate this, consider two steel balls of mass 4 kg which are in contact with each other and the earth's surface. From Newton's law of gravitational attraction, the weight-force of each ball becomes 39.2 newtons, whereas the force of mutual attraction existing between them is only 0.000 001 043 newtons.

Since the earth is not a uniform sphere, the acceleration due to gravity varies slightly from place to place; however, in most dynamics' problems the value of g can be taken as 9.8 m/s^2. For a quick solution to a problem, g can be taken as 10, this approximation introducing an error of about 2% into your answer.

Table 9.1. Summary of Units Used in Kinematics

Quantity	Linear units	Rotational units
Time	second (s)	second (s)
Displacement	metre (m)	radian (rad)
	kilometre (km)	revolution (rev)
Velocity	metres per second (m/s)	radian per second (rad/s)
	kilometres per hour (km/h)*	revolutions per second (rev/s)
Acceleration	metres per second per second (m/s^2)	radians per second per second (rad/s^2)
	kilometres per hour per second (km/h/s)†	revolutions per second per second (rev/s^2)

*This is not an SI unit, but is often used in practice. To convert km/h to m/s, remember that 36 km/h = 10 m/s.
†A very clumsy unit that must be converted into m/s^2.

EQUATIONS DESCRIBING STRAIGHT-LINE (RECTILINEAR) MOTION

Suppose that a body travelling in a straight-line path has an initial velocity, u. It now undergoes a uniform acceleration, a, for a length of time, t, and its velocity changes to v.

Since
$$\text{acceleration} = \frac{\text{change in velocity}}{\text{time taken}}$$

$$\therefore a = \frac{v - u}{t}$$

and,
$$v = u + at \tag{1}$$

Because the body undergoes uniform acceleration, the average velocity over the time, t, is

$$v \text{ (average)} = \frac{u + v}{2} \tag{2}$$

The distance, s, covered by the body in the time interval, t, is therefore

$$s = \text{average velocity} \times \text{time taken}$$

$$s = \frac{u + v}{2} \times t$$

$$\therefore s = \frac{u + (u + at)}{2} \times t, \quad \text{since } v = u + at$$

and thus,
$$s = ut + \tfrac{1}{2} at^2 \tag{3}$$

If $a = \dfrac{v - u}{t}$, then $t = \dfrac{v - u}{a}$

Substituting $t = \dfrac{v - u}{a}$ into $s = \dfrac{u + v}{2} \times t$

we obtain,
$$s = \frac{u + v}{2} \times \frac{v - u}{a}$$

$$\therefore s = \frac{v^2 - u^2}{2a}$$

and thus,
$$v^2 = u^2 + 2as \tag{4}$$

From the above calculations, it is apparent that the four equations which are most useful in solving problems of linear motion involving uniform accelerations are as follows:

$$v \text{ (average)} = \frac{u + v}{2}$$
$$v = u + at$$
$$v^2 = u^2 + 2as$$
$$s = ut + \tfrac{1}{2} at^2$$

where: u = initial velocity
v = final velocity
a = uniform acceleration
t = time interval
s = displacement.

Sign Convention

Any suitable sign convention can be used for the solution of problems, provided that the same convention is used throughout all parts of the same problem. Probably the most useful convention is to consider the direction of initial displacement as being positive, and thus all velocities, accelerations and displacements measured in this direction will be positive, while those measured in the opposite direction will be negative.

An exception to this convention is in the case of g, the acceleration due to gravity (or the acceleration of free fall) which is always positive vertically down and negative vertically upward.

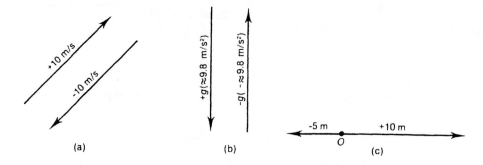

Figure 9.1 The application of a consistent sign convention for vectors: (a) shows two velocities; (b) shows *g*, the acceleration of free fall, acting positively down and negatively upward; and (c) shows positive and negative displacements in relation to the starting point *0*.

PROBLEM SOLVING

The most important distinction to be made when starting any problem is to separate the known information from the unknown. For example, there are five fundamental values in kinematic equations *s, t, u, v, a* and lists should be made of those which are known and those which are unknown before attempting to solve a particular problem. Often some values are implied rather than given directly. For instance, the words "falling freely" imply that the body is under the influence of the earth's gravitational pull and is thus being subjected to a downward acceleration of *g* (≈ 9.8 m/s^2).

Most problems can be solved by more than one method and graphical techniques often lead to a saving in time and effort.

SAMPLE PROBLEM 9/6

A car starts from rest with a constant acceleration of 5 m/s^2. Determine
(i) the speed at the end of the sixth second;
(ii) the average speed for the six-second interval;
(iii) the distance covered in six seconds.

Calculation:

(i)
$$v = u + at$$
$$v = 0 + (5 \times 6)$$
$$= 30 \text{ m/s}$$

where: $v = ?$
$u = 0$
$a = 5 \text{ m/s}^2$
$t = 6 \text{ } s$

(ii)
$$v \text{ (average)} = \frac{u + v}{2}$$
$$= \frac{0 + 30}{2}$$
$$= 15 \text{ m/s}$$

(iii)

$$s = ut + \tfrac{1}{2} at^2$$
$$= (0 \times 6) + (\tfrac{1}{2} \times 5 \times 6^2)$$
$$= 0 + (2.5 \times 36)$$
$$= 90 \text{ m}$$

Alternative solution:
$$s = v \text{ (average)} \times t$$
$$= 15 \times 6$$
$$= 90 \text{ m}$$

Result:
The speed at the end of the sixth second is 30 metres per second, the average speed for the first six seconds is 15 metres per second, and the distance covered is 90 metres. ∎

SAMPLE PROBLEM 9/7

A bullet is fired vertically upwards from a rifle having a muzzle velocity of 392 m/s. Determine the time taken for the bullet to reach its maximum height. Could the bullet hit a balloon having an altitude of 4500 metres?

Analysis:
When the bullet reaches its maximum height its velocity will be zero and while travelling vertically upward its acceleration is $-g$ (-9.8 m/s²)

Calculation:

$$v = u + at$$
$$0 = 392 - 9.8t$$
$$\therefore t = \frac{392}{9.8}$$
$$t = 40 \text{ seconds}$$

where: $v = 0$
$u = 392$ m/s
$a = 9.8$ m/s²
$t = ?$

The maximum height reached is found by substituting in
$$s = ut + \tfrac{1}{2}at^2 \qquad \text{(where } t = 40s)$$
$$\therefore s = (392 \times 40) + (\tfrac{1}{2} \times -9.8 \times 40^2)$$
$$\therefore s = 7840 \text{ m}$$
(Note that an alternative method for determining maximum height reached would be to use $v^2 = u^2 + 2as$.)

Result:
The bullet takes 40 seconds to reach its maximum height of 7840 metres above the earth's surface. It could therefore hit the balloon at the altitude of 4500 metres. ∎

GRAPHICAL SOLUTIONS IN KINEMATICS

Many problems in dynamics are readily solved by graphical methods, the answers being obtained without the need for lengthy calculations. Since graphical methods generally give less accurate answers, the greatest care should be used in the selection of scales and the plotting of the given data.

Displacement–Time Graphs

Suppose that a man on a bicycle starts from rest and is timed as he passes points along a straight road, these points being 10 metres apart. The resultant data are shown below.

Distance (metres)	10	20	30	40	50	60	70	80	90	100
Time from start (seconds)	2	4	6	8	10	12	14	16	18	20

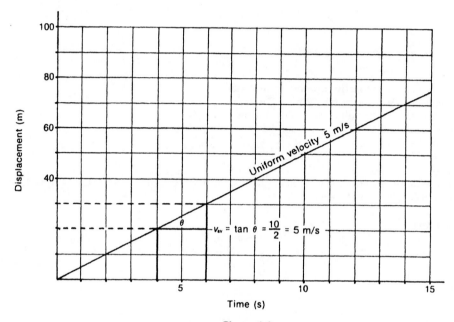

Figure 9.2

Using these data, the displacement–time graph shown in Figure 9.2 can be plotted if suitable scales for displacement and time are selected. From this graph the average velocity of the bicycle can be obtained for any given interval of time. For example, for the period from fourth to sixth second:

$$\text{Average velocity} = \frac{\text{displacement}}{\text{time}}$$

$$= \frac{30 - 20}{6 - 4}$$

$$= \frac{10}{2}$$

$$= 5 \text{ m/s}$$

This average velocity is, in fact, equal to the slope of the graph (expressed as the tangent of the angle θ) and, since the graph is a straight line, the velocity is uniform throughout the motion.

SAMPLE PROBLEM 9/8

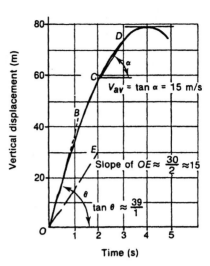

Problem 9/8

The graph shown here represents the displacement–time graph of the motion of a ball thrown from the earth with an unknown initial velocity. This graph was plotted from the data tabulated below.

Determine from this graph
 (i) the initial velocity of the ball;
 (ii) the instantaneous velocity after 4 seconds;
 (iii) the average velocity for the third second of motion.

Time from launch (seconds)	0	1	2	3	4	5
Displacement (metres)	0	34.3	58.8	73.5	78.4	73.5

Analysis and Calculation:
 (i) The initial velocity is equal to the slope of the tangent to the curve at time $= 0$ seconds. This tangent is represented by the line OB which cuts the time grid $t = 1$ at an interval of ≈ 39 units. That is, the slope of this tangent is $\approx 39/1$. Thus, initial velocity is ≈ 39 m/s.

 (ii) The instantaneous velocity after 4 seconds is obviously zero since the tangent to the curve at $t = 4$ seconds has a slope of $0/4$ or zero.

 (iii) The average velocity for the third second is represented by the slope of the cord CD. The slope of this cord is easily found if a line is drawn through the origin parallel to CD. This line is represented by OE which cuts the time grid $t = 2$ at ≈ 30. Thus, the average velocity for the third second is ≈ 15 m/s (i.e. $\tan \alpha = 15$).

Result:
The ball was thrown with an initial upward velocity of ≈ 39 m/s, the velocity after 4 seconds was zero, and the average velocity during the third second was ≈ 15 m/s. ■

Velocity–Time Graphs

Perhaps the most useful graph in kinematics is the velocity–time graph on which velocity (or speed) is plotted on the vertical axis and time is plotted on the horizontal axis. Figure 9.3 shows three possible velocity–time graphs:

(a) for an object having a uniform velocity of 4 m/s;

(b) for an object having a uniform acceleration of 3 m/s²;

(c) for an object having irregular velocity, that is, an acceleration varying with time.

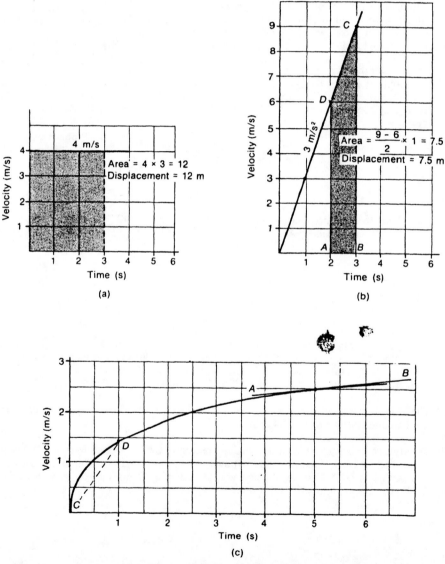

Figure 9.3 Velocity-time graphs: (a) shows constant velocity; (b) shows constant acceleration; (c) shows variable acceleration such as the motion of a projectile.

From these three velocity–time graphs the following significant features are readily apparent.

(i) Uniform velocity appears as a horizontal line.

(ii) Uniform acceleration appears as a straight oblique line which shows that the velocity is increasing by equal amounts per unit time (i.e. each second). Also the steeper the line, the greater is the uniform acceleration.

(iii) Varying acceleration is represented by some type of curved line from which the instantaneous velocity can be read off at any particular instant.

(iv) The displacement of the body whose motion is represented by a particular velocity–time graph is equal to the area under the graph. This is because displacement is equal to the product of velocity and time ($s = vt$).

For the uniform velocity shown in Figure 9.3 (a) the displacement for the first three seconds is equal to the area of a rectangle with sides 4 and 3 units respectively; that is, displacement for three seconds of motion is equal to 12 metres.

Similarly, the displacement for the third second of the motion shown in Figure 9.3 (b) is the area of the figure *ABCD* (shaded) which is equal to 7.5 metres.

The acceleration occurring during any part of the motion is found by the slope of the graph if it is a straight-line graph, or by the slope of the appropriate chord or tangent to the curve. For example, on the velocity–time graph shown in Figure 9.3 (c) the instantaneous acceleration at $t = 5$ seconds is represented by the slope of the tangent *AB* while the average acceleration for the period of the first second is represented by the slope of the chord *CD*.

SAMPLE PROBLEM 9/9

A conveyer is uniformly accelerated from rest to a velocity of 3 m/s in 5 seconds, travels at this constant velocity for a further 6 seconds, and is brought to rest by a uniform deceleration in a further 8 seconds. From a velocity–time graph, determine

(i) the acceleration of the first and last seconds of motion;
(ii) the displacement of the 13th second;
(iii) the total displacement.

Analysis and Calculation:

Using a suitable scale, construct a velocity–time graph as shown.

From the velocity–time graph:

$$\text{Uniform acceleration from } A \text{ to } B = \text{slope of the line } AB$$
$$= \frac{3}{5}$$
$$= 0.6 \text{ m/s}^2$$
$$\text{Uniform acceleration from } C \text{ to } D = \text{slope of the line } CD$$
$$= -\frac{3}{8}$$
$$= 0.375 \text{ m/s}^2$$

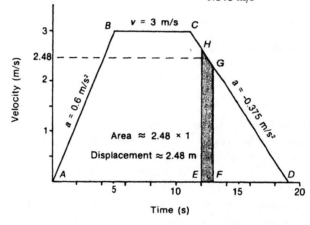

Problem 9/9

$$\text{Displacement of the 13th second} = \text{area of shaded area } EFGH$$
$$= 2.48 \text{ m}$$
$$\text{Total displacement} = \text{area of } ABCD$$
$$= (\tfrac{1}{2} \times 5 \times 3) + (6 \times 3) + (\tfrac{1}{2} \times 8 \times 3)$$
$$= 37.5 \text{ m}$$

Result:

Uniform acceleration during the first second was 0.6 m/s², deceleration during the last second was 0.375 m/s², the displacement during the thirteenth second was 2.48 m and the total displacement was 37.5 m. ∎

EXAMPLES OF MORE COMPLEX PROBLEMS

The following six problems are designed to illustrate how more complex problems of kinematics can be solved. Generally, the methods used to solve more complex problems involve *the resolution of the vectors operating in the direction of motion.* Always draw a simplified diagram of the problem (a free-body diagram) in order to assist you to visualise the nature of the problem and the vectors operating.

Three important categories of problems deserve special mention:

(a) those in which bodies move up or down inclined planes and friction is assumed to be absent. In these cases, the component of the acceleration due to gravity acting parallel to the inclined plane must be obtained by vector resolution. In practice, of course, friction does exist and such problems are considered in Chapter 10;

(b) those involving objects dropped from aircraft or balloons which may be moving horizontally, ascending or descending. In these cases it is often difficult to decide upon the initial motion of the falling object itself. A free-body diagram showing all the vectors operating at the instant of release will clarify this issue. For example, an object dropped from an ascending helicopter moves vertically upward before falling under the influence of gravity;

(c) those problems involving projectiles launched from earth or an aircraft. The flight of any projectile, such as a shell fired from a gun or a rocket launched from earth or an aircraft, is determined by its motion at the instant of launch, the acceleration due to gravity, and air resistance. Other minor influences could be wind gusting, or barrel rifling in the case of a shell fired from a gun. Since the effects of air resistance are complex and differ with the size and shape of the projectile, they will not be considered in any of the subsequent problems. In this book, a projectile will be considered to be any body whose motion in space can be regarded as being determined by gravity and the initial motion of the body itself.

SAMPLE PROBLEM 9/10

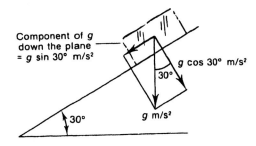

A block of ice slides down a polished metal shute in an iceworks. If the shute makes an angle of 30° with the horizontal, find the velocity of the block of ice after it has slid for 3 metres, and also find the time taken to travel this distance.

Problem 9/10: Diagram showing resolution of the acceleration of free fall, *g*, into components parallel and perpendicular to the inclined plane.

Analysis:
Since frictional forces will be negligible, the only vector acting is the component of *g*, the acceleration of free fall, acting down the plane.

Calculation:

$$a = g \sin 30°$$
$$= \tfrac{1}{2} g$$
$$\therefore a = 4.9 \text{ m/s}^2$$

From

$$s = ut + \tfrac{1}{2} at^2$$
$$3 = 0 + (\tfrac{1}{2} \times 4.9 \times t^2)$$
$$\therefore t = 1.1 \text{ s}$$

where: $s = 3$ m
$u = 0$
$a = 4.9$ m/s^2
$t = ?$

Substitute in

$$v = u + at$$
$$\therefore v = 0 + (4.9 \times 1.1)$$
$$\therefore v = 5.4 \text{ m/s}$$

Result:
The block of ice takes 1.1 seconds to travel down the shute and leaves the shute with a velocity of 5.4 m/s. ■

SAMPLE PROBLEM 9/11

A 25-mm diameter hardened steel ball is projected up a very smooth slide inclined at 30° to the horizontal with an initial velocity of 14.7 m/s. Calculate

(i) the time the ball takes to return to its starting point;

(ii) the distance it moves up the slope.

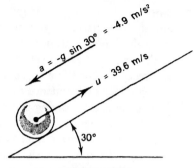

Problem 9/11

Analysis:

Assume frictional forces to be negligible. Let the initial velocity up the plane be positive. The component of g acting down the plane will be negative, i.e. acceleration down the plane = $-g \sin 30° = -4.9$ m/s².

Thus two vectors act on the ball, the initial velocity of $+14.7$ m/s up the slope, and the acceleration of -4.9 m/s² down the slope. Also, when the ball returns to its starting position, the displacement will be zero.

Calculation:

(i)

$$s = ut + \tfrac{1}{2}at^2 \qquad \text{where: } s = 0$$
$$\therefore 0 = 14.7t - (\tfrac{1}{2} \times 4.9 \times t^2) \qquad u = 14.7 \text{ m/s}$$
$$\therefore 0 = 6t - t^2 \qquad a = -4.9 \text{ m/s}^2$$
$$\therefore t(6 - t) = 0 \qquad t = ?$$

$\therefore$ either $t = 0$, or $t = 6$.

Thus, the ball takes 6 seconds to return to its starting position on the slide.

(ii) When the ball reaches its highest point, $v = 0$, and 3 seconds have elapsed (since time up = time down).

$$s = ut + \tfrac{1}{2}at^2 \qquad \text{where: } s = ?$$
$$s = t(u + \tfrac{1}{2}at) \qquad u = 14.7 \text{ m/s}$$
$$\therefore s = 3(14.7 - \tfrac{1}{2} \times 4.9 \times 3) \qquad a = -4.9 \text{ m/s}^2$$
$$\therefore s = 22.05 \text{ m} \qquad t = 3.$$

Note: $v^2 = u^2 + 2as$ could also be used to solve part (ii).

Result:

The ball takes 6 seconds to return to its starting position on the slide and travels 22.05 metres up the slide. ■

SAMPLE PROBLEM 9/12

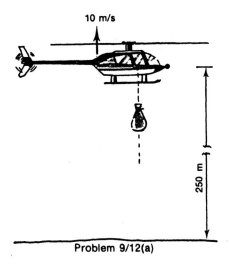

A weighted message bag is dropped from a helicopter which is 250 metres above the ground and ascending vertically at 10 m/s. If the ground troops see the message bag at the instant of release, how long do they have to wait for it to reach them?

Problem 9/12(a)

Analysis:

At the instant of release, the message bag has the same initial upward velocity as the helicopter and is also being acted upon by *g*. Consider $g \approx 10$ m/s² (see below for reason for this) and neglect air resistance.

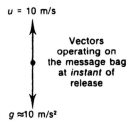

Problem 9/12(b)

Calculation:

$$s = ut + \tfrac{1}{2}at^2 \qquad \text{where: } s = +250 \text{ m}$$
$$\therefore 250 = -10t + 5t^2 \qquad u = -10 \text{ m/s}$$
$$50 = -2t + t^2 \qquad a = g = 10 \text{ m/s}^2$$
$$\therefore t^2 - 2t - 50 = 0 \qquad t = ?$$

Solving the above quadratic equation,

$$t = 8.14 \quad \text{or} - 6.14 \text{ seconds}$$

Result:

The message bag takes about 8 seconds to reach the ground. (The value -6.14 is obviously not applicable, since time is a positive value measured from the instant of release.)

The value $g \approx 10$ m/s² introduces an error of about 2% which is not significant since the answer itself is best-approximated to about 8 seconds. ∎

SAMPLE PROBLEM 9/13

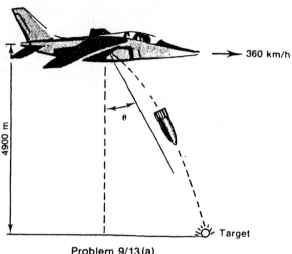

A high-altitude bomber lines up on his target 4900 metres below. If the bomb scores a direct hit, calculate the angle between the vertical and a line joining the bomber and its target at the instant of release. The bomber's velocity was 360 km/h horizontally at the time of release.

Problem 9/13(a)

Analysis:

Draw a free-body diagram showing all vectors and known displacements. The motion of the *bomb* can now be considered as the vector sum of a horizontal motion and a vertical motion.

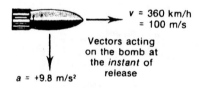

Vectors acting
on the bomb at
the *instant* of
release

Problem 9/13(b)

Calculation:

Consider the vertical motion only. Since the bomb falls a vertical distance of 4900 m under the influence of gravity

$$s = ut + \tfrac{1}{2} at^2$$
$$\therefore 4900 = 0 + (\tfrac{1}{2} \times 9.8t^2)$$
$$t^2 = \frac{4900}{4.9}$$
$$\therefore t^2 = 1000$$
$$t \approx 31.5 \, s$$

where: $s = 4900$ m
$u = 0$
$a = 9.8$ m/s²
$t = ?$

Consider the horizontal motion of the bomb. The horizontal distance travelled by the bomb in 31.5 seconds is found from

$$s = vt$$
$$\therefore s = 31.5 \times 100$$
$$= 3150 \text{ metres}$$

where: $v = 360$ km/h
$= 100$ m/s
$t = 31.5$ s

Now,
$$\tan \theta = \frac{3150}{4900} = 0.643$$
$$\therefore \theta \approx 32.75°$$

Result:

The angle required is $\approx 32.75°$. This is the angle that the bombardier in the aircraft would need to allow for at the instant of release. In most modern aircraft a minicomputer would be coupled to the bombsight to make such corrections automatically. ∎

SAMPLE PROBLEM 9/14

A shell is fired horizontally from the top of a hill. If the muzzle velocity is 300 m/s and the hill is 122.5 metres high, how long does the shell take to hit the ground and how far does it travel horizontally?

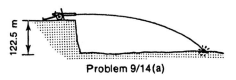

Problem 9/14(a)

Analysis:
The two influences determining the motion of the shell at any instant during the flight are its horizontal velocity and the acceleration of free fall which acts vertically down. Since these two vectors act at 90° to each other, the horizontal and vertical motions can be considered separately. (*Note:* The actual flight-path of the shell is a parabola as shown.)

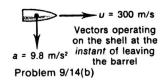

Vectors operating
on the shell at the
instant of leaving
the barrel
Problem 9/14(b)

Calculation:
Consider vertical motion of the shell

$$s = ut + \tfrac{1}{2}at^2$$
$$\therefore 122.5 = 0 + \tfrac{1}{2} \times 9.8t^2$$
$$\therefore 4.9t^2 = 122.5$$
$$\therefore t^2 = 25$$
$$\therefore t = 5s$$

where: $s = 122.5$ m
$u = 0$
$a = 9.8$ m/s^2
$t = ?$

Thus the shell will be in flight for 5 seconds.
Consider the horizontal motion of the shell

$$s = ut + \tfrac{1}{2}at^2$$
$$\therefore s = (300 \times 5) + 0$$
$$\therefore s = 1500 \text{ m}$$

where: $u = 300$ m/s
$t = 5s$
$a = 0$

Result:
The shell travels a horizontal distance of 1500 metres and hits the ground in 5 seconds. ■

SAMPLE PROBLEM 9/15

A projectile is fired from a gun and leaves the barrel with a velocity of 500 m/s. If the barrel is inclined at 30° to the horizontal, how high does the shell rise?

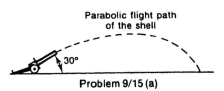

Parabolic flight path
of the shell

30°

Problem 9/15 (a)

Analysis:
The flight of the projectile is determined by the initial velocity and the effects of *g*, the acceleration of free fall. If the initial velocity is resolved into its horizontal and vertical components, the horizontal and vertical motions of the projectile can be considered separately.

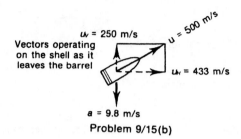

a = 9.8 m/s

Problem 9/15(b)

Calculation:

Resolve the initial velocity into its horizontal and vertical components.

$$u_H = 500 \cos 30° = 433 \text{ m/s}$$
$$u_V = 500 \sin 30° = 250 \text{ m/s}$$

Consider vertical motion only. The vectors operating are the initial velocity of 250 m/s and the downward acceleration of free fall − 9.8 m/s. At its maximum height, the vertical component of the velocity is zero. Let vertically down vectors be positive, i.e. $\downarrow +$

$$v^2 = u^2 + 2as \qquad\qquad \text{where: } v = 0$$
$$0 = (250)^2 + (2 \times -9.8 \times s) \qquad\qquad u = 250 \text{ m/s}$$
$$0 = 250^2 - 19.6s \qquad\qquad a = -9.8 \text{ m/s}^2$$
$$\therefore s = \frac{250^2}{19.6} \qquad\qquad s = ?$$
$$\therefore s \approx 3190 \text{ m}$$

Result:

The shell rises to a maximum height of ≈ 3190 metres. ·· ∎

(*Note:* The same answer could have been found by using $v = u + at$ to find the time of flight, then substituting the time into $s = ut + \frac{1}{2}at^2$ to find the maximum height. This would be a better method, if the horizontal distance of travel of the shell was also required.)

REVIEW PROBLEMS

(Use $g = 9.8 \text{ m/s}^2$ unless otherwise specified.)

9/16
A man walks for 35 minutes at 6.5 km/h and then 45 minutes.at 4.5 km/h. Determine his average speed (in m/s).

9/17
A motorcycle travelling along a straight road accelerates from 18 km/h to 108 km/h in 10 seconds. Calculate its average acceleration in m/s².

9/18
A car's speed increases uniformly from 36 km/h to 108 km/h in 30 seconds. Calculate
  (i) the average speed (in m/s) for the 30-second period;
 (ii) the acceleration operating (in m/s²);
(iii) the distance travelled during the 30-second period.

9/19
A car travelling at 50 km/h has its brakes applied and slows to 40 km/h in 10 metres. Given the same rate of braking, how much further (in metres) will the car run before stopping?

9/20
A jet fighter touches down on an airstrip and decelerates at 4 m/s^2 for 20 seconds before coming to rest. If the pilot touches down 1000 m from the trees at the end of the runway, will he crash?

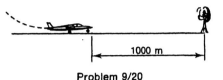

Problem 9/20

9/21
An aircraft travelling at 400 km/h is climbing at an angle of 30 degrees to the horizontal.
 (i) What is its ground speed?
 (ii) How high will it climb in 30 seconds assuming its rate of climb and speed remain constant?

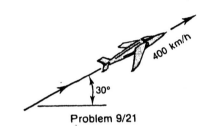

Problem 9/21

9/22
A boy on a snow-sled starts from rest down a run which has a uniform gradient. If the sled travels 10 metres in 4 seconds how long will it take to reach a velocity of 25 m/s down the plane?

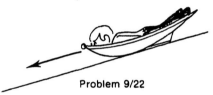

Problem 9/22

9/23
An astronaut making repairs to his space vehicle drops a wrench from a height of 15 metres. With what velocity does the wrench strike the surface if the space vehicle is standing on (i) the earth and (ii) the moon? Take $g = 1.7$ m/s^2 on the moon.

9/24
A stone thrown vertically down from a cliff with an initial velocity of 8 m/s hits the ground in 2 seconds. What is the height of the cliff and what is the striking velocity of the stone?

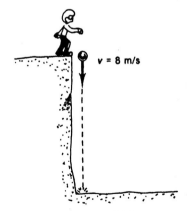

Problem 9/24

9/25
Fred's favourite football team, the Roosters, call for an "up and under". Fairfax obliges with a punt that gives the ball a velocity of 25 m/s. How much time have the players to get under the ball before it returns to the ground?

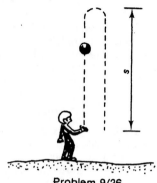

9/26
A ball thrown vertically upwards returns to the thrower in three seconds. Find its initial upward velocity and the height to which it rises.

Problem 9/26

9/27
A rocket is fired vertically and ascends with a constant resultant vertical acceleration of 9.8 m/s² for 30 seconds, after which its fuel is exhausted. Assuming that the rocket continues to rise vertically, calculate the maximum height reached and the time taken for the flight.

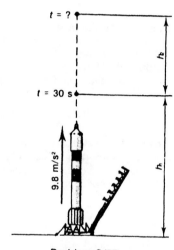

Problem 9/27

9/28
A man standing behind a gun hears the report of the gun and sees the shell-burst at the same instant. The shell, gun and man are in the same straight line, the man being 2000 metres behind the gun and the shell-burst occurring 5000 metres in front of the gun.

Given that sound travels at 335 m/s, determine the horizontal velocity of the shell.

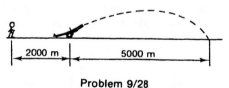

Problem 9/28

9/29
A jet of water issues from a small hole in the vertical pipe. If the initial velocity of the jet is 5 m/s horizontally, determine the vertical distance through which it will have fallen when its horizontal distance from the pipe is 2.5 m.

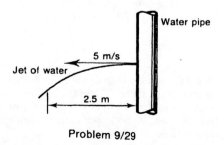

Problem 9/29

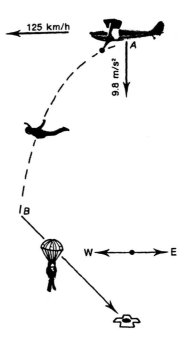

9/30

A parachutist exits from an aircraft that is in level flight at an altitude of 1000 metres and free-falls in a flat stable attitude for 10 seconds. If the aircraft is travelling with a ground speed of 125 km/h and is directly over the target at the time of exit, determine the position of the parachutist relative to the target at the end of his free fall. Assume that it is a windless day.

Problem 9/30

9/31

After his 10-second free fall, the sport parachutist in Problem 9/30 pulls his ripcord and opens his parachute at point *B*. The canopy is designed to expel air from the rear, so providing the parachutist with a maximum horizontal air-speed of 7 m/s.

If his vertical rate of descent is a uniform 5 m/s, is it possible for him to land on the target
 (i) on a windless day?
(ii) if the wind blows from the east at 3 m/s?

9/32

The graph shown illustrates the relationship between distance and time for a certain moving body. From the graph determine
 (i) the velocity during the first 5 seconds;
 (ii) the velocity for the 7th second;
(iii) the velocity during the 15th second.

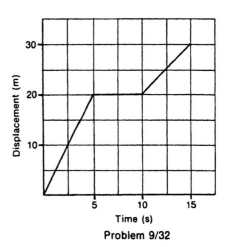

Problem 9/32

9/33

Draw a velocity–time graph for a machine element which has the following motion:

(i) uniform acceleration from rest to a velocity of 1.5 m/s in 3 seconds;

(ii) travels at 1.5 m/s velocity for the next 5 seconds;

(iii) decelerates at a uniform rate for 6 seconds, the final velocity being 1 m/s.

From this graph, determine

(a) the displacement in the first 2 seconds;

(b) the displacement in the 12th second;

(c) the total displacement;

(d) the acceleration in the first 3 seconds;

(e) the acceleration in the last second.

9/34

A conveyer belt in a factory carries partly completed sub-assemblies from one operator to the next in 20 seconds. The conveyer accelerates for the first 6 seconds, travels at uniform velocity for the next 10 seconds, then decelerates to rest for the next 4 seconds. If the operators are 10 metres apart, what is the uniform velocity during the 10-second period?

9/35

An electric car travelling between two stations 500 metres apart is uniformly accelerated for the first 10 seconds and covers 40 metres. It then runs with constant velocity for a certain time, after which it is uniformly decelerated and brought to a stop in the last 20 metres.

Draw a velocity–time graph for the motion of the electric car and determine

(i) the maximum velocity of the electric car;

(ii) the total time taken to cover the 500 metres.

9/36

A car is travelling at 28 m/s when the brakes are applied. The relationship that occurs between velocity and time is set out below. Draw the velocity–time graph and determine

(i) the approximate displacement that occurred during braking;

(ii) the average deceleration during the 6th second.

Time (s)	0	1	2	3	4	5	6	7	8	9	10
Velocity (m/s)	28	21	16.8	13.8	11.6	10	8.8	7.6	6.8	6.2	5.5

9/37

One cycle of the motion of a machine member is represented by the velocity–time graph shown. From the graph determine

(i) the acceleration during the first 2 seconds;

(ii) the maximum velocity;

(iii) the approximate acceleration at 3 seconds;

(iv) the approximate net displacement during the cycle.

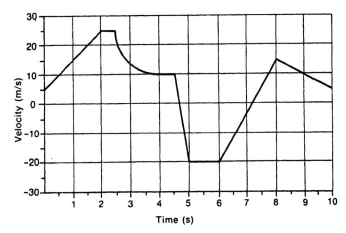

Problem 9/37

9/38

A mine cage starts from the 50-m gallery level and descends to the 150-m gallery level. It then ascends to the surface. What was its displacement

(i) after descending?

(ii) after ascending?

(*Note:* A suitable sign convention is essential in this problem.)

9/39

An aircraft flying at 200 km/h in level flight due north encounters a 60 km/h wind blowing from the SW direction.

(i) If no corrections are made to the aircraft heading, what will the actual flight path be?

(ii) What correction would the pilot need to make to his heading in order to maintain his original flight path?

9/40

Two ships leave a port at the same time. Ship *A* is set on a northeasterly course at 20 knots. Ship *B* steams due east at 25 knots. (1 knot = 1 nautical mile/hour.)

(i) At what speed are the ships moving away from each other?

(ii) In what direction is ship *A* moving relative to ship *B*?

9/41

An aeroplane is 500 nautical miles northwest of the airport which it has to reach in one hour. If the pilot observes that the wind is blowing from the east at 60 nautical miles per hour

(i) on what course will he set the plane?

(ii) at what *ground speed* must he fly, to arrive at his destination on time?

(iii) at what *air speed* must he fly, to arrive at his destination on time?

The first stage of the Saturn V rocket. The five F-1 engines provide power at the rate of 200 gigawatts (160 million horsepower) and develop a thrust of 33 meganewtons. The transporter is a converted tank retriever having a maximum speed of 6.5 km/h.

10

Forces and Motion

$$F = ma$$

Unlike kinematics which is concerned only with the "quality" or "geometry" of motion and the concepts of displacement, time, velocity, and acceleration, kinetics is concerned with those forces which determine the motion of an object.

In most instances when an object is in motion more than one force is operating. Some of these forces tend to cause motion, while others, like friction, tend to oppose motion. It is the resultant of all the external forces operating on an object that determines the motion of the object under consideration. If there is no resultant force, then the body will not accelerate; therefore, kinetics is concerned with motion resulting from the application of an *unbalanced* (i.e. non-equilibrium) force system to an object.

As an example, consider a motor car in motion along a straight level road.

(i) If the motive force transmitted from the engine to the rear wheels exceeds the frictional drag and air resistance, the car will accelerate.

(ii) If the motive force is less than the frictional drag and air resistance, the car will decelerate.

(iii) When the car reaches top speed, the maximum motive force is exactly equal to the frictional drag and air resistance, thus the force system acting on the car is balanced and no further acceleration is possible. It is important to note that this is not an instance of "no force therefore no acceleration", but rather one of "a balanced force system therefore no resultant acceleration" ($\Sigma F = 0$; $\therefore a = 0$).

A clear understanding of the concepts of force, weight, and mass is essential to the study of kinetics, and the following is a brief summary of the important points that have already been discussed in earlier chapters in relation to these concepts.

Force is simply a push or a pull in a particular direction. Forces are generated by the action of one body on another and motion always results from the application of an unbalanced force system to an object. Viewed in this way, statics problems are only particular types of dynamics problems where acceleration equals zero. The fundamental unit of force is the *newton* (N).

Weight is a force. The weight-force or gravity-force of an object within the earth's gravitational field can be calculated according to Newton's law of gravitation* and is found to be equal to $W = mg$, where m equals the mass of the body and g equals the acceleration due to gravity. If the term *weight* is used correctly, its units will naturally be those of force†. To avoid confusion, the terms *weight-force* or *gravity-force* are used in this book instead of the more common term, weight.

Mass is the measure of the inertia of a body and is directly related to the amount of substance present. The mass of an object can only be altered if substance is either added to or removed from the original object. The fundamental unit of mass is the *kilogram* (kg).

It is important to note the distinction between *mass* and *weight*. A kilogram of potatoes will remain a mass of one kilogram whether it is measured in London, Moscow, Sydney or on the moon. However, the weight (or gravity-force) exerted by the potatoes will be different in each of these places because it depends upon the acceleration due to gravity at each place.

---

*See Chapter 2, page 19.

†In everyday speech, "weight" may be expressed in kilograms; for example, the "weight" of a man could be 60 kg. This is, however, a confusion of terms resulting from the usual method of "weighing" a body to obtain its mass. The term "weight" in this book refers to the gravity-force of an object found, as stated above, from $W = mg$ (kg $\times$ m/s^2 = N).

SAMPLE PROBLEM 10/1

A car of mass 1 tonne is stationary on a level road.
What total force does this car exert on the road?
Assume $g = 9.8$ m/s^2.

Analysis:
The total force exerted on the road is the weight-force or gravity-force of the car.

Calculation:

$$W = mg$$
$$= 1000 \times 9.8$$
$$= 9800\,\text{N}$$
$$= 9.8\,\text{kN}$$

where: W = weight-force
m = 1 tonne = 1000 kg
g = 9.8 m/s^2

Result:
The total force exerted by the car on the road is 9.8 kilonewtons. ■

SAMPLE PROBLEM 10/2

What is the mass of a body having a weight of 980 newtons?
Assume $g = 9.8$ m/s^2.

Calculation:

$$W = mg$$
$$m = \frac{W}{g}$$
$$m = \frac{980}{9.8}$$
$$= 100\,\text{kg}$$

where: W = 980 N
m = ?
g = 9.8 m/s^2

Result:
The body has a mass of 100 kilograms. ■

NEWTON'S FIRST LAW AND INERTIA

According to Newton's first law of motion,
> *a body will remain at rest or in uniform motion in a straight line unless acted upon by an external unbalanced force.*

This law therefore suggests that a body cannot change its state of motion of its own accord. The property of a body which is responsible for its resistance to a change in motion is known as its *mass* or *inertia*. Two inertial states are recognised by this law, the obvious one when the body is at rest, and the not so obvious one when the body is travelling at constant velocity in a straight line (that is, when acceleration = 0).

Consider the following as a practical illustration of inertia. A small speedboat can increase its speed very quickly if the throttle settings are advanced; however, a large ocean liner will only increase its speed very slowly even though a tremendous increase in force can be applied by increasing the engine revolutions from "slow" to "full ahead". Similarly, while the speedboat can be quickly slowed down if the throttle settings are reduced, the ocean liner will take a long time to slow down appreciably once the engine revolutions are again reduced to "slow". The reason for this difference in response is inertia; the ocean liner obviously has a much greater mass or inertia than the speedboat and thus will respond more slowly to changes in the external unbalanced forces causing motion.

NEWTON'S SECOND LAW AND ACCELERATION

Newton's second law* can be stated as follows:

a mass acted upon by an external unbalanced force will accelerate in proportion to the magnitude of that force in the direction in which the force acts.

The law as stated above is fundamental to the study of kinetics in that it provides a direct relationship between applied force, mass, and resultant acceleration. Expressed in equation form, this law becomes

$$F = kma$$

where F is the unbalanced force causing motion, m is the mass of the body, a is the resultant acceleration, and k is a constant whose value depends upon the units selected.

The truth of this law can be demonstrated by experiment. Imagine a body of mass m subjected to a force F_1. The acceleration a_1 is measured and the ratio F_1/a_1 is found. This experiment is now repeated many times using different forces F_2, F_3, etc., and the values of the resulting acceleration a_2, a_3, etc., are also measured and found to be different. However, the ratios F_1/a_1, F_2/a_2, F_3/a_3, etc., are all found to be equal.

The constant value of the ratio F/a is a measure of the property of the body which does not change; it is obviously related to the constant in the experiment which is the mass of the body. Thus, a body with a large mass will accelerate slowly and a body with a small mass will accelerate more rapidly if identical external unbalanced forces are applied to them.

If a consistent system of units is selected, the constant k in the equation $F = kma$ becomes unity (1) and the equation can be written in its most common form,

$$F = ma$$

It is thus essential, prior to the application of the formula $F = ma$, to convert
(a) all forces into newtons;
(b) all masses into kilograms (kg);
(c) all accelerations into metres per second squared (m/s^2).

Also, remember that it is only the unbalanced or resultant force that produces an acceleration. Therefore, vector addition of force systems is often necessary prior to

---

*Some interpretations of this law state that it is *the time-rate change of momentum of the body that varies proportionally to the external unbalanced force*. Both this and the above interpretation are equally correct but, for simplicity, the interpretation used here is preferred at this stage. Momentum is discussed in Chapter 11.

the application of $F = ma$ to a problem in order to determine the resultant unbalanced force causing motion.

SAMPLE PROBLEM 10/3

A stationary body of mass 10 kg is acted upon by a force of 20 newtons for 5 seconds. Determine
(i) the acceleration that occurs;
(ii) the velocity after 5 seconds.

Analysis:
Since 20 N is the only force acting, it is, in effect, the unbalanced force present and $F = ma$ can be used to calculate the resultant acceleration.

Calculation:
(i)

$$F = ma$$
$$20 = 10a$$
$$a = 2 \text{ m/s}^2$$

where: $F = 20$ N
$m = 10$ kg
$a = ?$

(ii)

$$v = u + at$$
$$= 0 + (2 \times 5)$$
$$v = 10 \text{ m/s}^2$$

where: $v = ?$
$u = 0$
$a = 2 \text{ m/s}^2$
$t = 5$ seconds

Result:
The acceleration produced is 2 m/s^2 and the body is travelling at 10 m/s after 5 seconds.   ■

SAMPLE PROBLEM 10/4

Calculate the mass of an object if a force of 20 newtons produces an acceleration of 5 m/s^2.

Analysis:
When an unbalanced force acts on a body, the resultant acceleration is proportional to the mass or inertia of that body, the relationship being that of $F = ma$. Therefore, if a known force, F, produces a known acceleration, a, the mass, m, of the body must be $m = F/a$.

Calculation:

$$F = ma$$
$$\therefore m = \frac{F}{a}$$
$$= \frac{20}{5}$$
$$= 4 \text{ kg}$$

where: $F = 20$ N
$a = 5$ m/s^2
$m = ?$

Result:
The body is of mass 4 kilograms.   ■

SAMPLE PROBLEM 10/5

A 1-tonne car is travelling at 108 km/h in a straight line. Determine the retarding force that the brakes must exert to stop it in 75 metres.

Analysis:
The retarding force of the brakes is the unbalanced force opposing motion. Once the rate of deceleration is found, the relationship $F = ma$ can be used to calculate the total retarding force operating.

Calculation:
Find the deceleration that would occur.

$$v^2 = u^2 + 2as$$
$$0 = 30^2 + (2 \times a \times 75)$$
$$900 = -150 a$$
$$a = -6 \text{ m/s}^2$$

where: $v = 0$
$u = 108$ km/h
$= 30$ m/s
$a = -?$

Apply

$$F = ma$$
$$F = 1000 \times -6$$
$$= -6000 \text{ N}$$
$$= -6 \text{ kN}$$

where: $F = ?$
$m = 1$ tonne
$= 1000$ kg
$a = -6$ m/s^2

Result:
The force applied by the brakes would be 6 kilonewtons, the negative sign indicating that it acts to oppose the original motion of the car. ∎

FURTHER WORKED EXAMPLES NOT INVOLVING FRICTION

While friction is present in most situations when motion is occurring, there are some types of motion where friction is either not relevant to the motion involved or is negligible and can be ignored. The following selection of problems represents the most common types where calculations need not, for one reason or another, involve frictional forces.

SAMPLE PROBLEM 10/6

A 50-kg wool bale hangs on the end of a rope. Determine the resultant motion of the bale if the tension in the cord is
  (i) 490 newtons;
 (ii) 240 newtons; and
(iii) 890 newtons.
Assume $g = 9.8 \text{ m/s}^2$.

Analysis and Calculation:
  (i) The force system consists of the tension $T = 490$ N acting upwards and the weight-force $W = mg = 50 \times 9.8 = 490$ N acting down. Since the resultant force acting vertically $= 0$, (i.e. no unbalanced force present) the bale remains stationary.

$T = 490$ N

$W = mg$
$= 490$ N

Problem 10/6

(ii) If the tension T is reduced, an unbalanced force, $F = W - T$, will be present and an acceleration vertically down will occur.

Unbalanced force $= W - T$
$$= 490 - 240$$
$$= 250 \text{ N}$$

$$F = ma$$
$$\therefore a = \frac{F}{m}$$
$$= \frac{250}{50}$$
$$= 5 \text{ m/s}^2 \text{ vertically down}$$

where: $F = 250$ N
$m = 50$ kg
$a = \,?$

T = 240 N

W = 490 N

Problem 10/6

(iii) If the tension is increased to 890 N, an unbalanced force, $F = T - W$, will be present, and the bale will be subjected to an acceleration acting vertically upwards.

Unbalanced force $= T - W$
$$= 890 - 490$$
$$= 400 \text{ N}$$

$$F = ma$$
$$\therefore a = \frac{F}{m}$$
$$= \frac{400}{50}$$
$$= 8 \text{ m/s}^2 \text{ vertically upward}$$

where: $F = 400$ N
$m = 50$ kg
$a = \,?$

T = 890 N

W = 490 N

Problem 10/6

Result:
The accelerations acting are (i) zero; (ii) 5 m/s² vertically down; and (iii) 8 m/s² vertically upward. ■

SAMPLE PROBLEM 10/7

A man of mass 80 kg is trapped on a ledge 10 metres above ground in a burning building. The only rope available has a breaking strength of 650 newtons. Calculate the minimum rate at which he must accelerate down the rope in order to not break it. Assume $g = 9.8 \text{ m/s}^2$.

Analysis:
Weight-force of man $= 80 \times 9.8 = 784$ N. Maximum force that rope can support $= 650$ N. Unbalanced force that must be used to produce the acceleration of the man must be equal to the difference between his weight-force and the breaking strength of the rope.

Calculation:

$$\text{Unbalanced force } F = W - T$$
$$= 784 - 650$$
$$= 134 \text{ N}$$
$$F = ma$$
$$a = \frac{F}{m}$$

where: $F = 134$ N
$m = 80$ kg
$a = ?$

$$= \frac{134}{80}$$
$$\therefore a \approx 1.7 \text{ m/s}^2$$

Result:
The man must slide down the rope with a minimum acceleration of ≈ 1.7 m/s² or it will break. ■

SAMPLE PROBLEM 10/8

A cord passes over a virtually frictionless pulley and has a 10-kg mass tied on one end and a 20-kg mass tied on the other. Calculate the approximate acceleration of the system and the tension in the cord, assuming $g = 9.8$ m/s².

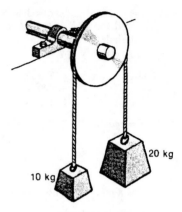

SPACE DIAGRAM

Problem 10/8

Analysis:
Since the masses are unequal, an acceleration will occur. Also since the string is continuous and friction is assumed to be zero, the string will have the same tension throughout its length.

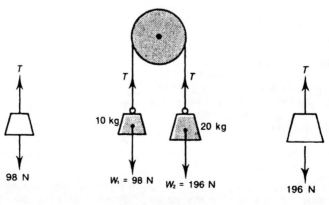

Problem 10/8

Calculation:
Consider the *left-hand side* of the system.

Unbalanced force F_1 causing motion $= T - 98$ N

$$F = ma$$
$$\therefore T - 98 = 10a$$

(i) $T = 10a + 98$

where: $F = T - 98$ N
$m = 10$ kg
$a = ?$

Consider the *right-hand side* of the system.
Unbalanced force F_2 causing the motion $= 196 - T$ N

$$F = ma$$
$$196 - T = 20a$$

(ii) $T = 196 - 20a$

where: $F = 196 - T$
$m = 20$ kg
$a = ?$

From Equations (i) and (i)

$$10a + 98 = 196 - 20a$$
$$30a = 98$$
$$a = 3.27 \text{ m/s}^2 \quad \text{(the 10-kg mass moving up)}$$

Substituting $a = 3.21$ m/s² in Equation (i)

$$T - 98 = 10 \times 3.27$$
$$T = 130.7 \text{ N}$$

Alternative solution for acceleration:
Consider the system as a whole.
Unbalanced force $= (20 - 10)\,9.8$ N $= 98$ N
Mass being accelerated $= 10 + 20 = 30$ kg

$$F = ma$$
$$\therefore a = \frac{F}{m}$$
$$= \frac{98}{30}$$
$$= 3.27 \text{ m/s}^2$$

where: $F = 98$ N
$m = 30$ kg
$a = ?$

Result:
Thus the tension in the cord is ≈ 131 N and the system accelerates at ≈ 3.3 m/s², the 20-kg weight moving downward. (*Note:* frictional forces will certainly be present, thus these approximates are suitable answers.) ■

SAMPLE PROBLEM 10/9

A crate of mass 30 kg rests on the floor of a goods hoist. Calculate the force exerted by the crate on the floor when
(a) the hoist is ascending with constant velocity;
(b) ascending with an acceleration of 1.2 m/s²; and
(c) descending with an acceleration of 1.2 m/s².
    Assume $g = 9.8$ m/s².

SPACE DIAGRAM

Problem 10/9

Analysis and Calculations:

(a) Since the lift is ascending with constant velocity, a = 0, and the reaction of the floor is $W = mg$.

$$\therefore W = 30 \times 9.8$$
$$= 294\,\text{N}$$

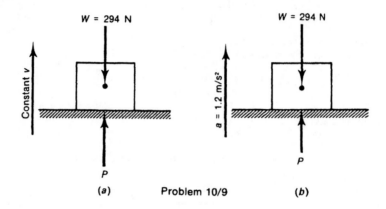

(a)   Problem 10/9   (b)

(b) The two forces acting on the crate are its weight-force downwards and the push, P, of the floor upwards.

Unbalanced force causing motion $= P - 294\,\text{N}$

$$F = ma$$
$$P - 294 = 30 \times 1.2$$
$$P = 294 + 36$$
$$P = 330\,\text{N}$$

where: $F = P - 294\,\text{N}$
$m = 30\,\text{kg}$
$a = 1.2\,\text{m/s}^2$

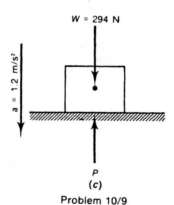

(c)

Problem 10/9

(c) Since the hoist is descending, the unbalanced force must be $294 - P$

$$F = ma$$
$$294 - P = 30a$$
$$P = 294 - (30 \times 1.2)$$
$$\therefore P = 258\,\text{N}$$

where: $F = (294 - P)\,\text{N}$
$m = 30\,\text{kg}$
$a = 1.2\,\text{m/s}^2.$

Result:

Thus the force exerted by the crate on the floor of the hoist is (a) 294 N; (b) 330 N; and (c) 258 N. ∎

SAMPLE PROBLEM 10/10

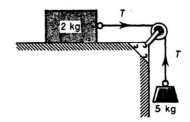

Consider the system of weights shown here. The plane that the 2-kg block slides upon is smooth (that is, can be considered frictionless) and the friction in the pulley bearing can be neglected. Determine the acceleration of the system and the tension in the cord. Assume $g = 9.8$ m/s^2.

SPACE DIAGRAM

Problem 10/10

Analysis and Calculations:
Since there is no friction, the 5-kg weight will move downwards. Consider the horizontal motion of the 2-kg weight.

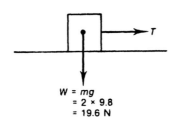

W = mg
= 2 × 9.8
= 19.6 N

Problem 10/10

Unbalanced horizontal force $= T$.
$$F = ma$$
$$\therefore T = 2a \qquad \text{(i)}$$

where: $F = T$
$m = 2$ kg
$a = ?$

Consider the vertical motion of the 5-kg weight.

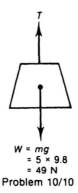

W = mg
= 5 × 9.8
= 49 N

Problem 10/10

Unbalanced vertical force $= 49 - T$.

$$F = ma$$
$$49 - T = 5a$$
$$T = 49 - 5a \qquad \text{(ii)}$$

where: $F = 49 - T$
$m = 5$ kg
$a = ?$

Solving Equations (i) and (ii) simultaneously
Thus,
$$49 - 5a = 2a$$
$$7a = 49$$
$$a = 7 \, \text{m/s}^2$$
Substitute $a = 7$ m/s^2 in Equation (i)
$$T = 2 \times 7$$
$$= 14 \, \text{N}$$

Result:
Thus, the tension in the cord is 14 N and the acceleration of the system is 7 m/s^2 (the 5-kg weight moving down). ∎

SAMPLE PROBLEM 10/11

The two weights of 10 kg and 20 kg respectively are attached by a cord which passes over the pulley as shown. The 20-kg weight hangs vertically and the 10-kg weight slides freely on the inclined plane. Neglecting friction, determine the tension in the cord and the acceleration of the system. . Assume $g = 9.8 \, \text{m/s}^2$.

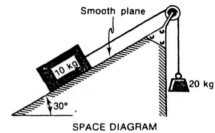

SPACE DIAGRAM
Problem 10/11

Analysis and Calculations:
Consider the forces acting on the 10-kg weight. These are
(a) the tension tending to pull it up the plane,
(b) the component of its own weight, $F_1 = mg \sin 30° = 49 \, N$, tending to move it down the plane,
(c) the normal reaction N, acting at 90° to the plane.

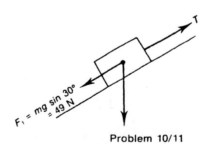

Problem 10/11

Assume that the 10-kg weight slides up the plane.
Unbalanced force causing motion $= T - 49$

$$F = ma$$
$$T - 49 = 10a$$
$$T = 10a + 49 \qquad \text{(i)}$$

where: $F = (T - 49)$ N
$m = 10$ kg

Consider the forces acting on the 20-kg weight.

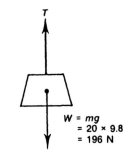

$W = mg$
$= 20 \times 9.8$
$= 196$ N

Problem 10/11

Unbalanced force $= 196 - T$

$$F = ma$$
$$196 - T = 20a$$
$$T = 196 - 20a \quad \text{(ii)}$$

where: $F = (196 - T)$ N
$m = 20$ kg

From Equations (i) and (ii)

$$196 - 20a = 10a + 49$$
$$30a = 147$$
$$a = 4.9 \text{ m/s}^2$$

The positive value for a indicates that the assumption that 10-kg weight moves up the plane was correct.

Substitute $a = 4.9$ m/s^2 in Equation (i)

$$T = 10a + 49$$
$$= 49 + 49$$
$$= 98 \text{ N}$$

Result:
The tension in the cord is 98 N and the system accelerates at 4.9 m/s^2, the 20-kg weight moving downwards. ■

(*Note:* Sample Problems 10/10 and 10/11 are unreal situations since friction would always be present on the flat surfaces. However, they are included because they illustrate the type of analysis necessary to solve more difficult problems.)

DYNAMIC FRICTION

In every instance of sliding or rolling motion, frictional forces are generated between the surfaces in contact. These frictional forces always act to oppose motion and thus they cause wear and loss of efficiency in machines. As an example of this, consider the concept of a "perpetual-motion machine", that is, a machine that, once started, would continue in motion forever. Such a machine is, of course, not possible because of frictional forces generated between its moving parts. Such frictional forces can be reduced by improving surface finish and by suitable lubrication, but they can never be completely eliminated. Therefore, a "perpetual-motion machine" cannot be built. However, frictional forces can be useful; for example, when walking or running, and when starting, stopping and turning corners in a motor car.

Friction has already been studied in Chapter 8 and the points listed below are a summary of what should already be known about frictional forces.

(1) Frictional forces always act to oppose motion and are tangential to the surfaces in contact.

(2) Frictional forces are dependent upon the relative roughness of the surfaces in contact.

(3) The maximum frictional force, called the *force of limiting friction*, which can be developed between two surfaces in contact just as motion is about to occur, is proportional to the normal force between the surfaces, and is found from the formula $F_r \text{ (lim)} = \mu_s N$, where μ_s = coefficient of static friction.

(4) Within wide limits, the frictional force is not dependent upon the area of apparent surface contact. (It is dependent, however, on the "true" area of contact.)

(5) While the two bodies in contact are in equilibrium with respect to each other, the frictional force is equal to the resultant of all other forces tending to cause motion.

Kinetic frictional forces are those that exist between surfaces that are in motion with respect to each other. Generally, once one object is set in motion over another, the force of the kinetic friction between the surfaces in contact is less than the force of limiting friction that existed when the first object was on the point of motion. Thus, the coefficient of kinetic friction, μ_k, is always less than the coefficient of static friction, μ_s. This is illustrated in Figure 10.1.

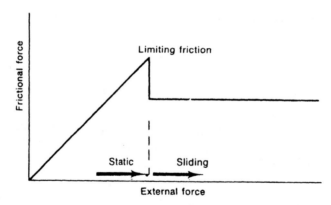

Figure 10.1   The graph shows the proportional increase of the frictional force with an increase of applied external force until a limit is reached when the object starts to slide. At this point, the coefficient of friction, μ_s, reduces to the significantly lower value, μ_k.

Only "dry friction" will be considered in this section, since lubricated surfaces have to be considered in quite a different way. In fact, when two surfaces are completely separated by a film of lubricant, it is the internal frictional resistance of the lubricant itself that becomes important, not the relative roughness of the bearing surfaces.

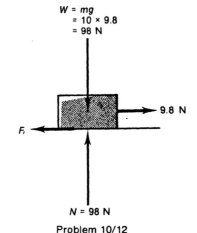

$W = mg$
$= 10 \times 9.8$
$= 98$ N

9.8 N

F_r

$N = 98$ N

Problem 10/12

SAMPLE PROBLEM 10/12

A block of mass 10 kg is moving with constant velocity along a rough horizontal surface under the influence of a 9.8 N force. Find the coefficient of sliding friction (μ) operating on this surface.

Analysis:
Determine all forces acting. These are
(i) The normal reaction, N, perpendicular to the surfaces in contact. $N = 10 \times 9.8 = 98$ N;
(ii) the force sustaining motion $= 9.8$ N;
(iii) the frictional force, $F_r = \mu N = \mu \times 98 = 98$
Since $a = 0$, this force system must be in equilibrium.

Calculation:
Consider $F = ma$ for horizontal motion.
Since $a = 0$, resultant horizontal force $= 0$
$$\therefore 9.8 - 98\mu = 0$$
$$\therefore 9.8 = 98\mu$$
$$\therefore \mu = \frac{9.8}{98}$$
$$\mu = 0.1$$

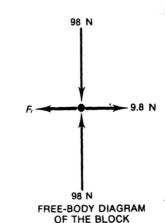

98 N

F_r 9.8 N

98 N
FREE-BODY DIAGRAM
OF THE BLOCK
Problem 10/12

Result:
The coefficient of sliding friction is 0.1.   ■

SAMPLE PROBLEM 10/13

A brick of mass 4 kg is at rest on a horizontal builder's plank. A horizontal force of 20 N is applied to the brick which accelerates away at 2 m/s². Calculate the coefficient of sliding friction present between the brick and the plank.

20 N   4 kg μ

Problem 10/13

Analysis:
Since an acceleration occurs, there must be an unbalanced horizontal force. This will be equal to the applied force of 20 N minus the opposing frictional force acting.

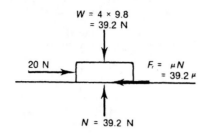

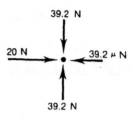

Unbalanced $F_H = 20 - F_r$ N

Problem 10/13

Calculation:
Let the coefficient of friction $= \mu$
Frictional force $F_r = \mu_s N = 39.2 \mu$
Sum of the horizontal forces, $\Sigma F_H = 20 - 39.2 \mu$

Apply $F = ma$
$$20 - 39.2 = 4 \times 2$$
$$39.2\mu = 20 - 8$$
$$\mu = \frac{12}{39.2}$$
$$\therefore \mu = 0.28$$

where: $F = 20 - 39.2\mu$
$m = 4\,\text{kg}$
$a = 2\,\text{m/s}^2$

Result:
The coefficient of sliding friction is 0.28.

■

SAMPLE PROBLEM 10/14

A brick of mass 4 kg rests on the middle of a builder's plank. One end of the plank is slowly raised until the angle of plank with the horizontal is 25° and the brick just begins to slide. After the brick is sliding freely, the plank is lowered until the angle becomes 16°, at which point the brick just ceases to slide freely. Calculate the coefficients of static and sliding friction operating.

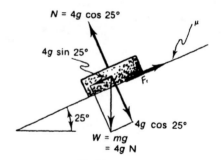

Problem 10/14

Analysis and Calculation:
When the angle is 25°, the force of limiting friction is present since motion just begins. Resolve forces parallel and perpendicular to the angle of the plank.

Force of limiting friction $F_r = \mu_s N = \mu_s \times 4g \cos 25°$
But $F_r =$ force down the plane, $\qquad F_r = 4g \sin 25°$
$$\mu_s \times 4g \cos 25° = 4g \sin 25°$$
$$\therefore \mu_s = \frac{4g \sin 25°}{4g \cos 25°}$$
$$= \tan 25°$$
$$\therefore \mu_s = 0.47$$

Therefore the coefficient of static friction is 0.47.

If the same method is applied to the second part of the problem, the coefficient of sliding friction will be found to be equal to tan 16°, that is 0.29.

Result:
The coefficients of friction are 0.47 (static) and 0.29 (sliding).

■

(*Note:* Problem 10/14 points out the important relationship between the frictional coefficient and the angle of the incline just sufficient to (i) initiate motion and (ii) sustain motion. Under either of these conditions the tangent of the angle of the incline or plane to the horizontal is equal to the appropriate frictional coefficient, and the mass of the body has no significance in the above calculations.)

SAMPLE PROBLEM 10/15

A parcel of mass 10 kg slides down a delivery shute 5 metres long. What will be its exit velocity if the shute is inclined at 30° to the horizontal and the coefficient of friction operating is 0.2?

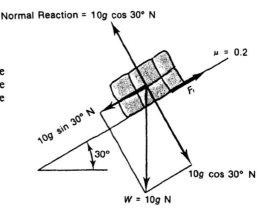

Normal Reaction = $10g \cos 30°$ N

$\mu = 0.2$

$10g \sin 30°$ N

$30°$

$10g \cos 30°$ N

$W = 10g$ N

Problem 10/15

Analysis and Calculation:
Draw a diagram showing all forces operating.
Now $F_r = \mu N = 0.2 \times 10g \cos 30°$

Resultant forces acting down the plane,

$$F = 10g \sin 30° - F_r$$
$$= 10g (\sin 30° - 0.2 \cos 30°)$$

Consider $F = ma$

$$10g (\sin 30° - 0.2 \cos 30°) = 10a$$
$$a = g (\sin 30° - 0.2 \cos 30°)$$
$$= 9.8 (0.5 - 0.173)$$
$$= 9.8 \times 0.327$$
$$\therefore a \approx 3.2 \text{ m/s}^2$$

Substitute $\quad a = 3.2 \text{ m/s}^2 \quad$ in $\quad v^2 = u^2 + 2as$
$$v^2 = 0 + 2 \times 3.2 \times 5$$
$$= 32$$
$$\therefore v \approx 5.65 \text{ m/s}$$

where: $v = ?$
$u = 0$
$a = 3.2 \text{ m/s}^2$
$s = 5 \text{ m}$

Result:
The exit velocity from the shute will be approximately 5.65 m/s. ■

SAMPLE PROBLEM 10/16

In a physics experiment, two weights of 5 kg and 2 kg are connected as shown. When the system is set in motion, an acceleration of 6 m/s² is measured for the 5-kg weight. Assuming no inertia for the pulley and no mass for the string, what coefficient of friction operated on the horizontal table, and what was the tension in the cord?

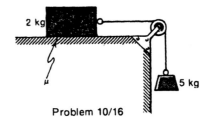

2 kg

μ

5 kg

Problem 10/16

Analysis and Calculation:
Draw a diagram showing all vectors operating.

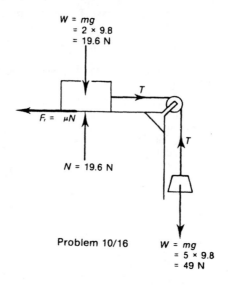

Problem 10/16

Consider horizontal motion.

$$F_r = \mu N$$
$$= 19.6\mu$$

Resultant horizontal force, $T - F_r = T - 19.6\mu$
Apply $F = ma$

$$T - 19.6\mu = 2 \times 6$$
$$\therefore \mu = \frac{T - 12}{19.6} \quad \text{(i)}$$

where: $F = T - 19.6$
$m = 2$ kg
$a = 6$ m/s^2

Now consider vertical motion.
Resultant vertical force, $W - T = 49 - T$
Apply $F = ma$

$$49 - T = 5 \times 6$$
$$T = 49 - 30$$
$$\therefore T = 19 \text{ N}$$

where: $F = 49 - T$
$m = 5$ kg
$a = 6$ m/s

Substitute $T = 19$ N in Equation (i)

$$\mu = \frac{19 - 12}{19.6}$$
$$= \frac{7}{19.6}$$
$$\therefore \mu = 0.36$$

Result:
The coefficient of friction between the horizontal surface and the 2-kg weight is 0.36 and the tension in the cord is 19 N. (Compare this answer with that for Sample Problem 10/10. Why is this a larger tension?) ∎

REVIEW PROBLEMS

(Use $g = 9.8\,\text{m/s}^2$ unless otherwise specified.)

10/17

A body of mass 20 kg is acted upon by a resultant force of 60 newtons. Determine the acceleration that occurs in the direction of that force.

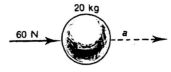

Problem 10/17

10/18

A horizontal force of 1 kN acts on a body and produces an acceleration of 10 m/s². What is the mass of the body?

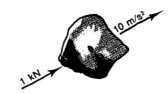

Problem 10/18

10/19

An air-to-air missile of mass 500 kg is fired from a fighter in level flight. If the missile accelerates at 90 m/s² relative to the fighter, what average thrust is provided by the rocket motor?

Problem 10/19

10/20

A rocket of mass 300 tonnes travelling horizontally at 3600 km/h has its velocity increased to 5400 km/h by a 3-second "burn" of its motor. What average thrust was exerted by the engine?

10/21

A car on a dragstrip accelerates at 10 m/s² from a standing start.
- (i) How long does it take to cover the 400 m course?
- (ii) What is the final velocity of the car?
- (iii) What average tractive force was exerted by the rear wheels on the concrete dragstrip if the car had a mass of 0.8 tonnes and the total force opposing motion remained constant at 500 N during the run?

Problem 10/21

10/22

A man on a rocket sled mounted on a straight track is travelling at 360 km/h when the retro-rockets are fired and the velocity is reduced to 36 km/h in 165 metres.

If the man has a mass of 80 kg, calculate the total force exerted by the man on his seat belts during deceleration. Express your answer in kilonewtons and as "g-force".

Problem 10/22

10/23

A balloon is drifting horizontally in still air when 10 kg of ballast is released. Determine the resultant vertical acceleration of the balloon if its mass is 1 tonne.

Problem 10/23

10/24

A 75-kg man is winched into a search-and-rescue helicopter which is hovering directly above the building in which the man was trapped. If the mass of the sling used is 10 kg, determine the tension in the cable when the man is ascending,

(i) with constant velocity, and

(ii) with a constant acceleration of 0.5 m/s^2.

(Neglect the mass of the winch cable.)

Problem 10/24

10/25

A machine of mass 4 tonnes is hauled vertically upwards by a cable and moves a distance of 3 metres in 4 seconds from rest. Assuming that the acceleration is constant, determine the tension in the cable.

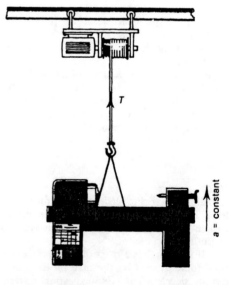

Problem 10/25

10/26

A surface-to-air missile is fired vertically from its launcher and its motor provides an average thrust of 5 kN over a 10-second period. If the missile has a mass of 102 kg, determine its acceleration and the altitude it reaches.

10/27

During a test, a thrust of 5 kN is developed for 15 seconds by a small rocket motor. If the minimum vertical acceleration of the rocket into which this motor is to be placed is 65 m/s^2, determine the maximum mass of the rocket. (Neglect the mass of the motor in your calculations.)

10/28

The Saturn V rocket had a mass of approximately 2.7×10^6 kg, of which 2.5×10^6 kg was the liquid oxygen/kerosene propellant. This produced a lift-off thrust of 33.5 MN.
  (i) What was the initial acceleration of the rocket?
  (ii) How does this compare with that of an average family car?
  (iii) How does the acceleration vary during the rocket launch and why?

Problem 10/28

10/29

During the last stage of its descent on to the moon's surface, a lunar module decelerated vertically at 1 m/s^2. Given that the module had a mass of 13×10^3 kg and that the acceleration due to gravity is 1.67 m/s^2 on the moon, determine the thrust exerted by the descent engine during this stage.

Problem 10/29

10/30

An aircraft climbs to an altitude of 1500 m at a constant angle of 30 degrees to the horizontal. If the total thrust of the jet engines is 500 kN and the total mass of the plane and its contents is 55 tonnes, determine the approximate time taken to reach the 1500 m altitude. Assume a ground speed at take-off of 200 km/h and an average air resistance of 40 kN.

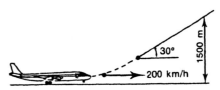

Problem 10/30

10/31

A body travelling at 30 m/s in a direction due east is acted upon by a force of 250 N acting due north for 2 seconds. If the body has a mass of 50 kg, determine its final velocity.

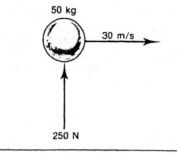

Problem 10/31

10/32

A motorless service elevator consists of a platform plus cage and a counterweight connected by a rope which passes over a small, free-running pulley. If the platform and the cage have a mass of 15 kg and the counterweight a mass of 14 kg, calculate the tension in the rope and the acceleration of the system if it is allowed to move freely when unloaded. Neglect friction, inertia and the weight of the rope.

10/33

In an experiment, the single-fixed, single-movable pulley system is loaded as shown. Neglecting friction, the mass of the cord, the mass of the pulley sheave blocks and the inertia of the pulleys, determine:

(i) the tensions in the cord at the points *A, B* and *C*;

(ii) the tension in the tie, *D*;

(iii) the accelerations of the two masses.

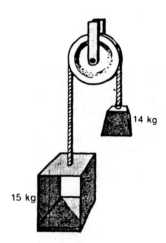

Problem 10/32

10/34

The fixed and movable pulleys and sheave blocks of the pulley system shown in Problem 10/33 are found to have a mass of 5 kg. For the same loads determine

(i) the tensions in the cord at the points *A, B* and *C*;

(ii) the tension in the tie, *D*;

(iii) the acceleration of the system.

(Neglect friction, the mass of the cord, and the inertia of the pulleys.)

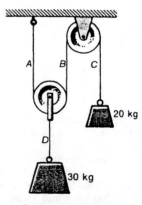

Problem 10/33

10/35

A trailer together with its load has a mass of 20 tonnes and is attached to a prime mover which can provide a maximum pull of 5.5 kN. If the total frictional force resisting the forward motion of the trailer and load is 3 kN, determine the resultant acceleration of the trailer and prime mover.

Problem 10/35

10/36

A motor vehicle of mass 1.5 tonnes starts from rest to climb a gradient of 1 in 10 against a frictional resistance of 180 N. If the tractive effort exerted by the rear wheels is 2.5 kN, determine
 (i) the acceleration of the vehicle up the gradient;
 (ii) the time taken to reach 60 km/h;
(iii) the total distance travelled.
(iv) What would be the acceleration on a horizontal stretch of road if the frictional resistance is unchanged?

Problem 10/36

10/37

A suburban train consists of 3 carriages each having a mass of 15 tonnes. The first carriage acts as an engine and exerts a driving force of 40 kN on the rails. The frictional drag of each carriage on the rails is 1 kN. Determine the acceleration of the train and the tension in the couplings between the carriages.

Problem 10/37

10/38

A barge of mass 10 tonnes is moved along a straight canal by two horizontal tow ropes attached to two prime movers on opposite sides of the canal. The prime movers operate so that the tow ropes always make angles of 30° with the direction of motion.

 If the barge moves from rest a distance of 40 metres in 20 seconds, and the average force resisting the forward motion of the barge is 180 N per tonne, determine
 (i) the average resultant force causing the forward motion of the barge;
 (ii) the tension in either tow rope.

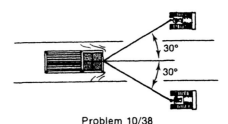

Problem 10/38

10/39

The engine of a motor car drives the rear wheels only. If the axle loads are 900 kg rear and 600 kg front, determine the maximum tractive force that the rear tyres can exert during acceleration. Take the coefficient of friction (skidding) between rubber and bitumen as 0.6.

Problem 10/39

10/40

Determine the tangential braking force *P* acting on the rotating railway carriage wheel when a pressure of 7 MPa exists between the cast iron brake shoes and the rim of the wheel. The coefficient of friction μ_k present is 0.2, and each brake shoe has an area of 30 000 square millimetres.

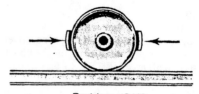

Problem 10/40

10/41

A bronze bush is drawn from the end of the shaft using the screw-operated puller as shown. If the radial pressure between the bush and shaft is 3.5 MPa and the coefficient of sliding friction is 0.2, determine the force required to remove the bush.

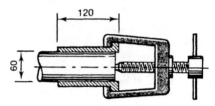

Problem 10/41

10/42

The start of a sand sled run has a constant slope of 45° to the horizontal. One competitor of mass 60 kg has a sled of mass 10 kg.

(i) What tension will he need to exert on the tow rope when pulling his sled up the sled run with constant velocity?

(ii) With what rate of acceleration will he come down this section of the sled run?

(iii) If this section of the sled run is 20 metres long, what will be his velocity after he covers that distance, assuming that his initial velocity was zero?

The coefficient of friction between sand and sled is 0.35.

10/43

A 100-kg fire door has a self-closing mechanism operated by the 25-kg counterweight as shown. Assuming negligible friction in the pulley, determine the approximate acceleration of the door when the restraining latch is released, if the coefficient of friction present in the overhead door track system is 0.2.

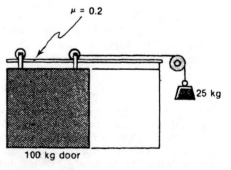

Problem 10/43

10/44

Part of a delivery system consists of a conveyer belt that carries 15-kg bundles of newspapers to the top of a metal shute 3 metres long and inclined at 25° to the horizontal.

At the bottom of the shute a pair of rubber flaps slow the bundles before they enter the storage area. If the bundles have an initial horizontal velocity of 0.5 m/s when they reach the slide and the coefficient of sliding friction present is 0.1, calculate the velocity with which the bundles hit these flaps.

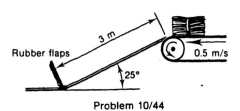

Problem 10/44

10/45

A constant push of 80 N applied at an angle of 30° to a crate of mass 15 kg gives the crate a horizontal velocity of 1 m/s in 3 seconds. Determine the coefficient of dynamic friction acting between the crate and the horizontal plane.

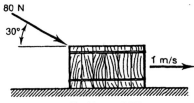

Problem 10/45

11

Impulse and Momentum

$I = Ft$
$M = mv$

All moving bodies possess momentum and the momentum of a body determines the difficulty of bringing that body to rest. For instance, it requires more effort to stop an ocean liner travelling at 18 kilometres per hour than it does to stop a light speedboat travelling at the same velocity, since the momentum of the ocean liner is very large compared to that of the much lighter speedboat.

Newton made frequent reference to the *quantity of motion* possessed by moving bodies during his discussion of the laws of motion. This quantity of motion, which we call *momentum*, was defined by Newton in the following way:

> *The quantity of motion is the measure of the same, arising from the velocity and quantity of matter conjointly.*

In other words, the momentum of a body is found from the product of its mass and its velocity at a given instant, that is

$$M = mv \qquad \text{where: } M = \text{momentum} \ (\text{kg m/s})$$
$$m = \text{mass}$$
$$v = \text{velocity}$$

Momentum is thus obviously a vector quantity since one of the quantities determining it—velocity—is a vector. Therefore, if momenta need to be added, the usual methods of vector addition apply.

SAMPLE PROBLEM 11/1

A car of mass 1 tonne is moving with a velocity of
36 kilometres per hour. Calculate its momentum
in kg m/s.

Analysis and Calculation:
Since the unit kg m/s is specified for the momentum, velocity must be converted into m/s and the
mass into kg.

$$Mass = 1 \text{ tonne} = 1000 \text{ kg}$$
$$Velocity = 36 \text{ km/h}$$
$$= \frac{36 \times 1000}{60 \times 60} \text{ m/s}$$
$$\therefore v = 10 \text{ m/s}$$
$$Momentum = 1000 \times 10$$
$$= 10^4 \text{ kg m/s}$$

Result:
The momentum of the car is 10^4 kg m/s. ■

IMPULSE

The concepts of impulse and momentum are very closely related, even though we
speak of the *impulse of a force* and the *momentum of a mass*. Forces for which impulses
are calculated usually act for only very short lengths of time; for example, the blow
of a hammer, an explosion, or the impact of a cue on a billiard ball.

The impulse of a force is calculated from the following relationship:

$$\text{Impulse} = \text{force acting} \times \text{time it acts, i.e.}$$
$$I = F \times t$$

The units of impulse will normally be those of force (newtons) multiplied by those of
time (seconds); that is, newton seconds (N s).

The relationship of the impulse of a force to momentum can be clearly seen from
the following example.

Consider a body of mass m kilograms moving with an initial velocity of u m/s.
A force of F newtons is applied for time t seconds and the velocity of the body changes
to v m/s.

Now: impulse of force operating, $I = Ft$ N s
initial momentum of body, $M_1 = mu$ kg m/s
final momentum of body, $M_2 = mv$ kg m/s
$\therefore$ change in momentum $= M_1 - M_2$
$= mv - mu$ kg m/s

Consider the acceleration of the body.

From $F = ma$, $$a = \frac{F}{m}$$

Substitute $a = F/m$ in the equation $v = u + at$, and v, the final velocity, becomes

$$v = u + \frac{F}{m} \times t$$

Multiply both sides of this equation by the mass, m, and thus

$$mv = mu + Ft$$
$$\therefore Ft = mv - mu$$

Therefore it is obvious that the *impulse of a force* causing the change in velocity is equal to *the change in momentum* of the body. Since these quantities are equal, they must also have the same units.

Unit for momentum is kg m/s, but unit for force, the newton, is kg m/s². Therefore, the unit for momentum can be rewritten as

$$\frac{\text{kg m}}{s^2} \times s = \text{N s, the same unit as for impulse of a force.}$$

(*Remember :* The impulse of the force acting is equal to the change in momentum of the body upon which that force acts.)

SAMPLE PROBLEM 11/2

A force of 20 N acts on a body of mass 500 g for 3 seconds. What change in momentum is produced in the body?

Analysis and Calculation :

$$\text{change in momentum} = \text{impulse of force}$$
$$I = Ft$$
$$= 20 \times 3$$
$$= 60 \text{ N s}$$

Result :
The change in momentum of the body is equal to 60 N s (or 60 kg m/s). ∎

SAMPLE PROBLEM 11/3

A rocket of mass 500 kg travelling horizontally at 3600 km/h is accelerated by an average force (thrust) of 250 kN obtained by a 3-second "burn" of its motor. Calculate the impulse of the force operating and the final momentum of the rocket.

Analysis :
The impulse of the thrust of the rocket motor is equal to the change in momentum of the rocket, and the final momentum of the rocket is the sum of its initial momentum and this impulse.

Calculation :

$$I = Ft$$
$$= 250 \times 10^3 \times 3$$
$$= 75 \times 10^4 \text{ N s (or } 75 \times 10^4 \text{ kg m/s)}$$

where: $F = 250 \text{ kN} = 250 \times 10^3 \text{ N}$
$t = 3 s$

original momentum $= mv = 500 \ \times 1000$
$$= 50 \times 10^4 \text{ kg m/s}$$

where: $m = 500 \text{ kg}$
$v = 3600 \text{ km/h} = 1000 \text{ m/s}$

final momentum $= (50 \times 10^4) + (75 \times 10^4)$
$$= 125 \times 10^4 \text{ kg m/s}$$

Result :
The impulse of the thrust was 75×10^4 N s and the final momentum was 125×10^4 N s (or kg m/s). ∎

CONSERVATION OF MOMENTUM

Newton's third law states that

To every action there is an equal and opposite reaction.

This means, in effect, that the action of one body on another produces an equal but opposite reaction on the first body.

As an example, consider two balls moving along the same path and in the same direction. If the second ball moves faster than the first, *a collision* will occur. In this collision, according to Newton's third law, the action or force of the first ball on the second will be equalled by the reaction or force of the second ball on the first. Both action and reaction will act for only the very short time that the two balls are actually in contact. Since the time interval is short, both action and reaction can be expressed as an impulse (refer to Figure 11.1).

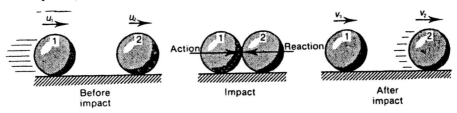

Figure 11.1

Again, because of Newton's third law, the impulses of the action and the reaction will be equal in magnitude but opposite in sign. However, the impulse of the action (force) of the first ball equals its change in momentum, and similarly, the impulse of the reaction (force) of the second ball equals its change in momentum. Thus,

$$I\ (\text{action}) = I\ (\text{reaction})$$
$$\therefore m_1\ v_1 - m_1\ u_1 = m_2\ v_2 - m_2\ u_2 \tag{1}$$

where: m_1 = mass of first ball

$\quad\quad v_1$ = final velocity of first ball

$\quad\quad u_1$ = initial velocity of first ball

$\quad\quad m_2$ = mass of second ball

$\quad\quad v_2$ = final velocity of second ball

$\quad\quad u_2$ = initial velocity of second ball

Rearranging Equation (1)

$$m_1\ u_1 + m_2\ u_2 = m_1\ v_1 + m_2\ v_2$$

But $\quad m_1\ u_1 + m_2\ u_2$ = total momentum before impact;

and $\quad m_1\ v_1 + m_2\ v_2$ = total momentum after impact.

Thus, *the total momentum of the two balls before impact equals the total momentum of the two balls after impact.* This is the *law of conservation of momentum.*

In applying the law of conservation of momentum, written as

$$m_1\ u_1 + m_2\ u_2 = m_1\ v_1 + m_2\ v_2$$

to the solution of problems, care must be taken to use a consistent sign convention.

Once a positive velocity direction is chosen, all other velocities must be assigned either a positive or negative direction consistent with this convention. If the direction of a velocity is unknown, assume it to be positive. If it turns out to be negative, then it is moving in a direction opposite to that which you initially chose as positive.

One final point is that, even though no loss of momentum occurs in a collision, there is usually a loss of energy involved* due to the generation of heat and sound. This makes the law of conservation of momentum very useful in solving problems involving collisions or the impact of objects.

SAMPLE PROBLEM 11/4

A block of wood of mass 2 kg is suspended on a long string. A bullet weighing 50 g is fired horizontally into the wooden block with a velocity of 400 m/s. Calculate the initial velocity of the block with the bullet embedded in it as they move off together as one mass.

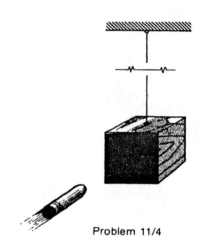

Problem 11/4

Analysis:
Since this is a problem involving the collision of two bodies, the law of conservation of momentum applies. Thus,

total momentum before impact = total momentum after impact

$$\therefore \frac{\text{momentum}}{\text{of bullet}} + \frac{\text{momentum}}{\text{of block}} = \text{momentum of block and bullet as they move off together}$$

Calculation:
Let v be the velocity of the block and bullet as they move off together.

$$\therefore \left(\frac{50}{1000} \times 400\right) + 0 = \left(2 + \frac{50}{1000}\right) \times v$$
$$\therefore 20 = 2.05\,v$$
$$\therefore v = 9.8\,\text{m/s}$$

Result:
The block with the bullet embedded in it moves off with a velocity of 9.8 m/s.   ■

---

*Refer to Chapter 14.

SAMPLE PROBLEM 11/5

A man driving a car of mass 1.5 tonnes sees a truck ahead and brakes hard. His wheels lock and he slides into the truck at 25 km/h. If the truck has a mass of 8 tonnes and was travelling at 10 km/h in the same direction as the car at the time, what would be the final velocity of the two vehicles if they locked bumber bars on impact?

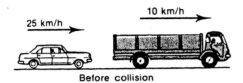

| Before collision | After collision |

Analysis and Calculation:
From the law of conservation of momentum, the total momentum before impact equals the total momentum after impact.

Let the final velocity of the car plus truck as they move off together be v.

$\therefore$ momentum of car + momentum of truck = momentum of car and truck

$$\therefore (1.5 \times 25) + (8 \times 10) = 9.5 \times v \quad \text{(momentum in t km/h)}$$

$$\frac{75}{2} + 80 = \frac{19}{2}v$$

$$\therefore 19v = 235$$

$$\therefore v = 12.4 \text{ km/h}$$

Result:
The two vehicles move off with a common velocity of 12.4 km/h after locking bumper bars. ■

    (*Note:* This problem illustrates the fact that it doesn't really matter what units are used for mass and velocity as long as they are used consistently for all parts of the calculation.)

SAMPLE PROBLEM 11/6

Problem 11/6

A 4-kg rifle is fired and the bullet leaves the barrel with a velocity of 750 m/s relative to the ground. If the bullet weighs 15 grams, calculate the initial recoil velocity of the rifle.

Analysis and Calculation:
Apply the law of conservation of momentum to the rifle and bullet. Let final recoil velocity of rifle be v.

| initial momentum of rifle | + | initial momentum of bullet | = | final momentum of rifle | + | final momentum of bullet |

$$0 + 0 = 4v + \left(\frac{15}{1000} \times 750\right)$$

$$\therefore 4v = -11.25$$

$$\therefore v = -2.8 \text{ m/s}^2$$

Result:
The initial recoil velocity of the rifle was 2.8 m/s. The negative sign in the answer simply means that this velocity is in a direction opposite to the velocity of the bullet. ■

ROCKET AND JET ENGINES

The thrust of a rocket or a jet engine depends upon the rate of change of momentum imparted to the gases consumed and expelled by the engine. The gases consumed are usually fuel and oxygen (collectively called "propellant") and the gases expelled are simply termed "exhaust gases". In point of fact it is really quite unnecessary to know the composition of any of the gases involved, since the calculation depends upon mass consumed and the relative velocities before and after combustion in the engine. Thus, three facts must be determined prior to calculation of thrust:

(a) the velocities of the gases upon intake, relative to the engine (or ground), in m/s;
(b) the velocity of the exhaust gases, relative to the engine (or ground), in m/s;
(c) the quantities (mass) of gases consumed by the engine in a given time, usually one second, in kg.

Calculations involving rocket engines are obviously simpler since the initial velocities of the gases consumed relative to the rocket engine are always zero.

SAMPLE PROBLEM 11/7

The rocket engine of a missile ejects 200 kg of exhaust gases per second at a speed of 1000 m/s relative to the engine. Calculate the thrust of the rocket engine when the missile is travelling at 300 m/s.

Analysis and Calculation:
The initial velocity of the fuel and oxygen relative to the rocket engine is zero. The final velocity of the exhaust gases is 1000 m/s. Therefore,

$$\text{change in momentum} = mv = mu$$
$$= (200 \times 1000) - (200 \times 0)$$
$$= 200 \times 10^3 \text{ kg m/s}$$

Let F be the thrust and I be its impulse (per second)
Now, $\hspace{3cm} I = Ft$
but $\hspace{3cm} t = 1 \text{ second}$
$\hspace{2cm} \therefore F = \text{change in momentum per second}$
$\hspace{3.5cm} = 200 \times 10^3 \text{ N}$
$\hspace{3.5cm} = 200 \text{ kN}$

Result:
The thrust of the rocket engine is 200 kN. ■

This problem illustrates the fundamental relationship between fuel (propellant) consumed per second, change in velocity of propellant, and thrust.

Thrust (N) = mass discharged per second (kg) × change in velocity of propellant (m/s)

(*Note:* This is not a new formula to be remembered as it is only one instance of the general rule.
$$I = mv - mu = m(v - u).)$$

As a contrast to the simpler rocket engine, consider the jet engine in Sample Problem 11/8.

SAMPLE PROBLEM 11/8

The engine of a jet fighter travelling at 720 km/h is consuming fuel at the rate of 1 kg per second and air at the rate of 100 kg per second. Calculate the thrust of the engine if the velocity of the exhaust gases is 1000 m/s relative to the aircraft.

Problem 11/8

Analysis and Calculation:
This problem has to be considered in two parts: (a) the thrust provided by the air; (b) the thrust provided by the fuel.

Thrust due to discharge of air

$$F_1 = \text{mass discharged/sec} \times \text{change in velocity}$$
$$= 100 \times (1000 - 200) \qquad\qquad (720 \text{ km/hr} = 200 \text{ m/s})$$
$$= 80 \times 10^3 \text{ N}$$
$$= 80 \text{ kN}$$

Thrust due to the discharge of fuel

$$F_2 = \text{mass discharged/sec} \times \text{change in velocity}$$
$$= 1 \times 1000$$
$$= 1 \text{ kN}$$

Result:
The total thrust of the jet is $F_1 + F_2$ and is equal to 81 kN. ■

ELASTIC COLLISIONS AND THE COEFFICIENT OF RESTITUTION

The problems involving collisions considered so far have avoided the issue of the *elasticity* of the bodies involved. For instance, consider a perfectly elastic ball dropped on to a hard, rigid surface which is also perfectly elastic. In this case, the ball would rebound up to its original height, and the collision would be termed perfectly elastic. This means that the velocity after impact was equal to the velocity immediately prior to impact. On the other hand, imagine a plastic mass such as a lump of putty dropped on to the same rigid slab. There would be no rebound and thus the velocity after impact would be zero. In practice, the rebounds of most bodies dropped in similar circumstances lie somewhere in between these two extremes.

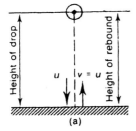
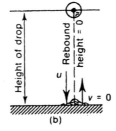
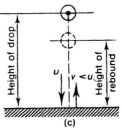

Figure 11.2 Types of collisions: (a) shows a perfect elastic collision in which the rebound velocity, *v*, is equal to the striking velocity, *u*; (b) is a plastic collision; and (c) is a normal elastic collision in which the rebound velocity is less than the velocity of impact.

The relationship governing relative velocities of two bodies before and after impact in an elastic collision can be stated as follows:

If two bodies collide, their relative velocity after impact is equal to their relative velocity before impact multiplied by their coefficient of restitution, C.

That is,

relative velocity after impact $= C \times$ relative velocity before impact

or
$$C = \frac{\text{relative velocity after impact}}{\text{relative velocity before impact}}$$

During the first part of the impact, the two bodies deform. If they possess elasticity, they will then tend to return to their original shapes. This latter part of the impact is known as the *restitution* of the bodies. The coefficient of restitution, C, is a measure of the abilities of the bodies involved to regain their original shapes (i.e. it is a measure of their relative elasticities).

Refer back to the previous example. Let the ball be a hardened steel ball bearing dropped from a height of 2 metres. If the height of rebound is 1.75 metres, then the coefficient of restitution is found from the relationship

$$C = \frac{\text{relative velocity after impact}}{\text{relative velocity before impact}}$$

Now, since the slab does not move (and it is reasonable to assume that the earth does not move in the impact),

(i) relative velocity between ball and surface is the velocity of the ball on impact and can be found from $v^2 = u^2 + 2\,as.$

$$v^2 = u^2 + 2\,as$$
$$= 0 + 2 \times 9.8 \times 2$$
$$\therefore v^2 = 39.2$$
$$v = 6.26 \text{ m/s}$$

Relative velocity on impact $= +6.26$ m/s.

(ii) Relative velocity between ball and surface after impact equals the velocity of the ball after impact. This will be the same as the velocity of a ball dropped from a height of 1.75 metres and can be found from $v^2 = u^2 + 2\,as.$

$$v^2 = u^2 + 2\,as$$
$$= 0 + 2 \times 9.8 \times 1.75$$
$$\therefore v^2 = 34.3$$
$$v = 5.86 \text{ m/s}$$

Relative velocity after impact $= -5.86$ m/s.

(*Note:* The negative sign indicates that the direction is opposite to velocity on impact.)

Substituting these values into our equation involving the coefficient of restitution:

$$C = \frac{\text{relative velocity after impact}}{\text{relative velocity before impact}}$$
$$\therefore C = \frac{-5.86}{+6.26}$$
$$\therefore C = -0.93$$

Note that the coefficient of restitution is a negative ratio; this is because of the change in direction of the relative velocities before and after impact. It is obvious from the above that C, the coefficient of restitution, varies from -1 for perfectly elastic collisions to 0 for perfectly plastic collisions in which the bodies move off together after collision.

SAMPLE PROBLEM 11/9

An engine of mass 60 tonnes used for shunting in a goods yard is rolling at 20 km/h along a straight track when it runs into an empty waggon of mass 6 tonnes which was moving in the same direction at a speed of 5 km/h. If the couplings fail to lock together, at what velocities will the engine and waggon move off when the coefficient of restitution is -0.4.

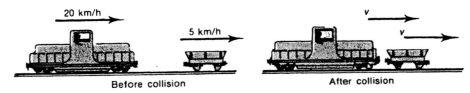

Before collision   After collision

Problem 11/9

Analysis and Calculation:
Let the velocity of the engine be v_1, and that of the waggon be v_2 after the impact. Now,

$$C = \frac{\text{relative velocity after impact}}{\text{relative velocity before impact}}$$

$$\therefore -0.4 = \frac{v_1 - v_2}{20 - 5}$$

$$v_1 - v_2 = -(0.4 \times 15)$$
$$v_1 - v_2 = -6 \tag{i}$$

Conservation of momentum

$$m_1 u_1 + m_2 u_2 = m_1 v_1 + m_2 v_2$$
$$(60 \times 20) + (6 \times 5) = (60\, v_1) + (6\, v_2)$$
$$205 = 10 v_1 + v_2 \tag{ii}$$

Adding Equations (i) and (ii)

$$10\, v_1 + v_2 = 205$$
$$v_1 - v_2 = -6$$
$$11\, v_1 = 199$$
$$\therefore v_1 = 18 \,\text{km/h}$$

From (i),

$$v_1 + v_2 = -6$$
$$v_2 = v_1 + 6$$
$$v_2 = 18 + 6$$
$$= 24 \,\text{km/h}$$

Result:
After the collision the engine moves off at 18 km/h and the waggon at 24 km/h.   ■

General Note Concerning Elastic Collisions

We have seen that elastic collisions between two or more objects that do not move off as one mass after the collision involve the use of a coefficient of restitution to determine relative velocities after impact. In these and all other types of collisions, except the theoretical case of a perfectly elastic collision, some energy is "lost" in the collision. This "lost" energy can be readily calculated knowing the initial and final velocities, and is in fact used up in the collision to make noise and heat. A discussion of such energy losses is given in Chapter 14, and the effects of these energy losses will not be considered before Chapter 14.

REVIEW PROBLEMS

(Use $g = 9.8 \text{ m/s}^2$ unless otherwise specified.)

11/10

Calculate the momentum (in kg m/s) of the following bodies:

(i) a motorcycle and rider, combined mass 400 kg moving at 72 km/h;

(ii) a truck, mass 8 tonnes, moving at 54 km/h;

(iii) a bullet, mass 15 g, moving at 280 m/s;

(iv) the Saturn V rocket of mass 2.7×10^6 kg when it is travelling at the "escape velocity" of 11 000 m/s.

11/11

A 6-gram bullet is fired horizontally into a 10-kg block of wood. The block and bullet move off with an initial velocity of 0.4 m/s. Determine the impact velocity of the bullet.

11/12

An empty 5-tonne skip loader is coasting at 4 m/s along a horizontal section of track. As it passes under a loader, 3 tonnes of sinter are suddenly dumped into it. Assuming that the sinter had no horizontal component to its velocity, determine the resultant velocity of the loader plus sinter.

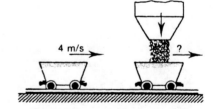

Problem 11/12

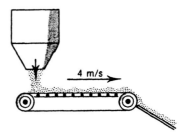

11/13

Sand drops at the rate of 50 kg per second from an overhead hopper on to a horizontal conveyer belt which moves at 4 m/s. Determine the additional force required to drive the conveyer belt, due to its load of sand.

Problem 11/13

11/14

An agitator truck full of concrete has a total mass of 19 tonnes. It is travelling at 60 km/h when it collides head-on with a small sedan of mass 1 tonne travelling at 100 km/h. If the wreckage moves off as one mass, determine its velocity immediately after impact.

11/15

A car and driver of combined mass 1 tonne are moving due east through an intersection at 36 km/h when struck centrally by a motorcycle plus rider moving due north at 72 km/h. If the combined mass of the motorcycle and rider is 300 kg, determine the velocity after impact if the car driver, motorcycle and rider become entangled and move off as one mass.

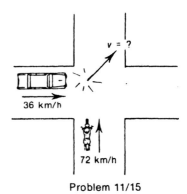

Problem 11/15

11/16

An 80-kg man dives off the end of a pier with an initial velocity of 2.5 m/s in the direction shown. Determine the horizontal and vertical components of the force exerted on the pier by the diver if he takes 0.75 seconds to leave the pier.

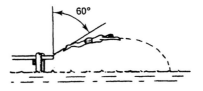

Problem 11/16

11/17

A sledge hammer of mass 6 kg has a velocity of 5 m/s when it hits squarely on to the end of a wedge. It rebounds with an initial velocity of 0.5 m/s. If the blow occurs in 0.01 seconds, what average force is exerted on the wedge?

11/18

A batsman strikes a 0.25-kg ball and causes the ball to reverse its direction of motion. If the initial velocity of the ball was 10 m/s and it moved off at 18 m/s in the opposite direction after the impact, determine the average force exerted on the ball if the duration of the impact was 0.01 seconds.

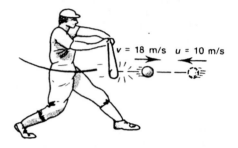

Problem 11/18

11/19

A man standing on a flatcar provides additional forward propulsion by hurling bricks off the back of the car.

   At a certain time the forward velocity of the flatcar together with its load of the man plus 10 bricks is 0.5 m/s. The flatcar plus man has a mass of 200 kg and each brick a mass of 4 kg. Calculate the velocity of the flatcar after the next brick is thrown, if it leaves the man's hand at 5 m/s as shown. (Ignore frictional losses.)

Problem 11/19

11/20

A high energy-rate forming press has a hammer of mass 400 kg. During the operation of the machine the hammer is accelerated vertically downwards and hits the metal slug in the die with a velocity of 150 m/s. If, during a particular forming operation, the hammer is brought to rest in 0.02 seconds, determine the average force exerted by the hammer.

11/21

A submarine of displacement 2000 tonnes fires a torpedo of mass 2 tonnes. If the torpedo is fired from a rear tube with an initial velocity of 30 km/h what momentary increase occurs in the forward velocity of the submarine?

Problem 11/21

11/22

A water jet from a horizontal pipe has a velocity of 15 m/s when it strikes the vertical wall of a mixing chamber. If 30 litres of water are released every second, determine the force that the water jet exerts on the wall. Note that
 (i) the water may be assumed to move parallel to the wall after impact;
(ii) the mass of 1 litre of water may be taken as 1 kg.

Problem 11/22

11/23

A gun of mass 250 kg fires a 1-kg projectile with a muzzle velocity of 500 m/s. Determine
 (i) the initial recoil velocity of the gun;
 (ii) the time taken for the recoil;
(iii) the distance travelled during recoil if the gun moves back against a constant resisting force of 2 kN.

11/24

A jet engine which is stationary in a test rig is consuming air at the rate of 100 kg per second and fuel at the rate of 1 kg per second. If the velocity of the exhaust gases relative to the engine is 600 m/s, what is the thrust of the engine?

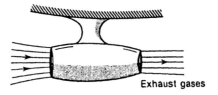

Problem 11/24

11/25

A marine jet engine in a surf-rescue craft travelling at 36 km/h is consuming 70 litres of water per second and discharging it with a velocity of 150 m/s relative to the boat. What thrust is the engine developing? Take the mass of 1 litre of water as 1 kg.

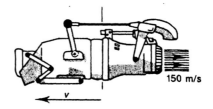

Problem 11/25

11/26

(a) A rocket ejects 20 kg of exhaust gases per second at a velocity of 600 m/s relative to the rocket. Calculate the propulsive force at this instant.

(b) If at a given time the mass of the rocket is 800 kg and it is moving vertically upwards, determine its acceleration.

11/27

While in level flight at an altitude of 3500 metres a jet aircraft scoops in 150 kg of air per second and discharges it with a velocity of 800 m/s relative to the aircraft. What thrust is produced by the engine if fuel consumption is 1.25 kg per second and the velocity of the aircraft is 900 km/h?

11/28

A ball is dropped from a height of 3 metres on to a concrete path and the height of rebound is measured as 1.8 metres. Determine the coefficient of restitution.

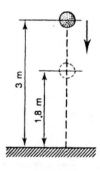

Problem 11/28

12

Circular Motion

$$s = r\theta = \omega_{av} \times t$$
$$v = r\omega$$
$$a = r\alpha$$

Chapters 9, 10, and 11 have been restricted to the motion of bodies of the following two types.
(a) Those bodies that move in a straight line (linear motion), such as a car on a straight road, a rocket moving vertically, or a piston moving in a cylinder.
(b) The motion of a projectile which follows a curvilinear path in space and whose motion can be analysed by the application of rectilinear equations to two directions of motion that are always at 90° to each other.

However, many objects move in circular paths; that is, they move so that they always remain a fixed distance from a given point known as *the axis of rotation*. This is illustrated in Figure 12.1.

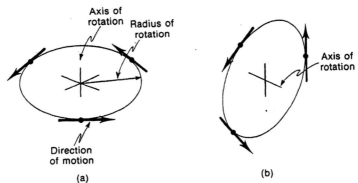

Figure 12.1 Two examples of circular motion show (a) motion in a horizontal plane; and (b) motion in a vertical plane.

Common examples of circular motion are flywheels on stationary engines, pulleys on shafts, satellites in circular orbits, a car driven around a curve of known radius, the centrifuge, and an aeroplane "banking" in a turn.

The previously discussed concepts of *displacement*, *velocity* and *acceleration* have their *angular equivalents* which are necessary for the solution of problems involving motion in a circular path.

ANGULAR DISPLACEMENT (θ, theta)

Consider the flywheel shown in Figure 12.2. When the point A on the rim of the flywheel has moved to position B, it has undergone a displacement. This displacement could be measured by the linear distance from AB around the circumference of the flywheel, or it could be measured as the *angular displacement*, which is equal to the angle θ. Consider the point C about halfway out from the centre of the flywheel. When it moves to position D its linear displacement (the arc CD) is obviously much less than the linear displacement of A (arc B) but its angular displacement is the same as the angular displacement of A (the angle θ). *Thus, it is obvious that all points on a rotating body such as this flywheel have the same angular displacements;* however their linear displacements are not equal but are proportional to their distances away from the axis of rotation.

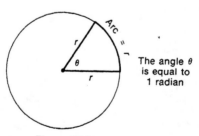

Figure 12.2

The unit of angular displacement is the *radian*, one radian being the angle subtended at the centre of the circle by an arc equal in length to the radius (refer to Figure 12.3).

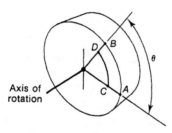

The angle θ is equal to 1 radian

Figure 12.3

Since the circumference of the circle is $2\pi r$, one revolution of a point around the circumference of a circle is equal to an angular displacement of $\dfrac{2\pi r}{r}$ or 2π radians.

(*Note:* 1 radian $\approx 57.3°$.)

Angular displacements can also be expressed in terms of revolutions, one revolution obviously being equal to 2π radians.

If a point moving in a circle undergoes an angular displacement of θ radians, then the linear distance moved by the particle around the circumference of the circle is equal to displacement in radians multiplied by the radius of rotation.

$$s = r\theta$$

To verify this, consider a particle that has moved through 1 revolution around a circle of radius r units. Its angular displacement is 2π radians, and the linear displacement is $2\pi r$ units, the circumference of the circle.

SAMPLE PROBLEM 12/1

The weight (or bob) of a 1-metre pendulum moves through an arc of 200 mm. Find its angular displacement in radians and degrees.

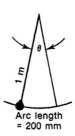

Arc length
= 200 mm

Problem 12/1

Calculation:

$$\theta \text{ radians} = \frac{\text{distance of arc}}{\text{radius}}$$

$$= \frac{200 \text{ (mm)}}{1000 \text{ (mm)}}$$

$$= 0.2 \text{ radians}$$

$$\theta \text{ degrees} = 0.2 \times \frac{360}{2\pi}$$

$$\approx 11.5°$$

where: arc = 200 mm
radius = 1 m
= 1000 mm

Result:
The bob moves 0.2 radians or $\approx 11.5°$. ∎

ANGULAR VELOCITY (ω, omega)

Angular velocity, denoted by the letter ω, is the time-rate change of angular displacement. Thus, the basic units of angular velocity are radians per second (rad/s), but other units in common use are revolutions per second (rev/s or rps) and revolutions per minute (rev/min or rpm).

Refer back to the flywheel in Figure 12.2. If it takes 0.1 second for point A to move through 1 revolution, its angular velocity can be expressed as:

$$\omega = \frac{\text{angular displacement}}{\text{time}}$$

$$= \frac{1 \text{ revolution}}{0.1 \text{ second}}$$

$$= 10 \text{ rev/s}$$

OR

$$\omega = \frac{2\pi \text{ radians}}{0.1 \text{ sec}}$$

$$= 20\pi \text{ rad/s}$$

$$\approx 62.8 \text{ rad/s}$$

(*Note:* To convert rev/s to rad/s, multiply by 2π).

Since all particles on the flywheel have equal angular displacements in any given time, their angular velocities will be the same.

However, the linear speed v of any particle on the flywheel depends upon its distance from the centre of rotation. Suppose point A on the flywheel moves through an angular displacement of θ radians in t seconds. If the radius of the flywheel is r units the distance moved by A around the circumference is $r\theta$ units. Therefore the linear speed of A is found from

$$v = \frac{\text{distance travelled}}{\text{time taken}}$$

$$= \frac{r\theta}{t}$$

But $\quad \dfrac{\theta}{t} = \omega$

Thus, $\quad v = r\omega$

The linear speed of a particle moving at an angular velocity of ω rad/s is found by multiplying ω by the radius of rotation.

SAMPLE PROBLEM 12/2

A pulley is turning at a constant rate of 700 revolutions per minute. Calculate
 (i) the angular velocity of any point on the pulley;
(ii) the linear speed of a point 50 mm from the axis of rotation on the pulley.

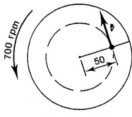

Problem 12/2

Calculation:
(i) $\qquad\qquad\qquad$ Angular velocity $\omega = 700$ rpm

$$\therefore \omega = \frac{700}{60} \text{ rps}$$

$$= \frac{700}{60} \times 2\pi \text{ rad/s}$$

$$\therefore \omega = 73.3 \text{ rad/s}$$

(ii) Linear speed of point.
$\qquad\qquad$ Angular velocity of the point $= 73.3$ rad/s
But, $\qquad\qquad\qquad\qquad v = r\omega$
$$= 0.05 \times 73.3 \quad \text{(since 50 mm} = 0.05 \text{ m)}$$
$$= 3.7 \text{ m/s}$$

Result:
The angular velocity of all points on the flywheel is 73.3 rad/s, and the linear speed of a point 50 mm from the axis of rotation is 3.7 m/s tangential to the circular path of the point. ■

ANGULAR ACCELERATION (α, alpha)

Angular acceleration is the time-rate change of angular velocity, the basic unit being radians per second *per second* (rad/s²). However, revolutions per second per second are also often used in engineering problems (rev/s²).

Consider a body moving with an initial angular velocity ω_0 which has its angular velocity uniformly increased to ω_t in time t seconds. The angular acceleration α in rad/s² can be found from

$$\alpha = \frac{\text{change in angular velocity}}{\text{time taken}}$$

$$\therefore \alpha = \frac{\omega_t - \omega_0}{t}$$

(*Note:* To convert α in rev/s² to rad/s² multiply by 2π.)

If a body is moving with a uniform angular acceleration α rad/s², its instantaneous linear acceleration tangential to its circular path is obviously also changing. To determine this tangential linear acceleration* at any instant, the relationship

$$a = r\alpha$$

can be used, where r is the radius of rotation.

SAMPLE PROBLEM 12/3

The spindle of an electric motor revolving at 2500 rpm slows down to 1500 rpm in the first 4 seconds after the power is turned off.
(i) What angular deceleration operated?
(ii) How long would it take the motor to stop after being switched off, assuming the same rate of deceleration?

Calculation: (i) Angular acceleration, $\alpha = \dfrac{\omega_t - \omega_0}{t}$

$$= \frac{1500 - 2500}{4}$$

$$= -250 \text{ rev/min/sec}$$

$$= \frac{-250}{60} \text{ rev/s}^2$$

$$= \frac{-250}{60} \times 2\pi \text{ rad/s}^2$$

$$\therefore \alpha = -26.2 \text{ rad/s}^2$$

(ii) Assuming the same rate of deceleration, let t seconds be the time taken

$$\therefore \alpha = \frac{\omega_t - \omega_0}{t} \qquad \text{where: } \alpha = 250 \text{ rev/min/s}$$
$$\omega_t = 0$$
$$\therefore -250 = \frac{0 - 2500}{t} \qquad \omega_0 = 2500 \text{ rpm}$$

$$\therefore t = 10 \text{ s}$$

Result:
The angular deceleration was 26.2 rad/s² and the time taken to stop was 10 seconds.   ■
(*Note:* This example shows that, provided one is consistent in their use, radians or revolutions can be used in the formulas to circular motion.)

---

*It is very important to note that there is another acceleration operating whenever a particle moves in a circular path. This is discussed fully in Chapter 13.

Relationships Between Linear and Angular Quantities

Note that the formulas for angular velocity and acceleration are identical in form to those for linear velocity and acceleration.

Angular	*Linear*
$\omega = \dfrac{\text{angular displacement}}{t} = \dfrac{\theta}{t}$	$v = \dfrac{\text{displacement}}{\text{time}} = \dfrac{s}{t}$
$\alpha = \dfrac{\text{change in angular velocity}}{\text{time}} = \dfrac{\omega_t - \omega_0}{t}$	$a = \dfrac{\text{change in velocity}}{\text{time}} = \dfrac{v - u}{t}$

This means that the same basic relationships hold equally well for angular and linear velocities and accelerations and two sets of formulas do not have to be remembered.

However, the formulas connecting linear and angular displacements, speeds and accelerations are new and must be memorised.

$s = r\theta$

$v = r\omega$

$a = r\alpha$

where:

s = linear displacement; θ = angular displacement

v = linear velocity; ω = angular velocity

a = linear acceleration; α = angular acceleration

and r = radius of rotation.

These are easily remembered since, in each case, the angular quantity is multiplied by the radius of rotation to find the linear quantity. (Note that these three formulas only apply when the basic quantity in the angular units is the radian.)

EQUATIONS OF ROTATION

The equations describing the "geometry" of rotational motion (with uniform acceleration) are identical in form to those derived in Chapter 9 for linear motion, except that the appropriate rotational symbols of θ, ω and α replace those of s, v, and a. The similarities of those equations are seen from the table below.

Linear	*Rotational*
$v = u + at$	$\omega_t = \omega_0 + \alpha t$
$s = ut + \frac{1}{2}at^2$	$\theta = \omega_0 t + \frac{1}{2}\alpha t^2$
$v^2 = u^2 + 2as$	$\omega_t^2 = \omega_0^2 + 2\alpha\theta$
$s = v_{av} \times t$	$\theta = \omega_{av} \times t$

where: s = linear displacement

u = initial velocity

v = final velocity

a = acceleration

where: θ = angular displacement

ω_0 = initial angular velocity

ω_t = final angular velocity

α = angular acceleration

Note: In most problems involving rotation, either radians or revolutions can be used as the basic unit of displacement. That is, velocities can be in radians per second, radians per minute, revolutions per second, or revolutions per minute, and accelerations in corresponding units. However, to obtain the correct solution to a particular problem, all units used in the calculation must be consistent.

For example, you could use
(a) rpm; rpm/min and minutes;
(b) rps; rps/s; and seconds;
(c) rad/min; rad/min^2; and minutes; or
(d) rad/s; rad/s^2; and seconds (this is the *true* SI usage).

The derivation of the rotational formulas is almost identical to the method used for the linear formulas in Chapter 9, thus it is not necessary to derive them from first principles. Also, due to the identical form of the two sets of equations, there is no need to memorise the formulas for rotation; simply remember the linear formulas with which you are already familiar and substitute the appropriate rotational values for them as required.

GRAPHICAL SOLUTIONS: MOTION DIAGRAMS

We have seen that motion diagrams or graphs are often useful aids when visualising the linear motion of an object. In the same way these diagrams often prove useful when considering rotational motion, particularly when variable velocities and accelerations are concerned.

Various units can be used when constructing rotational motion diagrams; however, *consistency must be used throughout the solution to any particular problem.*

Velocity–time graphs are perhaps the most informative and useful type of motion diagram, with rotational velocities being plotted in rps, rpm, rad/s or rad/m.

Displacement–time graphs are also very useful in problem-solving. When using these diagrams, it is often most convenient to plot displacement in radians and time in seconds so that the derived SI unit for angular velocity, the *rad/s*, is used for related velocities.

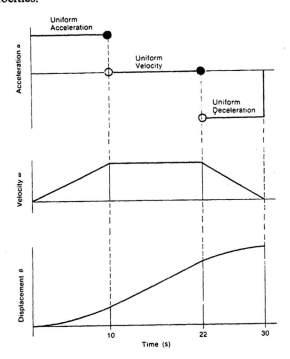

Figure 12.4 Graphs of the angular motion of a winding drum used to raise and lower a mine cage. These show clearly the relationships between displacement, velocity and acceleration, with respect to time.

SAMPLE PROBLEM 12/4

A pulley with an initial angular velocity of 40 rad/s is accelerated at 10 rad/s² for 6 seconds. What is its final angular velocity and what displacement occurred during the 6-second acceleration period?

Calculation:

$$\omega_t = \omega_0 + \alpha t \qquad \text{(Remember: } v = u + at\text{)}$$
$$\therefore \omega_t = 40 + (10 \times 6)$$
$$= 100 \text{ rad/s}$$

where: $\omega_t = ?$
$\omega_0 = 40$ rad/s

$$\theta = \omega_0 t + \tfrac{1}{2} a t^2 \qquad \text{(Remember: } s = ut + \tfrac{1}{2} a t^2\text{)}$$
$$= (40 \times 6) + (\tfrac{1}{2} \times 10 \times 6^2)$$
$$= 240 + 180$$
$$\therefore \theta = 420 \text{ radians}$$

$\alpha = 10$ rad/s
$t = 6$ s

Result:
The final angular velocity is 100 rad/s and the displacement during acceleration was 420 radians. ∎

SAMPLE PROBLEM 12/5

A flywheel rotating at 1800 rpm was slowed down to 1300 rpm in 10 minutes by friction in the motor bearing. It was then brought to rest by the application of the brake which caused an angular deceleration of 100 rpm per minute. Calculate the time taken for the flywheel to come to rest and the number of revolutions it makes in doing so.

Calculation:
(i) The time taken to come to rest is 10 minutes plus the time taken to slow from 1300 rpm.

$$\omega_t = \omega_0 + \alpha t$$
$$\therefore 0 = 1300 - 100t$$
$$\therefore t = 13 \text{ minutes}$$

where: $\omega_t = 0$
$\omega_0 = 1300$ rpm
$\alpha = -100$ rpm
$t = ?$ minutes

Total time to come to rest = 10 + 13 = 23 minutes.

(ii) Displacement.

In the first 10 minutes, when the deceleration, $\alpha = \dfrac{1800 - 1300}{10} = 50$ rpm/min

$$\theta = \omega_0 t + \tfrac{1}{2}\alpha t^2$$
$$= (1800 \times 10) + (\tfrac{1}{2} \times -50 \times 100)$$
$$= 15\,500 \text{ revs}$$

where: $\theta = ?$
$\omega_0 = 1800$ rpm
$\alpha = -50$ rpm/min
$t = 10$ s

In the last 13 minutes

$$\theta = \omega_0 t + \tfrac{1}{2}\alpha t^2$$
$$= (1300 \times 13) + (\tfrac{1}{2} \times -100 \times 13^2)$$
$$= 8450 \text{ revs}$$

where: $\theta = ?$
$\omega_0 = 1300$ rpm
$\alpha = -100$ rpm/m
$t = 13$ min

Total displacement = 15 500 + 8450 = 23 950 rpm.

Result:
The flywheel comes to rest in 23 minutes with a total displacement of 23 950 revolutions.

Alternative Graphical Solution:
Use the given data to plot a velocity–time graph. (Note that the deceleration of 100 rpm/m gives a loss of 1000 rpm in 10 minutes.)

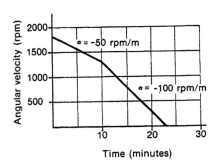

Problem 12/5

(i) Reading from the graph, total time taken by the flywheel to slow down is 23 minutes.

(ii)   Displacement = area under the graph
$$= (1300 \times 10) + (\tfrac{1}{2} \times 500 \times 10) + (\tfrac{1}{2} \times 1300 \times 13)$$
$$= 13\,000 + 2500 + 8450$$
$$= 23\,950 \text{ revs}$$

Result:
The flywheel comes to rest in 23 minutes with a total displacement of 23 950 revolutions. (Note the simplicity of the graphical solution to this problem.) ■

SAMPLE PROBLEM 12/6

Two balls of identical diameters are spinning about their vertical axes with opposing directions of rotation. Ball A was accelerated from rest at a constant rate of 100 rpm/min. Ball B, rotating at a constant velocity of 1400 rpm, begins to decelerate at 100 rpm/min at the same time that ball A begins to accelerate from rest. At what time can the two balls be touched together without "skidding" on each other? What is the angular velocity of each at this time?

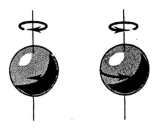

Problem 12/6(a)

Solution:

Construct a velocity–time graph from the given data. Note that

(i) if ball *A* accelerates from rest at + 100 rpm/m, its velocity after 5 minutes is 500 rpm;

(ii) if ball *B* decelerates from 1400 rpm at 100 rpm/m, its velocity after 5 minutes is 900 rpm.

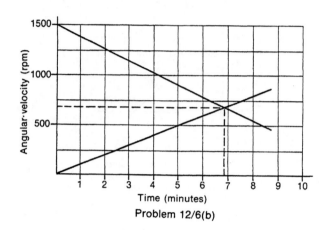

Problem 12/6(b)

Result:

6.8 minutes; 685 rpm.

REVIEW PROBLEMS

(Use $g = 10\,\text{m/s}^2$ unless otherwise specified.)

12/7

Express the following angles as radians:

(i) 90°;   (iii) 121°;

(ii) 16.5°;   (iv) 330°.

12/8

Express the following angles in degrees:

(i) 1 rad;

(ii) 19 rad;

(iii) 3π rad.

12/9
At certain different times during the operation of a machine, its flywheel is rotating with an angular velocity of
  (i) 200 rpm;
 (ii) 700 rpm;
(iii) 1440 rpm.
Express the angular velocity in each instance as radians per second.

12/10
Convert the following angular velocities to revolutions per minute:
  (i) 120 revolutions per second;
 (ii) 31 416 radians per minute;
(iii) 50 radians per second.

12/11
The Olympic Pool in Mexico City has an electronic timing device with a large public display incorporating a sweep second hand of 1.4 m effective radius. Determine the angular and linear velocities of the tip of the hand.

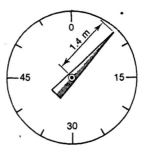

Problem 12/11

12/12
A flywheel of diameter 0.5 metres is rotating at 1000 rpm. Determine the linear speed of the point *A* on its circumference.

Problem 12/12

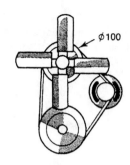

12/13

Determine linear speed (in m/s) of the fan belt of a motor vehicle, when the 100-mm diameter fan pulley is rotating at 4000 revolutions per minute.

Problem 12/13

12/14

A grinding wheel of 200 mm diameter is rotating at 1500 rpm when it begins to disintegrate. With what maximum linear velocity will pieces of the wheel be flung off?

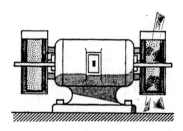

Problem 12/14

12/15

A water turbine starts from rest and accelerates at a uniform rate to its normal operating speed of 500 rpm in 5 minutes. Determine its angular acceleration (in rad/s²) and its total displacement during acceleration (in radians).

12/16

The elements of a circular saw are shown in the diagram. The driven pulley has a diameter of 50 mm and rotates at 4500 rpm. Given that the saw-blade diameter is 300 mm, determine the linear speed (in m/s) of the tip of one tooth of the saw blade.

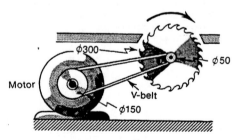

Problem 12/16

12/17

A pulley with an initial angular speed of 50 radians per second is accelerated at 10 radians per second squared for 10 seconds. What is its final angular velocity in rps, and what was its displacement during the 10-second period (in revolutions).

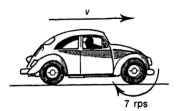

Problem 12/18

12/18

The wheels of a car have diameters of 560 mm. Determine the linear velocity of the car in m/s and km/h if the wheels revolve at 7 revolutions per second.

12/19

Determine the angular and linear velocities of the extremities of the hour and minute hands of the Town Hall clock, given that their respective lengths, as measured from their pivot points, are 1 and 1.3 metres. Use rad/s and m/s for your answers.

12/20

A boy riding his bicycle along a level road slows from 15 km/h to rest in 15 metres. If the bicycle wheels have diameters of 0.76 metres, determine
  (i) the average deceleration of the bicycle (in m/s^2);
 (ii) the angular retardation of the wheels (in rad/s^2).

12/21

The bascule bridge is opened by winding a pinion along a rack on each side of the bridge.
  (i) If the pinion is 250 mm in diameter and revolves at 5 rpm, what is the angular velocity of the bridge deck?
 (ii) How long will it take to open the bridge through 60°?

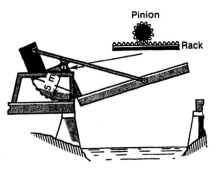

Problem 12/21

12/22

A bandsaw motor runs at 1425 rpm. The motor pulley has a diameter of 75 mm, the drive pulley has a diameter of 175 mm and the driving wheels have diameters of 400 mm each. Determine the linear speed of the saw blade as it passes the cutting table.

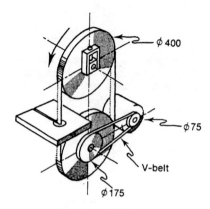

φ 400

φ 75

V-belt

φ 175

Problem 12/22

12/23

The spindrier of a washing machine accelerates from rest at 30 rpm per second, reaches its maximum velocity, and immediately begins to decelerate at 10 rpm per second. If the spin cycle takes 4 minutes, determine by using a suitable velocity–time graph the maximum velocity of the spindrier and the time taken to reach this velocity.

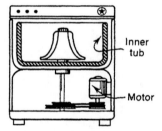

Inner tub

Motor

Problem 12/23

12/24

Two gears of equal diameters are rotating on adjoining shafts. One is rotating at a constant angular velocity of 140 rev/min in a clockwise direction. The other accelerates from rest at a constant angular acceleration of 2 rad/s² in an anti-clockwise direction. At what instant can the two gears be smoothly meshed together? Solve the problem analytically and graphically.

12/25

Pyrmont bridge has a centre section which opens in a horizontal arc in order to allow cargo vessels to enter the docks. The bridge takes $2\frac{1}{2}$ minutes to turn through 90°, its motion consisting of an initial constant angular acceleration for 60 seconds, a period of uniform angular velocity for 50 seconds, followed by deceleration to rest in the last 40 seconds.

Determine, with the aid of a graph of angular velocity against time,

 (i) the maximum angular velocity;
 (ii) the angular acceleration; and
(iii) the angular deceleration of this section of the bridge.

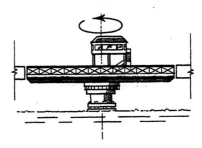

Problem 12/25

12/26

Friction in the shaft bearings brings a flywheel to rest in 3 minutes. If the angular speed of the flywheel was 1800 rpm, determine

 (i) the deceleration due to friction;
 (ii) the total number of revolutions turned during deceleration (that is, displacement);
(iii) the time taken to complete the first half of these revolutions.

This centrifuge was used in the training programme for the Apollo moonshots. When it is rotating, the astronauts in the three-man gondola at the end of the 15-metre arm experience inertia forces similar to those expected during lift-off and re-entry.

13

Rotational Motion

$$a_c = \frac{v^2}{r}$$

$$F_c = \frac{mv^2}{r} = m\omega^2 r$$

$$T = I\alpha$$

$$M = I\omega$$

Try this simple experiment. Attach one end of a piece of string to a relatively heavy object (say, a small block of wood) and whirl this object around in a horizontal circle while holding on to the other end of the string. You will find that, if you whirl the object with constant angular velocity, you need to apply a *restraining force* on the end of the string in order to maintain the original circular path of the object. This restraining force, F_c, acts radially inwards and is necessary to cause the continuous change in the *direction of* the linear velocity, V, of the object.

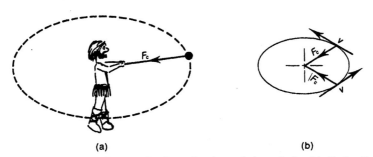

(a)   (b)

Figure 13.1 (a) An object moving in a circular path is restrained in that path by a radial force, F_c, which is directed towards the centre of rotation; (b) shows that this force F_c is necessary to change the direction of the tangential velocity, v.

However, Newton's second law ($F = ma$) states that the application of a force to a body causes an acceleration; thus, even though the object in our experiment was rotating with constant angular velocity, it must be subject to an acceleration caused by the restraining force F_c. In order to discover the magnitude and direction of this acceleration, it is necessary to draw a free-body diagram of the block showing all of the vectors operating upon it. This has been done in Figure 13.2, and now we can *analyse* the effects of these vectors. Since there is obviously an acceleration due to the force F_c, let the resultant acceleration acting on the block be represented by the vector a as shown in Figure 13.2 (a). This acceleration can be resolved into a tangential component and a radial component, as shown in Figure 13.2 (b).

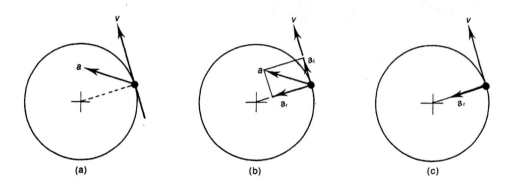

(a)   (b)   (c)

Figure 13.2 An object rotating in a circular path must be subjected to an acceleration, *a*, since its velocity, *v*, is constantly changing its direction: (a) shows this acceleration, *a*, in an assumed direction; (b) shows the tangential and radial components of the acceleration, *a*; and (c) shows that *a* is in fact directed radially inwards, the tangential component, a_t, always being zero.

If the object travels at *uniform linear speed*, then a_t must be zero, and thus the acceleration produced by the restraining force F_c must act radially inwards towards the centre of rotation as shown in Figure 13.2 (c). It is called the *centripetal acceleration* and the force producing it (F_c) is called the *centripetal force*.

CALCULATING CENTRIPETAL ACCELERATION AND CENTRIPETAL FORCE

Suppose that a body moves from A to B in time t around the horizontal circular path indicated in Figure 13.3 (a). If the body is moving at constant angular velocity ω, the magnitude of its tangential linear velocity v is constant, since $v = r\omega$, where r is the radius of rotation.

It has already been shown that, in moving from A to B, the body has a radially inward acceleration represented in Figure 13.3 (b) by the vector a_c. If the velocity diagram as shown in Figure 13.3 (b) is drawn, it is seen that the change in velocity from A to B is equal to the vector δv. If time t is considered to be very small, then the chord

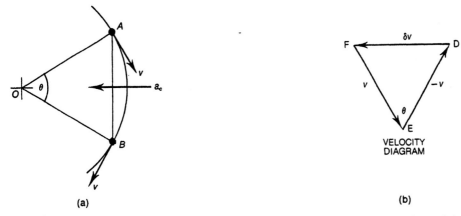

Figure 13.3 (a) A rotating mass in moving from *A* to *B* is subjected to the radial acceleration, a_c; (b) this radial acceleration produces the change in velocity *δv* shown in the velocity diagram.

AB becomes equal to the arc *AB* and the triangles *OAB* and *DEF* are similar (both have included angle θ and adjoining sides are proportional). Thus,

$$\frac{AO}{AB} = \frac{EF}{DF}$$

$$\therefore \frac{r}{vt} = \frac{v}{\delta v} \qquad \textit{Note: Side } AB = \text{speed} \times \text{time} = vt$$

$$\therefore \frac{\delta v}{t} = \frac{v^2}{r}$$

$$a_c = \frac{v^2}{r} \qquad \text{since } \frac{\delta v}{t} = \frac{\text{change in velocity}}{\text{time}} = \text{acceleration.}$$

Thus, the centripetal acceleration present for a constant angular velocity ω is equal to v^2/r or $\omega^2 r$ (since $v = r\omega$), where *v* is the linear speed and *r* is the radius of rotation.

The centripetal force F_c acting radially inwards can be calculated from Newton's second law and is obviously equal to the tension (restraining force) in the cord.

$$F = ma$$
$$\therefore F_c = ma_c$$
$$= m \times \frac{v^2}{r} \qquad \text{or} \qquad F_c = m \times \omega^2 r$$
$$\therefore F_c = \frac{mv^2}{r} \qquad \text{or} \qquad F_c = m\omega^2 r$$

CENTRIFUGAL FORCE

According to Newton's third law, to every action there is an equal but opposite reaction. The equal but opposite reaction to centripetal force is called centrifugal force; it obviously has a magnitude of mv^2/r (or $m\omega^2 r$) for constant angular velocity and is directed radially *away* from the centre of rotation.

Centrifugal force is a reactive or *inertia force*. It exists only as a reaction to the centripetal force. When an object which is rotating at constant angular velocity is released from its circular path it is not, as is commonly stated, "flung out by centrifugal force", *rather the object simply resumes linear motion* since the centripetal force causing the radial acceleration no longer exists.

Figure 13.4 When an object attached to one end of a cord is whirled in a circular path, the tension in the cord, (a), is the centripetal force, and the centrifugal force is the inertia force that balances this centripetal force, (b).

SAMPLE PROBLEM 13/1

A block of wood of mass 1 kg is tied to the end of a cord and whirled around in a horizontal circle of radius 1 metre at constant angular velocity of 3.5 revolutions per second. Calculate its linear speed, the centripetal acceleration, and the tension in the cord.

Calculation:

$$\text{Linear speed, } v = r\omega$$
$$= 1 \times 7\pi \text{ m/s}$$
$$= 22 \text{ m/s}$$

where: $\omega = 3.5$ rev/s
$= 7\pi$ rad/s
$r = 1$ m

$$\text{Centripetal acceleration, } a_c = \frac{v^2}{r}$$
$$= \frac{22^2}{1}$$
$$= 484 \text{ m/s}^2 \text{ towards centre of rotation}$$

where: $v = 22$ m/s
$r = 1$ m

$$\text{Tension in cord} = \text{centripetal force}$$
$$= ma_c \quad (\text{or } m\omega^2 r)$$
$$= 1 \times 484$$
$$= 484 \text{ N}$$

Result:
The body has a linear speed of 22 m/s, a centripetal acceleration of 484 m/s², and the tension in the cord is 484 N. ∎

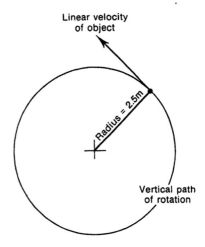

Linear velocity
of object

Radius = 2.5m

Vertical path
of rotation

Problem 13/2

SAMPLE PROBLEM 13/2

A 1-kg object is attached to a cord and whirled around in a vertical path of radius 2.5 metres. What minimum speed must it possess at the top of the circle so that the cord remains taut?

Analysis:

Draw a free-body diagram for the top and bottom position. At the top position, two forces act to pull the body down, these being its own weight-force and the tension T in the cord. When the cord is on the point of slackening, the resultant of these two forces is equal to the centripetal force necessary to keep the body in its circular path.

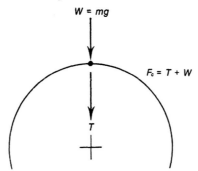

$W = mg$

$F_c = T + W$

T

Problem 13/2

Calculation:

Centripetal force, $F_c = T_1 + W$

$$\therefore \frac{mv^2}{r} = T_1 + mg \qquad \text{where: } m = \text{mass of the object}$$

But, when the cord begins to slacken, $T_1 = 0$

$$\therefore \frac{mv^2}{r} = mg$$

$$\therefore v^2 = rg \qquad\qquad \text{where: } r = 2.5 \text{ m}$$
$$= 2.5 \times 9.8 \qquad\qquad\qquad g = 9.8 \text{ m/s}^2$$

$$\therefore v = \sqrt{24.5}$$
$$\therefore v \approx 4.9 \text{ m/s}$$

Result:

The minimum speed at the top of the circular path must be ≈ 4.9 m/s for the cord to remain taut. (It is interesting to observe that this velocity is independent of the mass of the object.) ■

SAMPLE PROBLEM 13/3

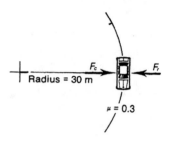

Radius = 30 m

$\mu = 0.3$

At what maximum speed can an automobile of mass 1 tonne be driven around an unbanked curve of radius 30 metres if the coefficient of friction between tyres and wet road is 0.3?

Problem 13/3

Analysis:

Consider the forces acting on the car. At the maximum speed, the outward reactive centrifugal force (inertia force) of the car must equal the total inward frictional force between tyres and the road. If the total inward frictional force between the tyres and the road is exceeded, the car will skid outwards.

Calculation:

Frictional force, $F_r = \mu N = \mu\,mg$, where m = mass of car.

Centrifugal force, F_c = centripetal force, $F_c = \dfrac{mv^2}{r}$.

But, on the point of skidding,

$$F_r = F_c$$
$$\therefore \mu N = \frac{mv^2}{r}$$
$$\therefore \mu\,mg = \frac{mv^2}{r}$$
$$\therefore \mu g = \frac{v^2}{r}$$
$$\therefore v^2 = \mu \times g \times r \qquad \text{where: } \mu = 0.3$$
$$\therefore v^2 = 0.3 \times 9.8 \times 30 \qquad\qquad g = 9.8 \text{ m/s}^2$$
$$= 88.2 \qquad\qquad\qquad r = 30 \text{ m}$$
$$\therefore v = \sqrt{88.2}$$
$$\approx 9.4 \text{ m/s}$$

Result:

The car can round the 30 m curve at about 9.4 m/s without skidding on the wet road. (Note that the mass of the car is immaterial and the width of the tyres also has no effect. Why?) ∎

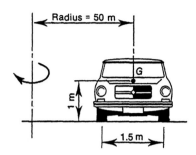

Radius = 50 m

SAMPLE PROBLEM 13/4

A car rounds a level unbanked highway curve of radius 50 metres. The wheel track of the car is 1.5 metres and its centre of gravity is 1 metre above the road. Assuming that side friction and roughness prevent skidding, at what speed will the car tend to tip over?

Analysis:
Draw a free-body diagram showing all of the forces on the car. When the car is on the point of tipping, there will be no reaction on the inner two wheels since they will rise slightly from the roadway.

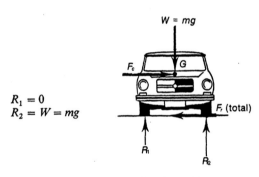

$R_1 = 0$
$R_2 = W = mg$

Problem 13/4

Also, if no sliding occurs, the total frictional force, F_r, between the tyres and roadway must be equal to F_c, the reactive centrifugal force, i.e.

$$F_r = F_c = \frac{mv^2}{r}$$

Calculation:
Take moments about the centre of gravity, G, when the car is on the verge of tipping. Assume $\oplus$
$\Sigma M = 0$ About G.

$$\therefore 1 \times F_r - 0.75\, R_2 = 0 \qquad \text{(Note: } R_1 = 0; R_2 = mg)$$

$$1 \times \frac{mv^2}{r} - 0.75\, mg = 0$$

$$\therefore v^2 = 0.75\, gr \qquad\qquad \text{where: } g = 9.8 \text{ m/s}^2$$
$$= 0.75 \times 9.8 \times 50 \qquad\qquad\qquad r = 50 \text{ m}$$
$$= 367.5$$
$$\therefore v = 19.2 \text{ m/s}^2$$

Result:
The car will tip over when its speed around the curve exceeds 19.2 m/s or $\approx$ 69 km/h. ∎

SAMPLE PROBLEM 13/5

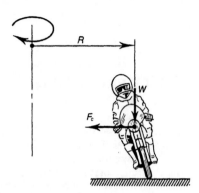

A motor cyclist rounding an unbanked corner of 100 metres radius leans at an angle of 6° from the vertical to maintain equilibrium. What is his speed in km/h?

Analysis:

The cyclist leans over to such an angle that will allow the horizontal component of his own weight-force to provide sufficient centripetal force to enable him to round the corner in equilibrium. Draw a free-body diagram of the forces acting on the cycle and rider, considering them as one mass. The combined weight (W) of the cyclist and machine acting through their common centre of gravity can be resolved into two components, one at 6° to the vertical (cb) and the other a horizontal component ab.

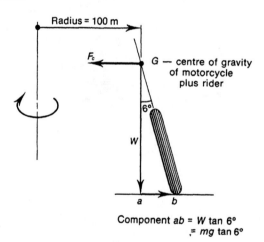

Component $ab = W \tan 6°$
$= mg \tan 6°$

Problem 13/5

Calculation:

The component ab of the weight-force W must equal the centripetal force F_c necessary to allow the bike and rider to negotiate the curve.

$$F_c = \text{component } ab$$

$$\frac{mv^2}{r} = mg \tan 6°$$

$$\therefore v^2 = rg \tan 6°$$
$$\therefore v^2 = 100 \times 9.8 \times 0.1051$$
$$\therefore v^2 = 103$$
$$\therefore v = 10.15 \text{ m/s}$$
$$\therefore v \approx 36.5 \text{ km/h}$$

where: $r = 100$ m
$g = 9.8$ m/s^2
$\tan 6° = 0.1051$

Result:

The bike and rider rounded the corner at a speed of 36.5 km/h.

SUPER-ELEVATION OF ROADS AND RAILWAY CURVES

A practical demonstration of the outward "pull" of centrifugal inertia (or reaction) force is felt by a passenger in a car rounding an unbanked curve on a highway. This pull acting on the car tends to cause the car to skid and can lead to roll-over of the vehicle depending on the relative movement of its centre of gravity (refer back to Sample Problem 13/4).

In order to compensate for the effects of centrifugal force in this situation, the outer edge of the roadway is elevated in relation to the inner edge. The relative amount of elevation is termed the *super-elevation* of the highway and may be measured as the angle (θ) between the road surface and the horizontal, or as the height, h, shown in Figure 13.5 (b).

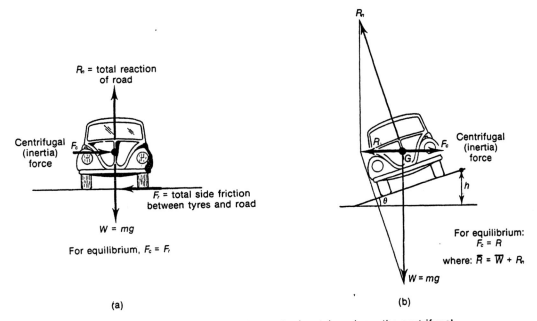

Figure 13.5 A car rounding a curve: (a) on a horizontal roadway, the centrifugal force, F_c, is balanced by the total side friction, F_r, between the tyres and the road; (b) however, on a properly banked (super-elevated) road, the resultant R of the weight-force, W, and the normal reaction, R_n, equals the centrifugal force, F_c, and side friction is thus zero.

When a car rounds a curve at the exact speed for which the curve was super-elevated, no side friction is developed between the tyres and the road surface. However, if the approach speed is greater or less than this equilibrium speed, side friction will operate either inwards or outwards in order to establish equilibrium.

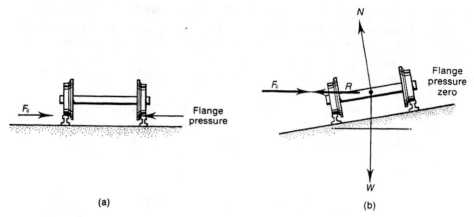

(a)   (b)

Figure 13.6 (a) On the unbanked railway curve, the flange pressure on the outer rail balances the centrifugal force, F_c; however, on a correctly banked curve (b) the resultant force, R, of the weight-force, W, and the normal reaction, N, balance the inertia force, F_c, and there is no flange pressure at the equilibrium speed.

When considering the super-elevation of railway lines, the same general principles apply, except that flange pressure replaces friction in the calculations. Thus, when a train rounds a curved track, no flange pressure occurs if the speed of the train is the exact speed for which the track was super-elevated.

SAMPLE PROBLEM 13/6

A diesel locomotive of mass 200 tonnes rounds a level curve with a radius of 150 metres at a speed of 18 km/h. Calculate the total flange pressure against the outer rails.

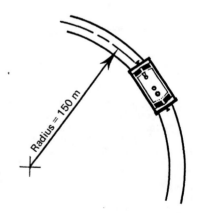

Analysis:
Draw a free-body diagram showing all forces acting. Obviously the centrifugal force, F_c, present must be opposed by an equal force of flange pressure, P, otherwise the engine will leave the tracks.

Forces on the locomotive

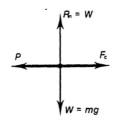

FREE-BODY DIAGRAM

Problem 13/6

Calculation:

$$P = F_c = \frac{mv^2}{r}$$

$$= \frac{200 \times 10^3 \times 5^2}{150}$$

$$= 33.33 \times 10^3 \text{ N}$$

$$= 33.33 \text{ kN}$$

where: $m = 200 \times 10^3$ kg
$v = 18$ km/h $= 5$ m/s
$r = 150$ metres

Result:
The total pressure exerted by the outer wheels on the rails is 33.33 kN. ■

SAMPLE PROBLEM 13/7

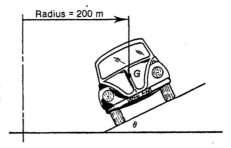

A curve of 200 metres radius on an expressway is to be banked so that an automobile can round the curve at 25 m/s without depending upon the frictional grip of the tyres. What must be the "bank" of the curve?

Analysis:
Draw a free-body diagram showing the forces acting on the car. From this diagram it is obvious that the resultant of N, the reaction of the road on the car, and W, the weight-force of the car, must be equal to F_c, the centripetal force required to hold the car in its circular path.

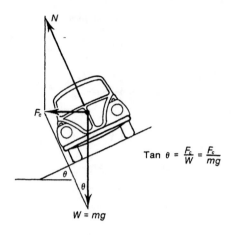

$$\text{Tan } \theta = \frac{F_c}{W} = \frac{F_c}{mg}$$

$W = mg$

Forces on the car
FREE-BODY DIAGRAM
Problem 13/7

Calculation:
From the free-body diagram,

$$\tan \theta = \frac{F_c}{mg}$$

$$= \frac{(mv^2/r)}{mg}$$

$$= \frac{v^2}{rg}$$

$$= \frac{25^2}{200 \times 9.8}$$

$$\therefore \tan \theta = 0.319$$

$$\therefore \theta = 17.7°$$

where: F_c = centripetal force
$v = 25$ m/s
$r = 200$ metres
$g = 9.8$ m/s²

Result:
The roadway would have to be banked to $\approx 18°$ from the horizontal.   ∎

TORQUE AND ANGULAR ACCELERATION

Consider the flywheel mounted on an axle as shown in Figure 13.7 (a). If a tangential force, F, is applied to this flywheel (say, by a tangential push of the hand), then the wheel will begin to rotate with a certain initial acceleration, α. The effectiveness of the force F in producing the angular acceleration α is measured by the *torque* of the force, *the torque being the product of the force itself and its perpendicular distance from the axis of rotation, 0—0* (Figure 13.7 (b)). In other words, the torque (or turning effect) of F is found from the relationship

$$T = Fr$$

where r is the distance from the axis of rotation.

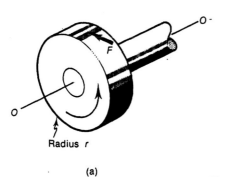

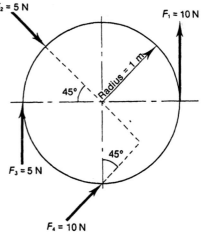

Torque T = F_r
about centre
of rotation O

(a)   (b)

Figure 13.7

Note that the torque is essentially the same as the moment of the force about the centre of rotation. The unit of torque is also the same as that of moment; that is, newton metre (N m).*

SAMPLE PROBLEM 13/8

A wheel of radius 1 metre has forces applied to it as shown. Find the torque applied by each force, and the resultant torque acting on the wheel.

$F_2 = 5$ N

$F_1 = 10$ N

45°

Radius = 1 m

45°

$F_3 = 5$ N

$F_4 = 10$ N

Problem 13/8

Analysis:
The torque applied by each individual force is found from the relationship $T = Fr$. The resultant torque is the vector sum of all of the torques present.

Calculation:
Assume anti-clockwise torque as being positive, that is, $\overset{+}{\circlearrowleft}$

$$\text{Torque of } F_1 = 10 \times 1 = 10 \text{ N m}$$
$$\text{Torque of } F_2 = 5 \times 0 = 0 \text{ N m}$$
$$\text{Torque of } F_3 = -5 \times 1 = -5 \text{ N m}$$
$$\text{Torque of } F_4 = 10 \times \sin 45° = 7.1 \text{ N m}$$

Total resultant torque $= 10 + 0 - 5 + 7.1$
$$= 12.1 \text{ N m}$$

∎

---

*Refer to Chapter 4.

Referring back to the flywheel shown in Figure 13.7, the problem now is to determine the relationship between the rate of acceleration and the applied torque ($T = Fr$). The main difficulty is that the various parts of the flywheel are moving with different tangential accelerations, since tangential acceleration is directly proportional to the distance from the axis of rotation (that is, $a = r\alpha$).

To overcome this difficulty, consider a simpler situation. Figure 13.8 shows a heavy but small "point" mass, m, rotating at the end of a thin arm of length r units whose mass can be considered as negligible.

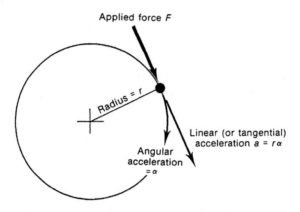

Figure 13.8

If the force F is applied, the mass m is subjected to the torque Fr which produces an angular acceleration α. The tangential linear acceleration a of the mass is equal to $r\alpha$ ($a = r\alpha$). Now consider Newton's second law,

$$F = ma$$
$$F = mr\alpha \quad \text{(since } a = r\alpha\text{)}$$

Multiply each side by r

$$Fr = mr^2\alpha$$
$$\therefore T = mr^2\alpha$$

Thus, the applied torque and the angular acceleration are proportional to each other, the proportional constant being mr^2.

The value mr^2 in this equation is known as the *moment of inertia, I*, of the mass m and is constant for any given rotating body.

MOMENT OF INERTIA (I)

The moment of inertia of a rotating body is *the resistance that the body offers to a change in angular velocity*; it is sometimes called the *second moment of area of the mass* (the first moment of area being the product *mr* which has no significance in this section of dynamics). Thus, the moment of inertia is not only concerned with the magnitude of the mass of the rotating body, but also with its position or distribution in relation to the axis of rotation. The moment of inertia is of similar importance to angular motion as mass is to linear motion in that both are a measure of the inertia of the body. The unit for moment of inertia is obviously $kg\,m^2$ (to prove this refer back to the simple example above where $I = mr^2$).

Now return to our original flywheel in Figure 13.7. We can consider the flywheel to be made up of a number of masses similar to that in Figure 13.8, each mass having its own radius of rotation about the centre O of the flywheel. The sum of the moments of inertia of all of these masses is equal to the total moment of inertia of the flywheel. If I is the moment of inertia of a rotating body like a flywheel, then the relationship between applied torque T and resultant angular acceleration α is

$$T = I\alpha \qquad \text{where: } T = \text{torque}$$
$$I = \text{moment of inertia}$$
$$\alpha = \text{angular acceleration}$$

From the discussion so far, it is reasonable to say that the further the mass from the centre of rotation, the greater the moment of inertia that it possesses.

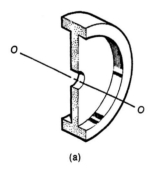
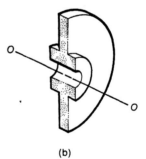

(a)   (b)

Figure 13.9   The mass of each wheel is the same, but the one on the left has a much greater moment of inertia than the one on the right.

Consider the two flywheels shown in Figure 13.9. It is obvious that flywheel (a) has a greater moment of inertia than flywheel (b) since the bulk of the mass of (a) is further out from the axis of rotation, $O\!-\!O$.

RADIUS OF GYRATION (k)

Consider that the entire mass of the flywheel shown in Figure 13.10 (a) can be concentrated into a very thin disc as shown in Figure 13.10 (b).

(a)   (b)

Figure 13.10   The radius of gyration, k, of a flywheel of radius r units. The moment of inertia, I, of this flywheel is found from $I = mk^2$, where m is the mass of the flywheel.

If the radius of rotation of this disc is k as shown, its moment of inertia would be mk^2 ($I = mr^2$ derived previously applies in this case) and is the same as the moment of inertia of the original flywheel. This radius k at which the entire mass of a rotating body is assumed to be concentrated is known as its *radius of gyration*.

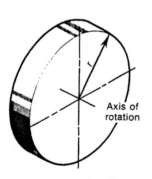
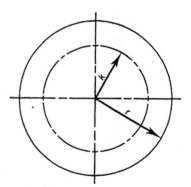

Axis of rotation

Figure 13.11

The moment of inertia of a flywheel of uniform section as shown in Figure 13.11 is found from the following relationship:

$$I = \frac{mr^2}{2}$$

where: m = mass of flywheel
r = radius

No proof is offered for this formula, since the determination of formulas for moments of inertia is outside the scope of this book. However, using the above formula, the radius of gyration of the uniform flywheel can be found as follows:

$$I = mk^2$$
$$\therefore \frac{mr^2}{2} = mk^2$$
$$\therefore k^2 = \frac{r^2}{2}$$
$$\therefore k = \frac{r}{\sqrt{2}}$$

This is a useful formula when flywheel or other rotating bodies of uniform section are involved.

For a practical illustration of the effects of the radius of gyration of a mass, consider a ballet dancer or a skater pirouetting; that is, spinning around on one toe or the tip of one skate. At the start of the motion, the dancer has her arms spread wide and her other leg held away from her body. Once motion begins, and the body is rotating with an initial angular velocity, the dancer brings her arms and leg in close to her body, thus decreasing her radius of gyration and hence her moment of inertia. Since the body's resistance to change in velocity is decreased (*I* is drastically reduced), the angular velocity increases dramatically and so does the dancer's spin. In order to slow down again, the dancer extends her arms, her moment of inertia increases, and consequently her angular velocity decreases.

You may care to try the following experiment. Sit in a swivel chair with a freely running bearing. Hold a shot put in each hand and stretch out your arms. Get a friend to give you a slight push to produce a slow initial angular velocity. Now bring your arms in to your sides. (*Caution:* Fix the chair to the floor and make sure that your centre of mass is exactly over the pivot of the chair, otherwise you will become unbalanced and leave the chair at a rate that could be quite sensational—for the observers!)

SAMPLE PROBLEM 13/9

A flywheel of mass 100 kg and radius of gyration 1 metre is rotating at 250 rpm. A brake is applied and supplies a retarding torque of 100 N m. Calculate the time taken to bring the wheel to rest.

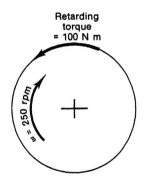

Problem 13/9

Analysis:
This problem is similar to a body being brought to rest by a retarding force, except that in this instance the resistance to retardation is measured by the moment of inertia of the flywheel.

Calculation:
Moment of inertia,

$$I = mk^2$$
$$= 100 \times 1$$
$$= 100 \text{ kg m}^2$$

where: $m = 100$ kg
$k = 1$ m

Now, since torque = moment of inertia × angular acceleration

$$T = I\alpha$$
$$-100 = 100\,\alpha$$
$$\therefore \alpha = -1 \text{ rad/s}^2$$

where: $I = 100 \cdot \text{kg m}^2$
$T = -100$ N m

Apply

$$\omega_t = \omega_0 + \alpha t$$
$$0 = (26.2) - (1 \times t)$$
$$\therefore t = 26.2 \text{ s}$$

where: $\omega_t = 0$
$\omega_0 = 250$ rpm ≈ 26.2 rad/s
$\alpha = -1 \text{ rad/s}^2$

Result:
The flywheel takes ≈ 26 seconds to come to a stop under a retarding torque of 100 N m. ■

SAMPLE PROBLEM 13/10

The wheel and axle shown in the diagram has a 10-kg mass attached to the free end of the cord which is wrapped around the axle. When released, the mass falls a vertical distance of 2 metres in 5 seconds. If the axle radius is 80 mm, calculate the moment of inertia of the wheel and axle.

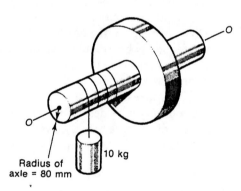

Radius of axle = 80 mm

10 kg

Problem 13/10

Analysis:
The linear acceleration of the 10-kg mass must be equivalent to the angular acceleration of the wheel and axle, and the tension in the cord must be the tangential force causing this angular acceleration. Once the tension is found, the relationship
                    torque = moment of inertia × angular acceleration
can be used to find the moment of inertia of the wheel and axle.

Calculation:
Linear acceleration, a, of the 10-kg mass is found from

$$s = ut + \tfrac{1}{2}at^2$$
$$2 = 0 + \tfrac{1}{2}a\,25$$
$$\therefore a = \frac{4}{25}$$
$$\therefore a = 0.16 \text{ m/s}^2$$

where: $s = 2$ m
$u = 0$
$t = 5$ s
$a = ?$

Angular acceleration of wheel and axle is found from

$$a = r\alpha$$
$$\therefore \alpha = \frac{a}{r}$$
$$\therefore \alpha = \frac{0.16}{0.08}$$
$$\therefore \alpha = 2 \text{ rad/s}^2$$

where: $a = 0.16$ m/s^2
$r = 80$ mm
$= 0.08$ m

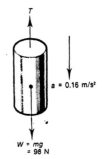

To find tension in the cord, apply $F = ma$ to 10-kg mass
Unbalanced force $F =$ weight-force $-$ tension (T)
$$\therefore 98 - T = 10 \times 0.16$$
$$T = 98 - 1.6$$
$$\therefore T = 96.4 \text{ N}$$
Thus, torque on wheel $= 96.4 \times r$
$$= 96.4 \times 0.08$$
$$\therefore T \approx 7.7 \text{ Nm}$$

Apply
$$T = I\alpha$$
$$\therefore I = \frac{T}{\alpha}$$
$$\therefore I = \frac{7.7}{2} \text{ kg m}^2$$
$$\therefore I = 3.9 \text{ kg m}^2$$

where: $T = 7.7$ Nm
$\alpha = 2$ rad/s^2
$I = ?$

Result:
The moment of inertia of the wheel and axle is 3.9 kg m^2.

■

SAMPLE PROBLEM 13/11

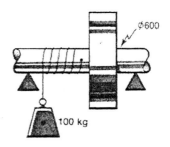

In an experiment on moment of inertia, a flywheel and shaft of combined mass 36 kg are fixed horizontally as shown. The diameter of the shaft is 600 mm, the radius of gyration of the shaft and flywheel combination is 400 mm and the load attached to the cord is 100 kg. Calculate the vertical acceleration of the load if the system is allowed to overhaul (that is, fall freely). You may assume that friction is negligible.

Problem 13/11

Analysis:
Since friction is negligible, the torque provided by the tension P in the cord must balance the inertia of the shaft and drum. The tension P can be found in terms of the vertical acceleration by analysing all of the vectors operating on the 100-kg load.

Calculation:
The unbalanced force causing the load to accelerate is
$$W - P = 980 - P$$
From
$$F = ma$$
$$980 - P = 100a$$
$$P = 980 - 100a \text{ N}$$
But $a = r\alpha$, where α is the angular acceleration of the flywheel.
$$P = 980 - 100\,r\alpha \text{ N}$$

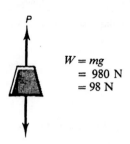

$$W = mg$$
$$= 980 \text{ N}$$
$$= 98 \text{ N}$$

Now consider the vectors on the flywheel:
The moment of inertia of the flywheel and shaft is found from
$$I = mk^2$$
$$= 36 \times 0.16 \text{ kg m}^2$$
$$\therefore I = 5.76 \text{ kg m}^2$$

where: $k = 0.4$ m
$m = 36$ kg

Inertia torque
$$T_1 = I\alpha$$
$$\therefore T_1 = 5.76\alpha \text{ N m}$$

Torque produced by tension P
$$T_2 = P \times 0.3 \text{ N m}$$
$$= 0.3 \ (980 - 100\,r\alpha) \text{ N m}$$

But for free fall
$$T_1 = T_2$$
$$5.76\alpha = 0.3 \ (980 - 100\,r\alpha)$$
$$5.76\alpha = 0.3 \ (980 - 100 \times 0.3 \times \alpha)$$
$$14.76\alpha = 294$$
$$\therefore \alpha = 19.9 \text{ rad/s}^2$$

where $r = 0.3$ m

But
$$\alpha = r\alpha$$
$$\cdot = 0.3 \times 19.9$$
$$\therefore \alpha = 5.98 \text{ m/s}^2$$

Result:
The vertical acceleration of the load is ~6 m/s².   ∎

ANGULAR MOMENTUM AND IMPULSE

In Chapter 11 we showed that a body of mass m moving with constant velocity v has a momentum equal to the product of its mass and its velocity, that is, $M = mv$.

Rotating bodies possess a similar angular momentum. However, we have already found that it is the moment of inertia, I, of a rotating body, and not its mass as such, that provides the inertia of the body.

Consider a flywheel having a moment of inertia, I, rotating at constant angular velocity ω. Since $M = mv$ for bodies moving at a constant linear velocity, the angular momentum of the flywheel must be

$$\text{momentum} = \text{moment of inertia} \times \text{angular velocity}$$
$$M \text{ (angular)} = I\omega$$

Now consider that the same flywheel moving with an initial angular velocity ω_0 is acted upon by an applied torque, T, for a length of time of t seconds. If the final velocity of the flywheel is ω_t, the angular acceleration, α, present is found from the relationship

$$\alpha = \frac{\omega_t - \omega_0}{t}$$

Since applied torque = moment of inertia × angular acceleration

that is $$T = I\alpha$$

$$\therefore T = I \times \frac{\omega_t - \omega_0}{t}$$

Now, in the same way that an applied force, F, acting for t seconds has an impulse equal to Ft, an applied torque, T, acting for t seconds has an *angular impulse* equal to the product of the applied torque and the time it acts.

$$\therefore \text{ angular impulse} = T \times t$$

But $$T = I \times \left(\frac{\omega_t - \omega_0}{t}\right)$$

$$\therefore \text{ angular impulse} = I \times \left(\frac{\omega_t - \omega_0}{t}\right) \times t$$

$$\therefore Tt = I\omega_t - I\omega_0$$

But $I\omega_t$ was the final angular momentum of the flywheel and $I\omega_0$ the initial angular momentum of the flywheel. Thus, *the impulse of the applied torque equals the change in angular momentum of the rotating body.* (*Note:* This is analogous to the relationship between impulse and change in momentum for the linear motion of a body.)

Units

The units for angular momentum and angular impulse are obviously the same. Since

$$\text{angular impulse} = T \times t$$

its unit must be equal to the product of the units of torque (N m) and time (s), that is

$$\text{unit} = \text{N m} \times \text{s}$$
$$= \text{kg m/s}^2 \times \text{m} \times \text{s} \quad (\text{since N} = \text{kg m/s}^2)$$
$$= \text{kg m}^2/\text{s}$$

SAMPLE PROBLEM 13/12

A shaft has a moment of inertia of 10 kg m² and is rotating clockwise at 100 rad/s. A second shaft, with a moment of inertia of 20 kg m², rotating at 200 rad/s anti-clockwise is engaged with the first shaft by means of a friction clutch.

After slipping in the clutch has ceased, both shafts rotate with a common angular velocity. Determine

(i) initial angular momentum of each shaft in kg m²/s;

(ii) their final common angular velocity, ω.

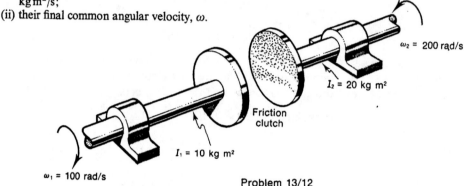

ω_2 = 200 rad/s

I_2 = 20 kg m²

Friction clutch

I_1 = 10 kg m²

ω_1 = 100 rad/s

Problem 13/12

Analysis:

The law of conservation of momentum,

$$\text{final momentum} = \text{initial momentum}$$

applies here, and enables the final common velocity, ω, to be determined. A suitable sign convention is essential in this problem.

Calculation:

(i) Initial momentum.

Let all anti-clockwise vectors be positive, i.e. ⊕

$$\text{momentum of first shaft} = I_1\,\omega_1$$
$$= 10 \times -100$$
$$= -1000 \text{ kg m}^2/\text{s}$$
$$\text{momentum of second shaft} = I_2\,\omega_2$$
$$= 20 \times 200$$
$$= +4000 \text{ kg m}^2/\text{s}$$
$$\text{total initial momentum} = 4000 + (-1000)$$
$$= +3000 \text{ kg m}^2/\text{s}$$

(ii) Final common velocity, ω

$$\text{initial momentum} = \text{final momentum}$$
$$+3000 = (10 + 20)\,\omega \quad \text{(since } I \text{ combined} = I_1 + I_2)$$
$$\therefore 30\,\omega = 3000$$
$$\therefore \omega = +100 \text{ rad/s}$$

(the + *ve* sign indicates ω is anti-clockwise).

Result:

The first shaft had an initial angular momentum of 1000 kg m²/s clockwise, the second shaft one of 4000 kg m²/s anti-clockwise, and they both had a final common velocity of 100 rad/s anti-clockwise. ∎

REVIEW PROBLEMS
(Use $g = 9.8 \, \text{m/s}^2$ unless otherwise specified.)

13/13

A flywheel of diameter 1.5 metres is rotating at 600 rpm. Determine the acceleration of any point on the outer rim of the flywheel.

13/14

A flywheel attached to a small engine is revolving at 1000 rpm. A crank pin attached to the face of the flywheel is located 150 mm from the centre of the flywheel. Determine the acceleration acting on this crank pin.

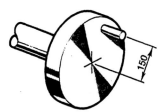

Problem 13/14

13/15

Part of a governor mechanism consists of a vertical shaft about which an arm of mass 2 kg rotates. Determine the force exerted on the shaft bearing when the arm rotates at 10 rad/s.

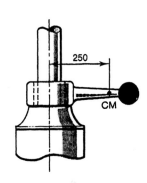

Problem 13/15

13/16

A lead weight is attached to the end of a piece of string and whirled around in a circle of diameter 2 metres. If the string remains horizontal and the lead has a mass of 0.2 kg, determine

(i) the tension, T, in the string when the angular velocity is 1 revolution per second;

(ii) the minimum angular velocity necessary to break the string, if the maximum tensile load it can sustain is 70 N.

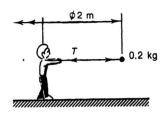

Problem 13/16

13/17

After being dynamically tested for balance, the front wheel of a car is found to need a 60-g balance weight fixed to the 500 mm diameter rim. What extra load would have been placed on the wheel bearings when the wheel rotated at 500 rpm if the wheel had not been balanced?

Problem 13/17

13/18

A girl of mass 25 kg is swinging on the end of a rope tied to the branch of a tree. The girl's centre of mass moves in a radius of 4 metres and at the bottom of each swing she moves with a horizontal velocity of 8 m/s. Determine the tension in the rope in this vertical position.

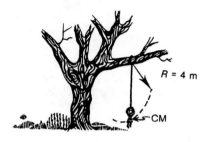

Problem 13/18

13/19

A plane is flying in a vertical circle (loop) of 200 metres radius. Determine the angular and linear velocities (in rad/s and m/s) of the plane at the highest point of the loop if the pilot experiences "weightlessness" at this stage of the manoeuvre. The pilot has a mass of 90 kg.

Problem 13/19

13/20

Pilots can withstand accelerations of up to 9 *g* for short periods without "blacking out" when wearing protective flying suits. Determine the minimum radius of curvature, *R*, with which a fighter pilot may turn his plane upward out of a dive if the speed of the plane is 700 km/h at the end of the dive.

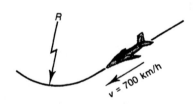

Problem 13/20

13/21

A ball of mass 0.5 kg is fastened to one end, *A*, of a piece of string of length 250 mm, the other end being attached to the fixed pivot, *B*. The ball rotates in a horizontal circle of radius *OA* so that the string is always inclined at 30° to the vertical as shown. Determine

  (i) the angular velocity of the ball;

  (ii) the tension in the string;

(iii) the total acceleration acting on the ball.

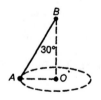

Problem 13/21

13/22

An automobile of mass 1.5 tonnes travels over the crest of a hill whose road-bed is part of a vertical circle of radius 40 metres.

 (i) Determine the maximum (linear) speed (in km/h) of the car over the crest if its wheels are not to leave the road.

(ii) If this speed is exceeded, how does the car move in relation to the crest of the road?

13/23
A car of mass 1.5 tonnes is travelling around a curve of radius 70 metres with a velocity of 20 m/s. Determine
(i) the coefficient of friction between the tyres and the road if the car is on the point of skidding.
(ii) the acceleration acting on the driver in this situation.

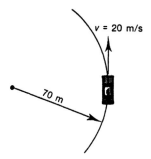

Problem 13/23

13/24
In the "wall of death" act at a sideshow, a stunt rider drives his motor bike in a horizontal circle around the inside of a cylindrical track. The cylindrical "wall of death" has an inside diameter of 11 metres and the bike plus rider a total mass of 250 kg. Determine the minimum linear speed (in km/h) that the rider must maintain in order to remain in this position, and the angle θ.

The coefficient of sliding friction between the wall and wheels is 0.6 and the centre of mass of the bike plus rider is 0.5 m radially from the wall as shown.

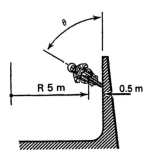

Problem 13/24

13/25
A car of mass 1.5 tonnes rounds a curve of radius 100 m. Determine its safe maximum speed (that is, speed at which skidding does not occur)
(i) if the roadway is horizontal;
(ii) if the roadway is super-elevated to an angle of 20°.
The coefficient of friction between the wheels and roadway is 0.58 (take $\tan^{-1} 0.58 = 30°$).

13/26
Determine the required super-elevation for a railway of standard gauge 1.435 metres if the equilibrium (normal) speed is 72 km/h and the radius of the curve is 600 metres. (*Hint:* Consider that, at equilibrium speed, the flange pressure between the outside wheels and the rails is zero.)

13/27
An electric trolley car with an effective track of 1.5 m turns a corner of radius 10 metres at a speed of 20 km/h. If the trolley way is level, determine how high the centre of mass can be without allowing the trolley to tip over.

13/28
Determine the torque produced when the 150-N force acts tangentially on the gear tooth as shown. The effective radius of the gear is 250 mm.

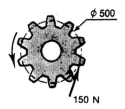

Problem 13/28

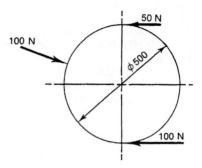

13/29
Determine the total torque on the flywheel when it is subjected to the three forces as shown.

Problem 13/29

13/30
A driving pulley attached to an electric motor has a diameter of 400 mm. When rotating at 1000 rpm, the belt tensions are 60 newtons on the slack side and 260 newtons on the tight side. What torque is being produced? (*Hint:* The available tangential force is equal to the difference between the taut and slack tensions of the belt.)

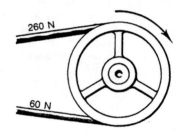

Problem 13/30

13/31
Determine the moment of inertia and the radius of gyration of the flywheel about its axis of rotation (*X—X*), if its mass is 50 kg.

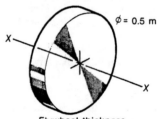

Flywheel thickness = 50 mm

Problem 13/31

13/32
An aeroplane propeller has a mass of 60 kg and a radius of gyration of 600 mm. Determine
 (i) its moment of inertia;
(ii) the unbalanced torque necessary to cause the propeller to revolve with an acceleration of 25 rad/s².

13/33
A shaft and flywheel of mass 100 kg is revolving at 500 rpm. The radius of gyration of this shaft and flywheel is 1 metre and the friction torque present in the bearings is 50 N m. How long does it take for the flywheel to come to rest?

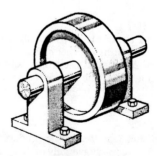

Problem 13/33

13/34
A wheel of diameter 0.25 m and moment of inertia 0.5 kg m^2 is acted on by a constant tangential force of 200 N. Determine
  (i) the resultant angular acceleration of the wheel;
  (ii) the angular speed (in rad/s) of the wheel after 4 seconds (from rest);
  (iii) the number of revolutions turned in the 4-second period.

Problem 13/34

13/35
Determine the angular momentum of a flywheel, with moment of inertia of 20 kg m^2, when it is revolving at
  (i) 100 rad/s;
  (ii) 500 rpm.

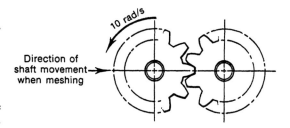

Problem 13/36

13/36
Two identical gear wheels on parallel shafts have moments of inertia of 1 kg m^2. At a given instant, one gear wheel is rotating at 10 rad/s clockwise when it is meshed with the other gear wheel. Determine the common rotational velocity of both gear wheels, if the second gear wheel was
  (i) stationary;
  (ii) rotating at 2 rad/s clockwise;
  (iii) rotating at 2 rad/s anti-clockwise at the time of meshing.

13/37
A wheel and axle is free to rotate about its horizontal axis. The 6-kg weight attached to the cord wrapped around the axle is allowed to fall freely a distance of 4 m from rest. Determine the moment of inertia of the wheel and axle if the weight takes 4 seconds to fall this distance. (Neglect friction.)

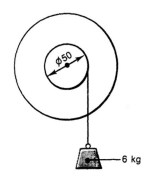

Problem 13/37

13/38
The pulley system has a moment of inertia of 30 kg m^2. If the weight A has a mass of 30 kg and the weight B a mass of 120 kg, determine the acceleration of the system.

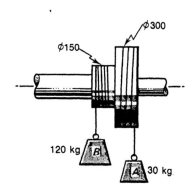

Problem 13/38

14

Work and Energy

$U = Fs$
$PE = mgh$
$KE = \frac{1}{2}mv^2$

In everyday situations we may *lift, carry, throw,* and *drag* various things and we may *walk, run, sit down,* and *stand up* at certain times. During every one of these and many other everyday activities we are *doing work* and we are also expending some of our available *energy*. For instance, a girl lifting a bag of groceries from the floor to the kitchen-bench top does *work* on that bag of groceries (or against gravity), while a gardener dragging a bag of fertiliser from the front gate to the back garden does *work* on that bag (and its contents). In each of these situations a *force* is involved; in the first instance, it is the weight-force of the bag of groceries that is involved, and in the second the work done by the gardener is done against the frictional resistance between the bag and the ground.

Work is always done when a force acts on a body and causes motion. Thus, *forces do work*, and the amount of *work done* is expressed as the product of the force acting and the distance through which its point of application moves.

Work done = force acting × distance moved in direction of force
$$U = Fs$$

Work can be done by a force or against a force. When work is done by a force, that is, when the body moves in the same direction as the acting force, positive work has been done. However, when the work is done against a force, that is, when the body moves in the opposite direction to that of the force, negative work has been done. Only external forces can do work, and thus internal reactive or inertia forces cannot do work. Three types of forces can be involved in calculations of work.

(1) *Applied forces*, such as accelerating or retarding forces acting on objects. Note that a retarding force applied to a moving object does negative work on the object, while an accelerating force does positive work.

(2) *Gravitational forces* or weight-forces that always act vertically downwards. Thus, when an object moves to a lower level, positive work has been done by its own weight-force, but when it moves to a higher level, negative work has been done against its own weight-force.

(3) *Friction forces*, which always do negative work, since they always act against the direction of motion.

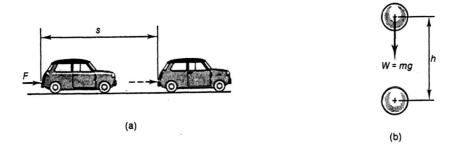

(a)

(b)

Figure 14.1 Work is done when a force moves its point of application: (a) The work done on the car by the motive force, *F*, is found from $U = Fs$; (b) the work done in raising the sphere of mass, *m*, through the vertical distance, *h*, is $U = Wh = mgh$.

It is important to note that the relationship $U = Fs$ can be used only where a constant force is acting. If the force has been variable, $U = Fs$ can only be used if only *F* is the average value of the variable force.

$$\text{Unit of work, } U = Fs$$
$$= \text{newton metre (N m)}$$
$$= \text{joule (J)}$$

By definition, *the work done when a force of one newton moves its point of application one metre along the line of action of the force is one joule (J)*. Thus, it is equally correct to express work done in newton metres (N m) or in joules (J).

SAMPLE PROBLEM 14/1

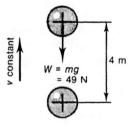

Problem 14/1

Calculate the work done in lifting a sphere of mass 5 kg through a vertical distance of 4 metres at constant velocity.

Analysis:
The force required to lift the sphere at constant velocity is equal in magnitude but opposite in direction to the weight-force of the sphere.

Calculation:

$$\text{Weight-force of sphere} = 5 \times 9.8$$
$$= 49 \text{ N}$$
$$\text{Work done: } U = Fs$$
$$\therefore U = 49 \times 4 \text{ N m}$$
$$= 196 \text{ J}$$

where: $F = 49$ N
$s = 4$ m

Result:
The work done in lifting the sphere was 196 joules. This is negative work since it is done against the gravitational force of the sphere. ∎

SAMPLE PROBLEM 14/2

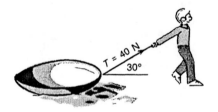

Problem 14/2

A boy pulls his sand sled a distance of 10 metres along a horizontal section of the beach, using a rope as shown. If the tension in the rope is 40 newtons, calculate the work done on the sled.

Analysis:
Work is done only by the component of the tension acting horizontally since the motion of the sled is horizontal. The vertical component of the tension does no work since there is no vertical motion.

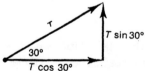

Problem 14/2

Calculation:

$$U = Fs$$
$$\therefore U = 40 \cos 30° \times 10 \text{ N m}$$
$$= 346.4 \text{ J}$$

where: $F = 40 \cos 30°$ N
$s = 10$ m

Result:
The work done on the sled was 346.4 J. ∎

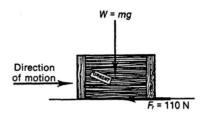

SAMPLE PROBLEM 14/3

Determine the amount of work done in sliding a crate of mass 50 kg across a horizontal floor if the frictional resistance is 110 newtons.

Problem 14/3

Analysis:
In this case work is done against friction, the amount of work done being equal to that done when a 110-N force moves a distance of 5 metres. The weight-force of the crate $(50 \times 9.8 = 490 \text{ N})$ does no work since it is perpendicular to the direction of motion.

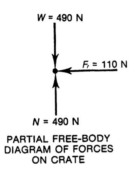

PARTIAL FREE-BODY
DIAGRAM OF FORCES
ON CRATE

Calculation:

$$U = Fs$$
$$\therefore U = 110 \times 5 \text{ N m}$$
$$= 550 \text{ J}$$

where: $F = 110$ N
$s = 5$ m

Result:
The work done against friction (negative work) was equal to 550 J. ■

WORK DONE BY A VARIABLE FORCE

The work done by a variable force is found from the product of the *average force acting* and the distance moved. Thus, when applying the relationship $U = Fs$, remember that, for variable forces, F is the average force acting for the given time.

Consider the following situation. You are standing on the roof of your house pulling up a rope which, when suspended from the guttering, did not quite reach the ground. When the first pull is made on this rope, the force acting is equal to the total weight-force of the rope. When all the rope is up, the force is, of course, zero. Thus, in this situation, the force being moved decreased from the full weight-force of the rope to zero. Therefore, the average force used to move the rope was half of its weight-force.

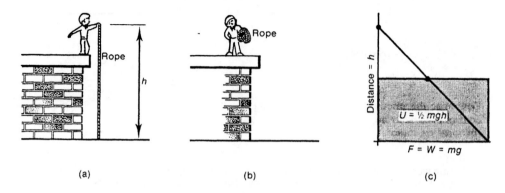

(a) (b) (c)

Figure 14.2 Pulling the rope on to the roof: (a) shows that when the rope is fully extended the force being moved is *W*, the weight-force of the rope; (b) shows that when all of the rope is up on the roof the force being moved is zero; and (c) shows that the work done by the boy on the rope is found from the product of half the weight-force and the distance moved ($U = \frac{1}{2} mgh$).

Another example of work done by a variable force is the work done in deforming a coil spring within its elastic range. *The force necessary to deform a coil spring by one unit length from its free length is called the spring modulus or stiffness*, and is related to the elastic modulus (or modulus of stiffness) of the material from which the spring is made. Deformation of springs will be treated as an example of strain energy later in this chapter.

SAMPLE PROBLEM 14/4

A force of 90 N stretches a spring 150 mm. (i) How much work must be done to stretch the same spring 200 mm? (ii) How much work must be done to stretch the spring another 50 mm; that is, from 200 mm to 250 mm? Assume that the elastic limit of the spring is not exceeded.

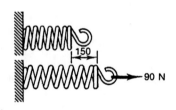

Problem 14/4

Analysis:
Since 90 N are required to stretch the spring 150 mm from its free length, the force required per millimetre of stretch is 90/150 = 0.6 N. Thus, the spring modulus or stiffness is 0.6 N/mm and the force required to deform the spring 200 mm must be equal to 0.6 × 200 = 120 N. Thus, in deforming the spring 200 mm the force increases uniformly from zero to 120 N; that is, the average force acting is 120/2 = 60 N.

Calculation:
(i)   Work done in extending spring 200 mm = *F* average × *s*
$$= 60 \times 0.2$$
$$= 12 \text{ J}$$

Similarly, the work done to extend the spring another 50 mm is the difference between the work done to extend it 200 mm and the work done to extend it 250 mm.

$$\text{Work done to extend spring to 250 mm} = F \text{ average} \times s$$
$$= \frac{0.6 \times 250}{2} \times 0.25$$
$$= 18.75 \text{ J}$$
$$\therefore \text{Work done to extend spring from 200 to 250 mm} = 18.75 - 12$$
$$= 6.75 \text{ J}$$

Result:
The work done in stretching the spring 200 mm from its free length is 12 joules, and the extra work done to stretch it another 50 mm is 6.75 joules. ■

SAMPLE PROBLEM 14/5

The coil spring in a spring buffer has a spring modulus of 10 newtons per millimetre. How much work must be done to compress the spring (i) 80 mm from its free length; and (ii) another 40 mm?

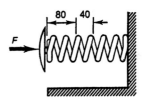

Problem 14/5

Analysis:
Since the spring modulus is 10 N/mm, the force required
 (i) to compress it 80 mm $= 10 \times 80$ $= 800$ N
 (ii) to compress it 120 mm $= 10 \times 120 = 1200$ N
   Thus, the average forces acting are
 (i) $\dfrac{800}{2} = 400$ N, during 80 mm compression

 (ii) $\dfrac{1200}{2} = 600$ N, during 120 mm compression

Calculation:
 (i)   Work done to compress the spring 80 mm $= 400 \times \dfrac{80}{1000}$ N m
$$= 32 \text{ J}$$
 (ii) Work done to compress the spring from 80 to 120 mm $= (600 \times \dfrac{120}{1000}) - 32$
$$= 72 - 32$$
$$= 40 \text{ J}$$

Result:
The work done in compressing the spring buffer 80 mm was 32 J and it required another 40 J to compress it a further 40 mm. ■

Some problems involving work done involve a constant force but variable distances. For example, when building a brick wall, the bricks for each successive row in the wall are lifted to progressively higher levels. Similarly, when digging a hole, the successive shovelfuls of soil are lifted progressively greater vertical distances. In these and similar instances, the work done is the product of the force acting and the *average distance* moved.

SAMPLE PROBLEM 14/6

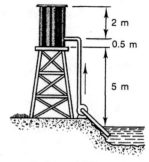

The water tank shown is filled by pumping water from a river, the water level of which is not affected during the pumping cycle. Neglecting frictional losses in the pipe, how much work must be done to fill the tank if it contains 15 000 litres of water? (*Note:* 1 litre of water has a mass of approximately 1 kg.)

Problem 14/6

Analysis:

Since the filler pipe enters the tank 0.5 metres from its base, all of the water below the filler pipe level must be raised 5.5 metres. However, the water in the upper 2 metres of the tank is raised an average distance between 5.5 metres and 7.5 metres; that is, 6.5 metres.

Calculation:

Work done to fill tank to filler pipe level $= \text{weight-force of water} \times 5.5 \text{ metres}$

$$= (15\,000 \times \frac{0.5}{2.5} \times 1 \times 9.8) \times 5.5$$

$$= 29\,400 \times 5.5 \text{ N m}$$

$$\approx 162 \times 10^3 \text{ J}$$

Work done to fill tank from filler pipe to top $= \text{weight-force of water} \times \text{average distance}$

$$= (15\,000 \times \frac{2.0}{2.5} \times 1 \times 9.8) \times 6.5$$

$$= 117\,600 \times 6.5 \text{ N m}$$

$$\approx 764 \times 10^3 \text{ J}$$

$$\text{Total work done} = (764 + 162) \times 10^3 \text{ J}$$

$$= 926 \times 10^3 \text{ J}$$

$$= 926 \text{ kJ}$$

Result:

The work done in filling the tank was 926 kilojoules. ■

ENERGY

The energy stored in an object is the measure of its capacity to do work. Energy exists in many forms, the most common being mechanical, chemical, heat, electrical, and atomic energy. Mechanical energy, with which we are concerned, can exist in one of the following forms.

Potential energy which is possessed by a body because of its position relative to a fixed reference position, often the earth's surface. Think back to the sphere shown in Figure 14.1 and recall that the work done to raise its centre of gravity a vertical distance *h* was equal to *mgh* joules. The work done, *mgh* joules, is stored in the body as *mgh* joules of potential energy, which is released if the body falls back to its original position.

$$PE = mgh$$

Unit of potential energy
$$
\begin{aligned}
mgh &= \mathrm{kg\,m\,s^{-2}\,m} \\
&= \mathrm{kg\,m^2\,s^{-2}} \quad (\text{or } \mathrm{kg\,m^2/s^2}) \\
&= \mathrm{N\,m} \quad (\text{since } \mathrm{N = kg\,m\,s^{-2}}) \\
&= \mathrm{J} \quad (\text{joules})
\end{aligned}
$$

Kinetic energy is energy possessed because an object is in motion, and again is a measure of the ability of the object to do work. The kinetic energy of a moving object is always equal to the amount of work that must be done to bring the object to rest.

Consider that a force F is applied to a body of mass m and that this force produces a velocity v after a displacement of s units from rest. Since

$$v^2 = u^2 + 2as$$
$$\therefore v^2 = 2as \quad (\text{since } u = 0)$$
$$\therefore a = \tfrac{1}{2}\frac{v^2}{s}$$

Thus, the force F produced an acceleration of $\tfrac{1}{2}\,v^2/s$ on the mass m.

From
$$F = ma$$
$$F = m \times \tfrac{1}{2}\frac{v^2}{s} \quad \left(\text{since } a = \tfrac{1}{2}\frac{v^2}{s}\right)$$
$$\therefore F = \tfrac{1}{2}\frac{mv^2}{s}$$

Since the force F moves the body of mass m, it does work on the body.

$$U = Fs$$
$$\therefore U = \tfrac{1}{2}\frac{mv^2}{s} \times s \quad \left(\text{since } F = \tfrac{1}{2}\frac{mv^2}{s} \text{ from above}\right)$$
$$\therefore U = \tfrac{1}{2}\,mv^2$$

Now, since work done always equals the energy generated, when the body of mass m is moving with a velocity of v, its kinetic energy is found from the relationship

$$KE = \tfrac{1}{2}\,mv^2$$

Unit of kinetic energy
$$
\begin{aligned}
= \tfrac{1}{2}\,mv^2 &= \mathrm{kg\,m^2\,s^{-2}} \quad (\text{or } \mathrm{kg\,m^2/s^2}) \\
&= \mathrm{N\,m} \quad (\text{since } \mathrm{N = kg\,m\,s^{-2}}) \\
&= \mathrm{J} \ (\text{joules})
\end{aligned}
$$

Strain energy, such as that existing in a spring or a bar of material deformed within its elastic limit.

Refer back to Sample Problem 14/4 which appears earlier in this chapter. We calculated that it took 12 joules of work to stretch this particular spring a distance of 200 millimetres from its free length. The work done, 12 joules, is stored in the spring as a special kind of potential energy called strain energy, and is available to do work when the spring is released. *Strain energy is stored in any piece of material deformed within its elastic limit, and does work when the deformation is released.**

---

*Strain energy in elastic materials is explained in detail in Chapter 16 where the relevant theory becomes more meaningful.

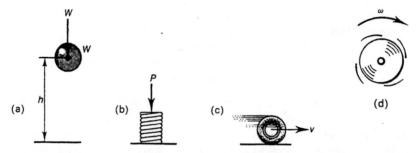

Figure 14.3 Forms of energy: (a) the sphere has potential energy by virtue of its height above the ground; (b) the coil spring possesses strain energy because it is compressed; (c) the wheel possesses kinetic energy because it is moving with velocity v; (d) the disc possesses rotary kinetic energy since it is rotating about its own axis with angular velocity ω.

CONSERVATION OF ENERGY

Energy can neither be created nor destroyed, but only transformed from one form to another.

In Newtonian mechanics this means that unless some external force is applied, or external resistance overcome, the sum of all of the energies possessed by a body at any instant will remain constant.* As an example, consider a car moving at constant velocity along a straight horizontal road. If the car is put into neutral, it gradually loses kinetic energy as it slows down. This *lost* kinetic energy is used to do work against the forces of frictional resistance such as those between the tyres and the road, those within wheel bearings, etc., and those produced by air resistance ("drag"). While this energy is *lost* for mechanical purposes, it reappears as heat and sound energies and the total energy at any instant is constant.

It is somewhat difficult to use the principle of conservation of energy for calculations since we often do not know how much energy is transformed into other forms such as heat and sound.

However, a most useful statement is the work-energy principle, which can be stated as

| Initial Kinetic Energy | + | Initial Potential Energy | ± | Work Done | = | Final Kinetic Energy | + | Final Potential Energy |

or

$$KE_1 + PE_1 \pm W = KE_2 + PE_2$$

If this principle is applied to a machine, it is readily seen that the power (work done per unit time) of any machine is used

---

*Note that, when considering the types of problems normally dealt with by Newtonian principles, it is not necessary to consider that mass can, under certain circumstances, be regarded as a form of energy. This facet of the theory of relativity, stated in the equation $E = mc^2$, where c is the velocity of light, is not considered in this text.

(a) to overcome external resistances or loads;
(b) to overcome frictional effects within the machine;
(c) to increase the energy of both the machine and its load.

This last use of the power of a machine is, of course, the desirable use of this power.

SAMPLE PROBLEM 14/7

Calculate the energies of the following bodies and state the type of energy involved in each case:
  (i) a ball of mass 10 kg on the roof of a building 10 metres high. (*Note:* Calculate energy relative to roof and the ground.);
  (ii) a car of mass 1 tonne moving at 36 km/h along a straight horizontal road;
  (iii) a compression spring, stiffness 1 newton per millimetre, compressed 50 mm from its free length.

Analysis and Calculation:
  (i) The energy of the ball is energy by virtue of position; that is, potential energy, which is calculated from
$$PE = mgh$$
Thus, the energy relative to the roof is zero, since h = zero. However, energy relative to ground is

$$PE = mgh$$
$$= 10 \times 9.8 \times 10$$
$$= 980 \text{ J}$$

where: $m = 10$ kg
$g = 9.8$ m/s^2
$h = 10$ m

  (ii) The car possesses energy by virtue of its motion; that is, kinetic energy, which is calculated from $KE = \frac{1}{2} mv^2$

$$\therefore KE = \frac{1}{2} \times 1000 \times 10^2$$
$$= 50\,000 \text{ joules}$$
$$= 50 \text{ kJ}$$

where: $m = 1t = 1000$ kg
$v = 36$ km/h $= 10$ m/s

  (iii) The energy stored in the spring is strain energy, a form of potential energy. It is equal to the work done in deforming the spring.

force to compress spring = stiffness $\times$ length of deformation
$$= 1 \times 50$$
$$= 50 \text{ N}$$

Now.   work done = average force exerted $\times$ amount of deformation (compression)
$$= \frac{50}{2} \times \frac{50}{1000}$$
$$= 1.25 \text{ J}$$

where: $F_{av} = \frac{50-0}{2}$;

compression $= \frac{50}{1000}$ m

But, energy stored in spring = work done in deforming it
$$= 1.25 \text{ J}$$

Results:
  (i) Potential energies are zero relative to roof and 980 joules relative to the ground.
  (ii) Kinetic energy of the car is 50 kilojoules.
  (iii) Strain energy stored in the spring is 1.25 joules.

■

SAMPLE PROBLEM 14/8

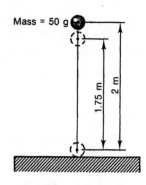

Problem 14/8

A steel ball bearing of mass 50 grams is dropped from a height of 2 metres on to a smooth, flat, rigid steel plate and rebounds to a height of 1.75 metres. Calculate
  (i) its original potential energy;
  (ii) its kinetic energy the instant before it hits;
  (iii) its kinetic energy the instant it starts to rebound;
  (iv) the velocity upon impact.
Also, how do you account for the apparently "lost" energy of the collision?

Analysis:
The original potential energy of the ball bearing is equal to its kinetic energy of impact, and the kinetic energy immediately after impact is equal to the potential energy possessed by the ball bearing when it has reached its full rebound height. Also, since original height $\neq$ height of rebound, the collision is not perfectly elastic.

Calculation:

(i)
$$\text{original PE} = mgh$$
$$= 0.05 \times 9.8 \times 2$$
$$= 0.98 \text{ J}$$
where: $m = 50 \text{ g} = 0.05 \text{ kg}$
$g = 9.8 \text{ m/s}^2$
$h = 2 \text{ m}$

(ii)
$$\therefore \text{ KE on impact} = 0.98 \text{ J}$$

(iii)
$$\text{KE after impact} = \text{PE when at full rebound height}$$
$$= mgh$$
$$= 0.05 \times 9.8 \times 1.75$$
$$\therefore \text{ KE after impact} = 0.86 \text{ J}$$
where: $m = 0.05 \text{ kg}$
$g = 9.8 \text{ m/s}^2$
$h = 1.75 \text{ m}$

(iv)
Now,
$$\text{KE on impact} = 0.98 \text{ J}$$
but
$$\text{KE} = \tfrac{1}{2} mv^2$$
$$\therefore v^2 = \frac{\text{KE}}{\tfrac{1}{2}m}$$
$$= \frac{2 \times 0.98}{0.05}$$
$$v^2 = 39.2$$
$$\therefore v = \sqrt{39.2}$$
$$\therefore v = 6.3 \text{ m/s}$$
where: $m = 0.05 \text{ kg}$
$\text{KE} = 0.98 \text{ J}$

(Note that v, the velocity upon impact, could have been found using $v^2 = u^2 + 2as$, since the ball falls freely. However, the method used employs the "energy principle" and is sometimes useful in more difficult problems.)

Result:
The original potential energy and the kinetic energy of impact are both 0.98 J, and the ball bearing hits with a velocity of 6.3 m/s. The kinetic energy immediately after impact was 0.86 J. It is obvious that the collision was not perfectly elastic and thus the "lost" energy $(0.98 - 0.86 = 0.12 \text{ J})$ was used in deforming the ball and plate and in producing sound and heat. (Refer back to Chapter 11, "Coefficient of Restitution", to revise work on elastic collisions.) ∎

SAMPLE PROBLEM 14/9

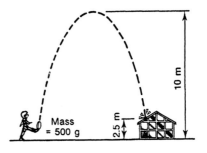

Problem 14/9

A boy kicks a football of mass 500 grams as shown. The ball rises to a maximum height of 10 metres. How much of its energy is used to smash the glasshouse roof if the velocity of the ball is 8 m/s after passing through the roof?

Analysis:
Apply the principle of work-energy conservation.

Calculation:
With the ball at its maximum elevation,

$$\therefore\ mgh_1 + \tfrac{1}{2}mv_1^2 - W = mgh_2 + \tfrac{1}{2}mv_2^2$$
$$PE_1 + KE_1 - W = PE_1 + KE_1$$

$$(0.5 \times 9.8 \times 10) + 0 - W = (0.5 \times 9.8 \times 2.5) + (0.5 \times 0.5 \times 8^2)$$
$$49 - W = 12.26 + 16$$
$$W = 20.75 \text{ J.}$$

where: $m = 0.5$ kg
$h_1 = 10$ m
$h_2 = 2.5$ m
$v_1 = 0\ \downarrow$
$v_2 = 8$ m/s $\downarrow$

Result:
It took 20.75 J of energy to break the roof of the glasshouse.

■

SAMPLE PROBLEM 14/10

How much work must be done to change the velocity of a car of mass 1.5 tonnes from 36 to 72 km/h? You may ignore frictional resistance in this problem.

Analysis:
The work done is equal to the energy change of the car. Thus, work done equals final KE − initial KE of the car.

Calculation:

$$\text{Final KE of car} = \tfrac{1}{2}mv^2$$
$$= \tfrac{1}{2} \times 1500 \times 400$$
$$= 300\ 000 \text{ J}$$

where: $m = 1500$ kg
$v = 72$ km/h
$= 20$ m/s

$$\text{Initial KE of car} = \tfrac{1}{2}mv^2$$
$$= \tfrac{1}{2} \times 1500 \times 100$$
$$= 75\ 000 \text{ J}$$

where: $v = 36$ km/h
$= 10$ m/s

$$\text{Work done} = 300\ 000 - 75\ 000$$
$$= 225\ 000\ \text{J}$$
$$= 225\ \text{kJ}$$

Result:

The work done was 225 kilojoules.

SAMPLE PROBLEM 14/11

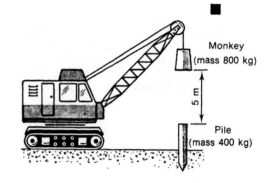

Monkey
(mass 800 kg)

5 m

Pile
(mass 400 kg)

A pile of mass 400 kg is driven into the ground by a
pile-driver monkey of mass 800 kg.
(i) What is the velocity of the pile immediately
after the blow, if the hammer drops 5 metres
and the hammer and pile move off together?
(ii) How far does the pile penetrate per blow of the
monkey, if the average soil resistance is 70 kN?

Problem 14/11

Analysis:

Velocities of bodies involved in collisions are best found by a consideration of momentum
before and after collision, since total initial momentum = total final momentum. The
penetration of the pile can easily be found by applying the work-energy principle to the
pile + driver after the impact.

Calculation:

(i) Impact velocity of pile-driver

$$v^2 = u^2 + 2as \qquad\qquad \text{where: } u = 0$$
$$\therefore v^2 = 2as \qquad\qquad\qquad\qquad a = 9.8\ \text{m/s}^2$$
$$= 2 \times 9.8 \times 5 \qquad\qquad\quad s = 5\ \text{m}$$
$$\therefore v^2 = 98$$
$$\therefore v \approx 9.9\ \text{m/s}$$

$$\underset{\text{of pile}}{\text{initial momentum}} + \underset{\text{of monkey}}{\text{initial momentum}} = \underset{\text{of pile and monkey}}{\text{final momentum}} \qquad \text{where: } v = \text{common velocity}$$

of pile plus monkey
after impact

$$(400 \times 0) + (800 \times 9.9) = (400 + 800)\,v$$
$$\therefore v = \frac{800 \times 9.9}{1200}$$
$$\therefore v = 6.6\ \text{m/s}$$

(ii) Apply the work-energy principle to the pile + driver. Let the distance the pile is driven
be x m.

$$\underset{\text{Pile + Driver}}{\text{Initial PE of}} + \underset{\text{of Driver}}{\text{Initial KE}} - \underset{\text{in Driving}}{\text{Work Done}} = \frac{\text{Final PE + Final KE}}{\text{of Pile + Driver}}$$

$$PE_1 + KE_1 - W = PE_2 + KE_2$$
$$\text{or, } mgh_1 + \tfrac{1}{2}mv_1{}^2 - F_x = mgh_2 + \tfrac{1}{2}mv_2{}^2 \qquad\qquad \text{where: } m = 1200\ \text{kg}$$
$$v_1 = 6.6\ \text{m/s}$$
$$\therefore (1200 \times 9.8 \times x) + (0.5 \times 1200 \times 6.6^2) - (70\ 000 \times x) = 0 + 0 \quad h_1 = x$$
$$58240\,x = 26136 \qquad\qquad\qquad\qquad v_2 = 0$$
$$x = 0.449\ \text{m}. \qquad\qquad\qquad\qquad h_2 = 0$$
$$F = 70\ 000\ \text{N}$$

Result:

The velocity of the pile is 6.6 m/s immediately after impact and the pile penetrates ~ 450 mm
with each blow of the monkey.

EFFICIENCY

Machines can either transform energy from one form to another or they can do work. As an example of energy transformation, consider an electric motor in which electrical energy is transformed into mechanical energy; and as an example of a machine doing work, consider a mechanical hoist or a pulley and belt system.

No machine is perfect in the sense that it is capable of putting out as much work, energy or power as is put into it in either a similar or a different form; in other words, *no machine can be 100% efficient.* Friction is the main reason why a machine never makes all of its theoretical power available as output, and frictional losses are usually observed in the form of heat which does no useful work.

The efficiency of a machine can be simply expressed as the ratio of *output* to *input*

$$\text{efficiency, } \eta = \frac{\text{output}}{\text{input}}$$

where output and input can be expressed in units of work, energy or power. The efficiency ratio is often expressed as a percentage.

The overall efficiency of a number of machines coupled together or working in series is the product of their individual efficiencies. For instance, if an electric motor of 85% efficiency is driving a compressor of 75% efficiency, the overall efficiency of the combined unit is equal to $85 \times 75 = 63.75\%$.

WORK DONE BY OR AGAINST A TORQUE

A torque acting on a rotating object does work as it moves about its centre of rotation in the same way that a force does work as it moves its point of application through a certain distance. The relationship $U = Fs$ has as its rotational equivalent the relationship

$$U = T\theta,$$

where T is the torque acting and θ is the angular displacement through which it moves. If torque is expressed in N m and θ in radians, then the work done is expressed in *joules*. (N m = J.)

Negative work can be done when a rotating body moves *against* a resisting torque, such as *friction* in a bearing.

The relationships $T = Fr$ and $T = I\alpha$, already discussed in Chapter 13, are often useful when considering rotary work and energy.

KINETIC ENERGY OF ROTATION

Just as a brick sliding down a plank possesses linear (or translational) kinetic energy, rotating bodies such as a flywheel or shaft possess kinetic energy of rotation (so-called "rotary KE").

An object rotating with an angular speed of ω radians per second and having a moment of inertia, I, has a rotary kinetic energy of $\frac{1}{2}I\omega^2$. Provided that I is expressed in kg m^2 and ω in radians per second, the rotary kinetic energy will be expressed in joules. Note that the proof of the equation

$$\text{KE (rotary)} = \tfrac{1}{2}I\omega^2$$

is similar to that given earlier for KE (linear) $= \frac{1}{2}mv^2$, except that linear units in the basic formulas used must be replaced by their rotational equivalents (that is, ω for v, T for F, I for m and α for a).

SAMPLE PROBLEM 14/12

A flywheel has a moment of inertia of 2250 kg m^2
and is rotating at a constant velocity of 500 rpm.
Calculate its kinetic energy.

Calculation:

$$\text{KE of flywheel} = \tfrac{1}{2}I\omega^2$$
$$\text{Now, } \omega = 500 \text{ rpm}$$
$$= \frac{500 \times 2\pi}{60} \text{ rad/s}$$
$$= 52.4 \text{ rad/s}$$

$$\therefore \text{KE} = \tfrac{1}{2} \times 2250 \times (52.4)^2 \text{ J}$$
$$= \tfrac{1}{2} \times 2250 \times 2745.8$$
$$\therefore \text{KE} \approx 3.09 \text{ MJ}$$

Result:
The flywheel has kinetic energy of 3.09 MJ by virtue of its rotary motion. ∎

SAMPLE PROBLEM 14/13

A uniform disc with a diameter of 0.8 m and mass
of 20 kg rolls along a horizontal path with a linear
velocity of 4 m/s. Determine the total kinetic
energy of the disc.

Analysis:
When a disc or wheel rolls, the total KE is made up of two parts:
 (i) the KE *of translation* of its centre of mass;
(ii) the KE *of rotation* about its centre of mass.
Total KE = $\tfrac{1}{2}mv^2 + \tfrac{1}{2}I\omega^2$, where v and ω are the linear and angular velocities of the wheel
respectively.

Calculation:
Moment of inertia of disc

$$I = \frac{mr^2}{2} \qquad \text{where: } m = 20 \text{ kg}$$
$$ \qquad\qquad\qquad r = 0.4 \text{ m}$$
$$= \frac{20 \times (0.4)^2}{2}$$
$$\therefore I = 1.6 \text{ kg m}^2$$

From $v = r\omega$,
$$\omega = \frac{v}{r} \qquad \text{where: } v = 4 \text{ m/s}$$
$$= \frac{4}{0.4}$$
$$\therefore \omega = 10 \text{ rad/s}$$

$$\text{Translational KE} = \tfrac{1}{2}mv^2$$
$$= \tfrac{1}{2} \times 20 \times 4^2$$
$$= 160 \text{ J}$$

$$\text{Rotary KE} = \tfrac{1}{2} I\omega^2$$
$$= \tfrac{1}{2} \times 1.6 \times 10^2$$
$$= 80 \text{ J}$$

where: $I = 1.6 \text{ kg m}^2$
$\omega = 10 \text{ rad/s}$

Result:
The total KE possessed by the rolling disc is $80 + 160 = 240$ J. ■

SAMPLE PROBLEM 14/14

A flywheel which is rotating at 500 rpm has a mass of 5 tonnes and a radius of gyration of 1 metre.
Determine
 (i) its kinetic energy
(ii) the time taken for it to come to rest if a resisting torque of 500 Nm is applied to the flywheel by means of a friction brake.

Analysis (i):
The moment of inertia of this flywheel can be found from $I = mk^2$ and the kinetic energy from $KE = \tfrac{1}{2} I\omega^2$.

Calculation:

$$I = mk^2$$
$$= 5 \times 10^3 \times 1 \text{ kg m}^2$$
$$\therefore I = 5 \times 10^3 \text{ kg m}^2$$

where: $m = 5$ tonnes
$= 5 \times 10^3$ kg
$k = 1$ m

$$KE = \tfrac{1}{2} I\omega^2$$
$$= \tfrac{1}{2} \times 5 \times 10^3 \times 52.4^2 \text{ kg m}^2/\text{s}$$
$$= 6864 \times 10^3 \text{ N/s}$$
$$\therefore KE \approx 6.9 \text{ MJ}$$

where: $I = 5 \times 10^3 \text{ kg m}^2$
$\omega = 500$ rpm
$= \dfrac{500 \times 2\pi}{60}$ rad/s
$= 52.4$ rad/s

Analysis: (ii)
This kinetic energy is expended by doing work against the frictional torque of 500 Nm. If θ is the angle through which the flywheel turns in coming to rest, then the work done against friction is equal to $T\theta$, where T is the applied friction torque.

Calculation:

$$\text{Work done against friction} = T\theta$$
$$= 500 \theta \text{ J}$$

where: $T = 500$ Nm

$$\text{Since KE of flywheel} = 6.9 \times 10^6 \text{ J}$$
$$\therefore 500 \theta = 6.9 \times 10^6$$
$$\therefore \theta = 13\,800 \text{ radians}$$

$$\text{Now, average speed during retardation} = \frac{52.4}{2} \text{ rad/s}$$
$$= 26.2 \text{ rad/s}$$

$$\text{Time taken to stop} = \frac{\text{angular displacement}}{\text{average speed}}$$
$$= \frac{13\,800}{26.2} \quad \left(\frac{\text{rad}}{\text{rad/s}} = \text{s}\right)$$
$$= 526.7 \text{ s}$$

Result:
The kinetic energy of the flywheel was 6.9 MJ and it took 8 minutes 46.7 seconds to come to rest under a braking torque of 500 Nm. ■

REVIEW PROBLEMS

(Use $g = 9.8 \, \text{m/s}^2$ unless otherwise specified.)

14/15
State the type of energy possessed by the following objects:
(i) a loose brick balanced on a wall of height 2 metres;
(ii) a bullet travelling at 300 m/s;
(iii) water in a dam that is used to drive a turbine;
(iv) a rotating flywheel.

14/16
(i) Determine the work done against gravity when a mass of 50 kg is lifted 6 metres vertically.
(ii) How much of this energy is lost when the mass falls 1 metre from this height?

14/17
A boy of mass 70 kg climbs a cliff 30 metres high.
(i) How much (useful) work has he done?
(ii) What is his potential energy in relation to his starting point?
(iii) If he dislodges a stone of mass 1 kg from the clifftop with what KE will it hit the ground?

14/18
(i) A cylindrical water tank, 2 m high, contains 20 000 litres of water when full. What is the potential energy of the water relative to an outlet point 20 m vertically below the bottom tank?
(ii) If the tank is refilled by means of a pump situated at the same level as the outlet, what is the useful work done by the pump?
Take the mass of 1 litre of water as 1 kg.

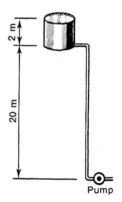

Problem 14/18

14/19
A tank, 20 m long by 8 m wide by 8 m deep, sits on level ground, and contains water to a depth of 4 metres. A second tank, 15 m long by 8 m deep by 8 m wide, sits on a stand so that the bottom of the tank is 15 m above ground. Find the work required to pump all the water from the first tank to the second tank. The density of water is 1000 kg/m³. Neglect frictional losses.

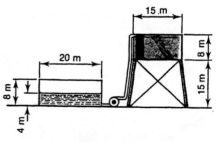

Problem 14/19

14/20

A rubber ball of mass 0.5 kg falls 4 metres vertically before striking a horizontal surface, from which it rebounds to a height of 3 metres. Determine
 (i) its potential energy before it falls;
 (ii) its kinetic energy at the instant of impact;
(iii) its kinetic energy at the instant of rebound.
Explain what happens to the energy "lost" during the impact.

14/21

A sled of mass 10 kg is pulled 10 metres along level ground. The tension in the tow rope is 100 N and it is inclined at 30° to the ground as shown. Determine
 (i) the total work done;
 (ii) the work done against friction, if the coefficient of friction between sled and ground is 0.3.

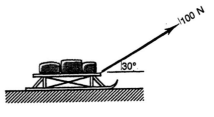

Problem 14/21

14/22

An archer pulls back a bowstring a distance of 500 mm and then releases his arrow which has a mass of 100 g. Determine the average force exerted on the arrow if its release velocity is 40 m/s. Solve this problem by the work-energy method.

14/23

The 6-kg pendulum of a Charpy notched-bar impact testing machine falls freely through a vertical height of 0.8 metres before it hits the specimen. Determine
 (i) its kinetic energy immediately before impact;
 (ii) its kinetic energy immediately after impact, if it rises 0.3 m vertically after striking the specimen;
(iii) the work done on the specimen by the impact.

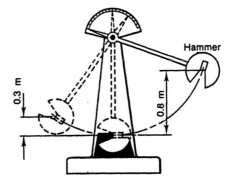

Problem 14/23

14/24

A concrete post of uniform cross-section has a mass 1 tonne and a length of 6 m. Determine the amount of work done on the pole when it is lifted from a horizontal to a vertical position.

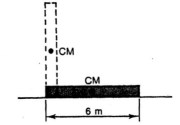

Problem 14/24

14/25

A man building a wall wishes to move a block of dimension stone measuring 300 mm by 300 mm by 1 m across a concrete path. Given that the block has a mass of 80 kg, will he use more energy if he slides it or rolls it?

The coefficient of friction between path and stone is 0.6. (*Hint:* Consider the energy used to displace the block by a distance equal to its 300-mm width by rolling and by sliding.)

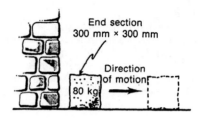

Problem 14/25

14/26

The high-jump world record was recently broken be Nee-Hi-Me at 2.4 m.

(i) If the athlete had a mass of 70 kg, and raised his centre of mass through 1.4 m from *A* to *B* during the jump, what was his gain in potential energy?

(ii) If he then fell on a pile of plastic mattress material at *C*, which compressed from 1.5 m down to 0.9 m at *D*, what was his kinetic energy at *C*, and how much energy altogether was absorbed by the plastic? Neglect any kinetic energy he may have had at *B*.

(iii) Assuming his deceleration from *C* to *D* was uniform, what was its value?

Assume throughout that his centre of mass was 150 mm above the bar at *B*, above the uncompressed mattress at *C*, and above the compressed mattress at *D*, and neglect all horizontal components of velocity.

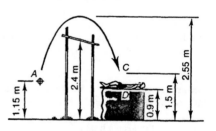

Problem 14/26

14/27

The planned excavation for the basement of a building is 20 m by 40 m by 5 m deep. If the spoil has a mass of 1800 kg per cubic metre, how much work will be done in raising the spoil to the original ground level? (*Note:* No allowance is to be made for heaping up the spoil once it is excavated from the hole.)

14/28

During the operation of a windlass, 30 metres of chain are to be wound on to the drum.

(i) Determine the work done on the chain if the chain has a mass of 0.5 kg/m.

(ii) A bucket containing spoil is attached to the end of the chain. If the bucket plus spoil has a mass of 25 kg, determine the work done in lifting the bucket through the 30 metres.

(iii) What is the total work done on both bucket and chain?

(iv) How much energy does the operator expend, if the windlass has an efficiency of 45%?

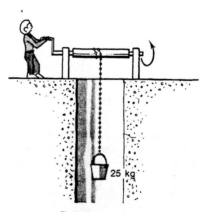

Problem 14/28

14/29
A mortar shell of mass 15 kg was fired from a mortar barrel of length 3 metres. If its velocity on leaving the barrel was 600 m/s, determine the average force exerted by the charge on the shell as it was fired.

14/30
A water tank 5 m long by 4 m wide by 4 m deep contains water to a depth of 3.5 m. A water turbine is in position 6 m below the bottom of the tank. Determine the output work of this turbine when all of the water in the tank flows through it, if the overall efficiency of the system is 60 %. The mass of 1 cubic metre of water is 1000 kg.

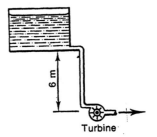

Problem 14/30

14/31
A car of mass 1 tonne travelling at 60 km/h has its speed uniformly reduced to 20 km/h in 5 seconds.
  (i) Assuming that the reduction in speed is solely due to the brakes being applied, calculate the energy absorbed by the brakes.
 (ii) Determine the average braking force acting on the car.

14/32
The mean force required to compress a given spring from its free length of 100 mm to 75 mm is 300 N. Determine the energy stored in it in the compressed state.

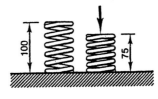

Problem 14/32

14/33
A spring has a free length of 100 mm. When a force of 40 N is applied, the spring compresses 10 mm; an additional force of 40 N causes the spring to compress a further 10 mm.
  (i) Assuming the spring behaves in accordance with Hooke's law, how much work would be done on it if it is compressed from its free length of 100 mm to a length of 50 mm?
 (ii) What would be the strain energy contained in the spring in the compressed state?

14/34
A crate of mass 50 kg is pushed 13 metres up an incline that rises 5 metres vertically over a horizontal distance of 12 metres. If the frictional resistance is constant at 120 N, calculate the work done on the crate.

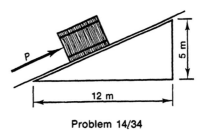

Problem 14/34

14/35
A diesel locomotive pulls a train of mass 1000
tonnes up a 1% grade at constant speed. If the
average resistance to motion is 20 N per tonne,
calculate
 (i) the drawbar pull supplied by the locomotive;
 (ii) the work done in moving the train 2 kilometres
   up the track.

14/36
A train of mass 300 tonnes, moving at 72 km/h
along a horizontal track begins to climb an incline
of 1% slope. During the climb the engine exerts a
constant tractive force (drawbar pull) of 30 kN
and the frictional resistances to motion remain
constant at 100 N per tonne. How far will the train
move up the incline?

14/37
A pile driving hammer (monkey) of mass 500 kg
falls 2.5 metres from rest onto a pile of mass 140 kg.
Assuming there is no rebound and the pile is driven
150 mm into the ground, determine the following:
 (i) the velocity of the hammer just before impact;
 (ii) the common velocity of the pile and hammer
   after impact;
 (iii) the work performed in sinking the pile 150 mm
   into the ground;
 (iv) the average resisting force of the ground in
   bringing the pile and driver to rest.

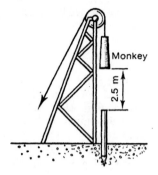

Problem 14/37

14/38
Two views of the basic units (disc and brake pads)
of a disc brake are shown in the diagram. The
mean effective radius of the disc is 100 mm, the
force on each brake pad is 100 N and the coefficient
of friction between the pads and disc is 0.4. If the
disc is rotating at 3000 rpm, determine:
 (i) the moment resisting the rotation of the disc;
 (ii) the energy absorbed through the brake pads if
   the disc takes 20 revolutions to come to rest.

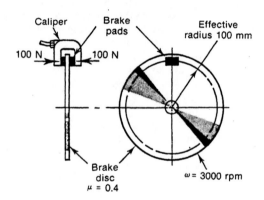

Problem 14/38

14/39
A rotating pulley and shaft has a moment of
inertia of 48 kg m². Given that the bearing friction
is equivalent to a resisting couple of 300 N m,
determine the torque necessary to accelerate the
shaft from rest to an angular speed of 600 rpm in
12 revolutions from rest. Solve by a work-energy
method and also by using kinetics of rotation.

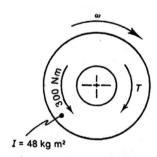

Problem 14/39

14/40

A flywheel has a radius of gyration of 1.5 m and a mass of 10 tonnes. Determine
(i) its kinetic energy when it revolves at 500 rpm;
(ii) the time taken to bring it to a rest when a resisting torque of 680 N m is applied to it.

14/41

A shaft and drum rotating at 720 rpm has a moment of inertia of 34 kg m². A brake block acting on the surface of the drum brings it to rest in 12 revolutions, the coefficient of friction present being 0.45. Determine the normal force, F, between the brake block and drum, given that the drum diameter is 1 metre.

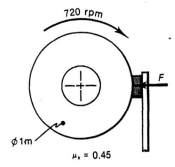

Problem 14/41

14/42

A shaft having a moment of inertia of 16 kg m² is accelerated from 1200 rpm to 1320 rpm during 2 revolutions. Determine
(i) the increase in kinetic energy of the shaft;
(ii) the torque required to provide this acceleration, given that a friction couple of 80 N m is acting to resist rotation.

14/43

In an experiment to determine the moment of inertia of a shaft and flywheel, the assembly was allowed to roll freely down an inclined plane formed by two highly polished parallel steel bars set the required distance apart. The flywheel and shaft rolled, without slipping, 1.5 metres down the incline and at the same time fell through a vertical distance of 500 mm, the time of motion being 5 seconds.

Neglecting friction, determine the approximate moment of inertia of the flywheel and shaft, given that its mass was 20 kg and the shaft diameter was 50 mm.

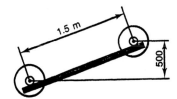

Problem 14/43

14/44

Fred's road roller has a total mass of 10 tonnes. The front roller assembly has a mass of 2 tonnes, a diameter of 1.2 m, and a radius of gyration of 0.45 m. The rear wheels and axle taken together have a mass of 2.5 tonnes and a radius of gyration of 0.6 m, the diameter of the rear wheels being 1.5 m. If the road roller is moving with a constant velocity of 10 km/h, determine
(i) the kinetic energy of rotation of the front roller assembly;
(ii) the kinetic energy of rotation of the rear wheels and axle;
(iii) the total kinetic energy of the road roller;
(iv) the braking force necessary to bring it to rest in 10 metres on a horizontal stretch of road.

Problem 14/44

15

Power

$$P = \frac{U}{t} = \frac{Fs}{t}$$

Machines such as the internal combustion engine, the steam engine, the water turbine and the electric motor are major sources of *power* both in industry and in everday life. We use the power provided by these and other machines in many different ways—for example, to transport goods and ourselves from one place to another, to operate generators in hydroelectric power plants and to run machinery such as lathes and drills in our manufacturing industries.

In all of these and similar situations the power available is being used to *do work*, and indeed the power of any engine or device is only a measure of the *rate* or speed at which it can *do work*. For any given machine, engine, or even a person, it is useful to know the maximum power output that can be sustained over a certain length of time and the efficiency present, that is, the percentage of usable power available at any given time.

When a known force moves its point of application through a known distance, the amount of work done is readily calculated from the previously discussed relationship, $U = Fs$. However, in many instances the time taken to perform this amount of work is often of equal importance to the actual amount of work done. Consider the following two examples:

(a) One car on a drag strip covers the distance in eleven seconds while another car covers it in 9.5 seconds. If the frictional resistance of both cars is equal, then obviously the engine of the second car has done work at a faster rate than the engine of the first car and thus its power was greater.

(b) A man walks up a flight of stairs in 30 seconds while another man of the same mass runs up the stairs in 5 seconds. Obviously, both men have done the same amount of work (work done = change in potential energy of each man), but the second man has done work at a much faster rate than the first, and consequently developed greater power.

Both of these examples illustrate that *the rate at which work is done* (or the rate at which energy is expended) is important in power calculations.

Power is the time-rate of doing work. The derived unit of power is the *watt* (W), one watt of power being equal to one joule of work done per second.

$$P = \frac{U}{t} = \frac{Fs}{t}$$

$$1 \text{ watt} = \frac{1 \text{ joule}}{\text{second}} = \frac{N\,m}{s} \quad \text{(since } 1 \text{ J} = 1 \text{ N m)}$$

More convenient units are the kilowatt (1 kW = 1000 W) and the megawatt (1 MW = 10^6 W). Because of the equivalence of work and energy, power can also be considered to be the time rate at which energy is either expended or increased.

SAMPLE PROBLEM 15/1

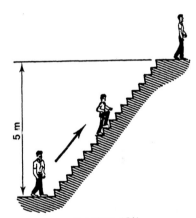

A man of mass 65 kg runs up a flight of stairs in 3.2 seconds. If the stairs have a total vertical rise of 5 metres, how much power did the man use?

Problem 15/1

Analysis:
Power is the time-rate of doing work. In this case the work done equals the change in potential energy of the man. Thus, the power used equals the time-rate change in potential energy.

Calculation:

$$\text{Power} = \frac{\text{change in PE}}{\text{time}}$$

$$= \frac{mgh}{t}$$

$$\therefore P = \frac{65 \times 9.8 \times 5}{3.2}$$

$$= 995.3 \text{ W}$$

where: $m = 65$ kg
$g = 9.8$ m/s^2
$h = 5$ m
$t = 3.2$ s

Result:
The power used by the man was 995.3 watts (or approximately 1 kW). ■

SAMPLE PROBLEM 15/2

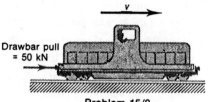

A diesel locomotive exerts a constant drawbar pull of 50 kilonewtons while pulling a train along a level track. During a certain distance the velocity of the train increases from 30 metres per second to 45 metres per second. Calculate the power delivered by the locomotive at each velocity.

Drawbar pull = 50 kN

Problem 15/2

Analysis:

$$\text{Power} = \frac{\text{work}}{\text{time}} = \frac{Fs}{t} = Fv \quad \left(\text{since } v = \frac{s}{t}\right)$$

Thus, if an object is moving at a known velocity, the power being developed can readily be calculated from $P = Fv$, where F is the constant force sustaining motion.

Calculation:
(i) Power at 30 m/s

$$P = Fv$$
$$= 50 \times 10^3 \times 30$$
$$= 1500 \times 10^3 \text{ W}$$
$$= 1.5 \text{ megawatts}$$

where: $F = 50 \times 10^3$ N
$v = 30$ m/s

(ii) Power at 45 m/s

$$P = Fv$$
$$= 50 \times 10^3 \times 45$$
$$= 2250 \times 10^3 \text{ W}$$
$$= 2.25 \text{ megawatts}$$

where: $F = 50 \times 10^3$ N
$v = 45$ m/s

Result:
The power developed by the locomotive at 30 m/s is 1.5 MW and at 45 m/s is 2.25 MW. ■

SAMPLE PROBLEM 15/3

Determine the power output of the engine of a car that is maintaining a constant velocity of 50 km/h against a combined resistance force (friction, wind, etc.) of 4 kN.

4 kN

Problem 15/3

Analysis:
The power of the engine does work against the total force resisting motion.

$$P = \frac{\text{work done}}{\text{time}}$$
$$= \frac{Fs}{t}$$
$$= Fv, \text{ since } v = \frac{s}{t}$$

Calculation:

$$P = Fv$$
$$P = 4 \times 10^3 \times 13.9 \ \text{Nm/s}$$
$$\therefore P = 55.6 \times 10^3 \ \text{J/s (or W)}$$
$$\therefore P = 55.6 \ \text{kW}$$

where: $F = 4 \times 10^3$ N
$$v = 50 \ \text{km/h}$$
$$= \frac{50 \times 10^3}{60 \times 60} \ \text{m/s}$$
$$= 13.9 \ \text{m/s}$$

Result:
The power output of the engine is equal to 55.6 kW.   ■

SAMPLE PROBLEM 15/4

A V-belt running in a pulley of diameter 0.25 metres exerts an effective pull of 500 newtons on the pulley. If the pulley is turning at 300 revolutions per minute, determine the power being transmitted by the belt.

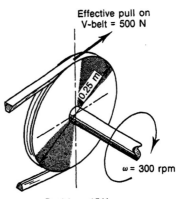

Effective pull on
V-belt = 500 N

ω = 300 rpm

Problem 15/4

Analysis:
In one revolution a point on the circumference of the pulley travels 0.25π metres. Thus the work done per revolution of the pulley, Fs, equals effective pull of belt × circumference of pulley and the power transmitted equals the work done per second.

Calculation:

Work done per revolution

$$W = Fs$$
$$= 500 \times 0.25\pi$$
$$= 392.7 \ \text{J}$$

where: $F = 500$ N
$$s = 0.25\pi \ m.$$

Work done per second = work done per revolution × revolutions per second
$$= 392.7 \times \frac{300}{60}$$
where: rps $= \frac{300}{60}$
$$= 1963.5 \ \text{J}$$
$$\text{Power} = \text{work done/second J}$$
$$= 1963.5 \ \text{W}$$

Result:
The power transmitted by the belt to the pulley is 1963.5 watts (or approximately 2 kW).   ■

POWER OF VEHICLES ON GRADIENTS

When an object is either resting on or moving up or down an inclined plane, its weight-force can be readily resolved into two perpendicular components, one acting parallel to the plane and the other acting normal to the plane. For instance, the block of mass m units shown in Figure 15.1 resting on the given inclined plane has two perpendicular components, $mg \sin \theta$ parallel to the plane and $mg \cos \theta$, perpendicular to the plane.

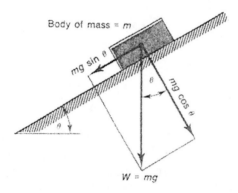

Body of mass = m

$mg \sin \theta$

θ

$mg \cos \theta$

$W = mg$

Figure 15.1

If the body moves up the plane, the component $mg \sin \theta$ opposes motion, but if the body moves down the plane, the component $mg \sin \theta$ assists motion to continue.

If inclined planes or gradients are expressed in terms of vertical unit rise per n units of movement up the slope (see Figure 15.2), the gradient $1/n$ becomes $\sin \theta$, where θ is the angle of inclination of the plane to the horizontal. The component of the weight-force acting parallel to the gradient, $mg \sin \theta$, therefore becomes mg/n.

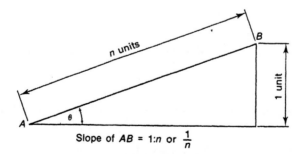

B

n units

1 unit

A

θ

Slope of AB = 1:n or $\dfrac{1}{n}$

Figure 15.2

In order to simplify problems involving gradients, all gradients will be considered *as the ratio of unit vertical rise to the distance, n, moved up the incline.*

SAMPLE PROBLEM 15/5

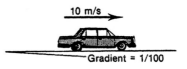

A car of mass 1.5 tonnes is being driven up a gradient of 1 in 100 and the frictional resistances are equal to 200 newtons per tonne. Calculate the power being developed by the engine of the car at the instant when the velocity of the car is 10 metres per second.

Problem 15/5

Analysis:

The total force moving the car up the plane must be the sum of the following two forces:

(i) the force necessary to overcome the frictional resistance, F_r

(ii) a force equal in magnitude but opposite in direction to the component of the weight-force of the car acting down the plane.

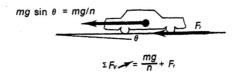

Problem 15/5

Calculation:

(i) Total frictional force

$$F_r = 200 \times 1.5 \text{ N}$$
$$= 300 \text{ N}$$

(ii) Component of weight-force acting down the plane

$$F = mg \sin \theta$$

$$= \frac{mg}{n}$$

$$= \frac{1500 \times 9.8}{100}$$

$$= 147 \text{ N}$$

where: θ = angle of gradient
$m = 1500$ kg
$g = 9.8$ m/s^2
$n = 100$

Thus, the total force necessary to maintain a constant velocity of 10 m/s up the slope

$$= 300 + 147$$
$$= 447 \text{ N}$$

From

$$P = Fv$$
$$\therefore P = 447 \times 10$$
$$= 4470 \text{ W}$$
$$= 4.47 \text{ kW}$$

where: $F = 447$ N
$v = 10$ m/s

Result:

The engine of the car must develop the power of 4.47 kW in order to achieve the velocity of 10 m/s up the incline. ∎

SAMPLE PROBLEM 15/6

A car of mass 1 tonne is using power at the rate of 10 kilowatts when travelling at 30 km/h on a level road. The gradient suddenly becomes 1 in 6 and the engine is turned off. How far up the incline will the car run before it stops? You may assume that frictional resistances remain constant.

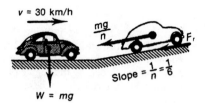

Problem 15/6

Analysis:

When the car begins to move up the incline, the kinetic energy of the car is used up as work done against the combined forces opposing motion up the plane. These combined forces are
 (i) the frictional resistance; and
(ii) the component of the weight-force of the car acting down the plane.

Calculation:

$$P = F_r v$$
$$\therefore F_r = \frac{P}{v}$$
$$\therefore F_r = \frac{10 \times 10^3}{8.3}$$
$$= 1206 \text{ N}$$

where: F_r = frictional force
$P = 10 \times 10^3$ W
$v = 30$ km/h $= 8.3$ m/s

Component of weight-force acting down the grade is
$$F = mg \sin \theta$$
$$= \frac{mg}{n}$$
$$\therefore F = \frac{1000 \times 9.8}{6}$$
$$= 1633 \text{ N}$$

where: θ = angle of gradient
$m = 1000$ kg
$n = 6$

Thus, total force opposing motion of the car up the incline
$$F_T = 1206 + 1633$$
$$= 2839 \text{ N}$$

KE of car at bottom of incline
$$KE = \tfrac{1}{2} mv^2$$
$$= \tfrac{1}{2} \times 1000 \times (8.3)^2$$
$$= 34\,445 \text{ J}$$

where: $m = 1000$ kg
$v = 8.3$ m/s

KE of car equals work done against friction
$$34\,445 = 2839\,s$$
$$s = 12.1 \text{ m}$$

where: s = distance car travels
up slope
$F_T = 2839$ N

Result:
The car will coast 12.1 metres up the 1 in 6 gradient before coming to rest. ■

POWER REQUIREMENTS OF HOISTS

Hoists are machines commonly used to raise or lower relatively heavy loads, hoist mechanisms forming the basis of lifts in buildings and mine-shafts, and cranes on construction sites. All hoists consist of three main parts:

(i) the load being lifted, which may be a lift cage, a bucket containing concrete, or some other relatively heavy mass;

(ii) a motor, which supplies the power necessary to operate the whole mechanism;

(iii) the *lift drum*, about which the cable (or cables) doing the lifting are wrapped. In some hoists a large pulley is used instead of, or as well as, a drum.

A simple hoist is shown in Figure 15.3 (a)

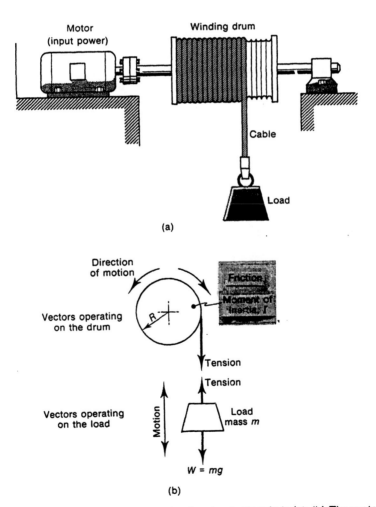

Figure 15.3 (a) The basic components of a simple electric hoist; (b) The vectors operating on the load and the winding drum when the hoist is raising the load.

Problems involving hoist systems can always be divided into three sections (refer to Figure 15.3):

(a) the vectors operating on the load, including the tension in the supporting cable;

(b) the power necessary to operate the hoist in a certain way, that is, to raise or lower the load at constant velocity or with uniform acceleration;

(c) vectors involved in the rotation of the winding drum. These are the *friction* in the bearings and the *moment of inertia* of the drum. They affect the efficiency of the hoist but may not need to be directly considered in calculations. This is shown in Sample Problems 15/7 and 15/8.

EFFICIENCY

The efficiency of machines is dealt with in more detail in Chapters 14 and 18. Efficiency is calculated from

$$\eta = \frac{\text{output}}{\text{input}}$$

where output and input can be expressed as work, energy, or power.

As the following sample problems show, no new concepts are involved in some problems concerned with the operation of hoists or similar lifting machines.

SAMPLE PROBLEM 15/7

An electric motor rated at 8 kW is coupled to a lifting machine which raises a 1-tonne load 3 metres in 5 seconds with uniform velocity. Determine the efficiency of the hoist.

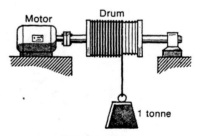

Problem 15/7

Analysis:
The work done by the hoist is equal to the change in energy of the load moved. Since the velocity is constant during the lift, there is no change in kinetic energy. However, there is an increase in the potential energy of the load. The time-rate change in potential energy is thus equal to the effective power of the hoist (that is, its output). Since the input is known to be 8 kW, its efficiency can now be easily determined.

Calculation:

$$\text{Change in PE of load} = mgh$$
$$= 1000 \times 9.8 \times 3 \text{ N m}$$
$$= 29\,400 \text{ J}$$

where: $m = 1$ tonne
$= 1000$ kg
$h = 3$ m

$$\text{Output of hoist} = \frac{\text{change in PE}}{\text{time}}$$

$$= \frac{29\,400}{5}$$

$$= 5880 \text{ W}$$

$$= 5.88 \text{ kW}$$

$$\text{Efficiency, } \eta = \frac{5.88,}{8} \times 100$$

$$= 73.5\%$$

Result:
The efficiency of the hoist is 73.5%.   ■

SAMPLE PROBLEM 15/8

A motor having an efficiency of 90% operates a crane having an efficiency of 35%. With what steady velocity does the crane lift a 400-kilogram load, if the power supplied to the motor is 5 kilowatts?

Analysis:
The overall efficiency of the coupled machines is the product of their individual efficiencies. Once this is known, the output can be calculated and application of $P = Fv$ will enable the velocity of the load to be determined.

Calculation:

$$\text{Overall efficiency} = 90 \times 35$$

$$= 31.5\%$$

$$\text{Output power} = \text{total efficiency} \times \text{power supplied to motor}$$

$$= \frac{31.5}{100} \times 5000$$

$$= 1575 \text{ W}$$

Now,
$$P = Fv$$
$$\therefore v = \frac{P}{F}$$

where: $P = 1575$ W
$F = 400 \times 9.8$ N
$v = ?$

$$= \frac{1575}{400 \times 9.8}$$

$$= 0.4 \text{ m/s}$$

Result:
The load moves upward with a velocity of 0.4 metres per second.   ■

ROTARY POWER

We have already seen that a body rotating with an angular velocity ω possesses rotary kinetic energy equal to $\frac{1}{2}I\omega^2$ where I is the moment of inertia of the body itself.

For example, consider a flywheel rotating at 500 rpm. Given that the moment of inertia of this flywheel is 200 kg m², then its rotary kinetic energy is

$$\text{KE (rotary)} = \frac{1}{2}I\omega^2 \qquad \text{where: } \omega = 500 \text{ rpm}$$

$$= \frac{1}{2} \times 200 \times 52.4^2 \qquad\qquad = \frac{500 \times 2\pi}{60} \text{ rad/s}$$

$$= 274\,576 \text{ J}$$

$$\approx 274.6 \text{ kJ} \qquad\qquad\qquad \approx 52.4 \text{ rad/s}$$

If a load is now applied, resulting in the flywheel being brought to a stop in 2 minutes, the power expended can be calculated from

$$P = \frac{\text{initial energy of flywheel} - \text{final energy of flywheel}}{\text{time taken}}$$

$$= \frac{(274\,576) - 0}{.2 \times 60} \text{ J/s}$$

$$= 2288 \text{ W}$$

$$\therefore P \approx 2.3 \text{ kW}$$

Thus, rotating bodies can deliver power, the amount of power available depending upon the time-rate loss of rotary kinetic energy.

The rotational equivalent of the linear relationship, $P = Fv$, which relates power with a force moving at constant velocity, is often useful in problem-solving and is written as

$$P = T\omega,$$

where T is the torque acting, and ω is the angular speed of the rotating body. If torque is expressed in N m and ω in rad/s, the units for power become watts (W).

Sample Problem 15/9 illustrates the calculations required when the moment of inertia of the winding drum of the hoist needs to be included in the power calculation. *The inertia torque always acts to oppose the acceleration or the deceleration of the drum.*

SAMPLE PROBLEM 15/9

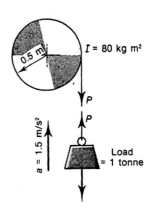

A hoist drum has a moment of inertia of 80 kg m²
and is used to raise a mass of 1 tonne with an
upward acceleration of 1.5 m/s². If the hoist drum
diameter is 1 metre, calculate
(i) the torque required at the drum, and
(ii) the power required at the end of the fifth
     second of motion.

Problem 15/9

Analysis and Calculation:
Since the lift is moving vertically upward there must be an unbalanced force, *F*, acting on the
1-tonne load.

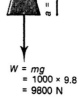

This unbalanced force, *F*, determines the vertical motion of the load.
Thus,
$$F = P - mg \text{ N}$$
$$\therefore F = P - 9800 \text{ N}$$

From
$$F = ma$$
$$P - 9800 = 1000 \times 1.5$$
$$\therefore P = 11\,300 \text{ N}$$

$W = mg$
$= 1000 \times 9.8$
$= 9800$ N

Problem 15/9

Now consider the vectors operating on the drum. The total torque necessary is made up of two
components:

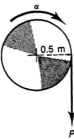

Problem 15/9

(i) the torque (T_1) necessary to counteract the effects of the tension, *P*, in the cable, which is
$$T_1 = P \times 0.5$$
$$= 11\,300 \times 0.5$$
$$= 5650 \text{ N m}$$

(ii) the torque (T_2) necessary to overcome the inertia of the drum itself. This is found from $T = I\alpha$, where I is the moment of inertia of the drum and α is its angular acceleration. Any point on the rim of the drum must have a linear acceleration of 1.5 m/s².

Since

$$a = r\alpha$$

$$\therefore \alpha = \frac{a}{r}$$

$$= \frac{1.5}{0.5}$$

$$\therefore \alpha = 3 \text{ rad/s}^2$$

Now,

$$T_2 = I\alpha \quad (\text{kg m}^2 \text{ rad/s}^2 = \text{kg m}^2/\text{s}^2$$
$$= 80 \times 3 \text{ N m} \quad (\text{since N} = \text{kg m/s}^2)$$
$$\therefore T_2 = 240 \text{ N m}$$

Total torque required

$$T \quad = T_1 + T_2$$
$$= 5650 + 240$$
$$= 5890 \text{ N m}$$

After accelerating 5 seconds from rest, the velocity, v, of the load can be found from

$$v = u + at \qquad\qquad \text{where: } u = 0$$
$$\therefore v = 0 + (1.5 \times 5) \qquad\qquad a = 1.5 \text{ m/s}^2$$
$$\therefore v = 7.5 \text{ m/s} \qquad\qquad t = 5 \text{ s}$$

Therefore, the angular velocity, ω, of the drum can be found from

$$v = r\omega \qquad\qquad \text{where: } v = 7.5 \text{ m/s}$$
$$\omega = \frac{v}{r} \qquad\qquad r = 0.5 \text{ m}$$
$$\omega = \frac{7.5}{0.5}$$
$$\omega = 15 \text{ rad/s}$$

Power required

$$P = T\omega \qquad\qquad \text{where: } T = 5890 \text{ N m}$$
$$= 5890 \times 15 \qquad\qquad \omega = 15 \text{ rad/s}$$
$$= 88350 \text{ N m/s}$$
$$\approx 88.4 \text{ kW} \qquad (\text{N m/s} = \text{J/s} = W)$$

Result:
The total torque necessary on the hoist drum was approximately 5.9 kN m and the power necessary was approximately 88.4 kW. ∎

It was assumed in this problem that there are no appreciable frictional forces acting on the hoist drum. If these are present, they are usually expressed as a *friction couple* (or torque), which must be added to the total torque required to cause rotation.

Suppose that a friction couple of 1 kN m is present in the hoist used in Sample Problem 15/9. Then the total torque required is increased from 5890 N m to 5890 N m + 1 kN m (6890 N m). This means that more power is required to operate the hoist.

Remember that friction couples always act to oppose the rotation of the drum.

SAMPLE PROBLEM 15/10

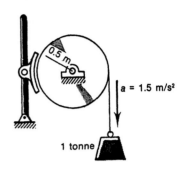

A hoist drum has a moment of inertia of 80 kg m²
and is used to lower a load of 1 tonne with an
acceleration of 1.5 m/s². If the hoist drum has a
diameter of 1 metre and a friction couple of
1.5 kN m acts in the bearings, determine
  (i) the braking torque required;
 (ii) the power absorbed in the brake after a drop
      of 10 metres.

Problem 15/10

Analysis:
Consider the vectors operating on the load, and let the tension in the cable be P.

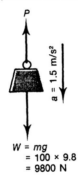

$W = mg$
$= 100 \times 9.8$
$= 9800$ N

Problem 15/10

The unbalanced force present must be $W - P$ since the load accelerates downwards.
From $F = ma$
$$9800 - P = 1000 \times 1.5$$
$$P = 8300 \text{ N}$$

Consider the vectors operating on the drum; these are

(a) The friction couple, F_r, which opposes motion, that is, assists braking
$$F_r = 1.5 \text{ kN m}$$
$$\therefore F_r = 1500 \text{ N m}$$

(b) The accelerating torque (T_1) provided by the cable tension, P
$$\text{i.e. } T_1 = 8300 \times 0.5 \text{ N m}$$
$$= 4150 \text{ N m}$$

(c) The inertia couple of the drum itself
$$T_2 = I\alpha$$
$$= 80 \times \frac{1.5}{0.5} \text{ kg m}^2 \text{ rad/s}^2$$
$$\therefore T_2 = 240 \text{ N m}$$

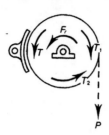

Problem 15/10

For rotation of the hoist drum, the accelerating torque, T_1, due to the rope tension must balance the sum of the decelerating torques. These are the inertia torque, T_2, the friction torque, F_r, and the unknown braking torque, T.

Thus,
$$T_1 = T_2 + F_r + T$$

Now, the braking torque, T, must be equal to $T_1 - T_2 - F_2$
$$\therefore T = 4150 - 240 - 1500$$
$$= 2410 \text{ N m}$$

After accelerating from rest for 10 metres, the velocity, v, of the load can be found from
$$v^2 = u^2 + 2a\,s \qquad \text{where: } u = 0$$
$$v^2 = 0 + (2 \times 1.5 \times 10) \qquad a = 1.5 \text{ m/s}^2$$
$$v^2 = 30 \qquad s = 10 \text{ m}$$
$$\therefore v \approx 5.5 \text{ m/s}$$

Therefore, the angular velocity of the hoist drum can be found from
$$v = r\omega$$
$$\therefore \omega = \frac{v}{r}$$
$$= \frac{5.5}{0.5}$$
$$\therefore \omega = 11 \text{ rad/s}$$

Power required
$$P = T\omega$$
$$= 2410 \times 11$$
$$\therefore P = 26\,510 \text{ W}$$
$$\therefore P \approx 26.5 \text{ W}$$

Result:
The braking torque required is about 2.4 kN m and after a drop of 10 metres the brake has absorbed 26.5 kW of power. ∎

FRICTIONAL LOSSES IN BEARINGS

One major source of power loss in machines occurs due to bearing friction, the "lost" power being used in heating up the bearing surfaces and assisting their mechanical destruction.

Even though lubricants are used to reduce these frictional losses, they are always significant, as Sample Problem 15/11 shows.

SAMPLE PROBLEM 15/11

A shaft of 50 mm diameter running at 400 rev/min
is supported by two bearings each having a coefficient of friction of 0.015. If the load on each bearing
is 2 tonnes, determine the power loss due to friction.

Analysis:
The power loss due to friction can be regarded as the work done per unit of time against the
friction force, F_r, generated within the bearing. Since the friction force F_r is tangential to the
bearing surface, the work done in overcoming it can be found from $W = F_s$, where s is the linear
distance moved per second by any point on the bearing surface.

Calculation:

$$F_r = \mu_k \, N$$
$$= 0.015 \times 2 \times 10^3 \times 9.8 \, N$$
$$\therefore F_r = 294 \, N$$

Total frictional force (2 bearings) = 588 N

$$\text{Linear distance } (s/\text{second}) = \pi d \times \frac{\text{rpm}}{60}$$
$$= \frac{22}{7} \times \frac{50}{1000} \times \frac{400}{60}$$
$$\therefore s = 1.047 \, m$$
$$\text{Power loss } (P) = \frac{Fs}{t}$$
$$= \frac{588 \times 1.047}{1}$$
$$= 615.6 \, W$$

Alternative Solution:
The relationship $P = T\omega$, where T is the torque acting and ω is the angular velocity, can also be
used to determine the power loss of this bearing, since in this case the torque, T, is the *friction
torque* present.

$$\text{Total friction torque, } T = F_r \times r$$
$$= 588 \times \frac{25}{1000} \, N\,m$$
$$\text{Angular velocity } (\omega) = \frac{400}{60} \times 2\pi \, \text{rad/s}$$
$$\therefore \omega = 13.33 \, \text{rad/s}$$
$$\text{Power loss } (P) = T\omega$$
$$= \frac{588 \times 25 \times 13.33\pi}{1000}$$
$$\therefore P = 615.6 \, W$$

Result:
The power loss due to bearing friction is about 616 W.

■

REVIEW PROBLEMS
(Use $g = 9.8 \, \text{m/s}^2$ unless otherwise specified.)

15/12
A jet engine can develop 25 kN of thrust when travelling at 650 km/h. Determine the power that the engine is delivering.

15/13
Determine the average power that would be necessary to raise a hopper of concrete through 30 metres in 2 minutes with constant velocity, given that the combined mass of hopper and concrete is 750 kg.

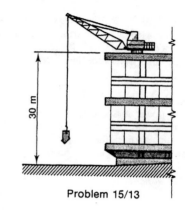

Problem 15/13

15/14
A man of mass 65 kg sits on a sled being pulled by a mule. The mule exerts a constant horizontal pull of 250 newtons and the sled moves with a constant velocity of 0.5 m/s. At what rate is (i) the man working? (ii) the mule working?

15/15
The output of a certain machine is equivalent to 15 kW. If the efficiency of the machine is 80%, what is the necessary input power?

Problem 15/14

15/16
A crane lifts a girder of mass 12 tonnes vertically upwards through 30 metres. During the first 5 m of lift, the load is accelerated uniformly to a velocity of 1 m/s. Thereafter the lift is continued at that velocity. Find:
  (i) the tension in the cable during the period of acceleration;
 (ii) the tension in the cable during the period of uniform velocity;
(iii) the capacity of the motor necessary to complete this operation, without overloading;
(iv) the total work done in the lifting operation.
Friction and inertial effects to be neglected for (iii) and (iv).

15/17
A centrifugal pump raises water through a vertical height of 20 metres. If the pump motor supplies 20 kilowatts of power and the efficiency of the pump is 80%, determine the quantity of water that can be pumped in one hour. Take the mass of 1 litre of water as 1 kg.

15/18

During a cutting stroke of a shaping machine it is found that the average force on the cutting tool is 1.5 kN. Given that the cutting speed is 0.3 m/s and the overall efficiency is 70%, determine the suitable power rating for its motor.

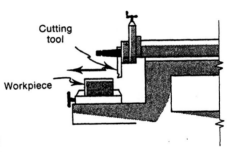

Cutting tool

Workpiece

Problem 15/18

15/19

The water used to drive a turbo-generator plant falls in a pipe through an effective height of 50 m to the turbine, from which it is released at a velocity of 1.3 m/s. If the combined losses in the pipe to the turbine and in the turbine itself are 25% of the potential power, what is the power available to drive the generator, when 60 000 litres of water pass through the turbine every minute? Take the mass of 1 litre of water as 1 kg.

Problem 15/19

15/20

A pile-driver hammer of mass 500 kg is raised vertically a distance of 4 m in 6 seconds. The hoisting equipment has an efficiency of 80%. Determine the power of the motor necessary to operate the hoist, assuming that the lift is at constant velocity and friction is negligible.

15/21

A pump raises 13 cubic metres of water per hour through a height of 100 metres. Calculate the power required to drive the pump if 60% of its power is used to do useful work. The mass of 1 cubic metre of water = 1 tonne.

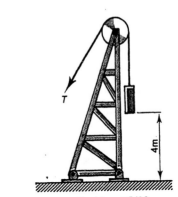

Problem 15/20

15/22

An engine delivering 4 kW is used to pump water from a shaft 30 m deep. How many litres will be raised in 24 hours.
 (i) if friction losses are negligible;
 (ii) if the overall efficiency of the system is 85%?
Take the mass of 1 litre of water as 1 kg.

15/23

In a test on a lathe cutting tool, it was observed that the power required for a certain size cut at an effective 150 mm radius and a speed of 200 rev/min was 4 kW. If power losses in the machine accounted for 1 kW, determine the tangential force on the cutting tool.

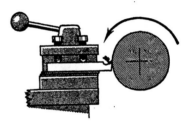

Problem 15/23

15/24

During a test on a motor vehicle with a total load of 5 tonnes, its road speed was steady at 50 km/h. The vehicle was travelling up an incline of 1 in 15 and instruments indicated that the engine was delivering 55 kW. Determine:

  (i) the power which produced no useful output;

  (ii) the power which did produce useful output;

(iii) the efficiency of the engine at this instant.

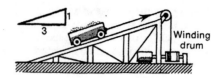

Problem 15/25

15/25

A steel cable is used to haul a truck up an incline as shown. The mass of the truck and load is 1 tonne and the frictional resistance to motion is 104 N. Determine the power necessary to haul the truck up the slope at 10 km/h.

15/26

An electrically powered conveyer carries 30 000 packages per hour a distance of 25 m up an incline of 1 in 10. Each package has a mass of 30 kg and the power absorbed by friction in the drive is 2 kW.

  (i) Determine the power of the motor driving this conveyer.

  (ii) What is the efficiency of the conveyer?

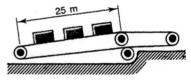

Problem 15/26

15/27

A diesel locomotive can deliver a maximum of 3 MW of power to the drawbar. Given that rolling resistances are 50 N per tonne, how many 60-tonne freight cars can the locomotive pull

  (i) along a level track at 100 km/h;

  (ii) up a 2% grade at the same speed?

15/28

A cylindrical bearing 100 mm in diameter is subject to a load of 8 tonnes. Determine the power absorbed by friction between the bearing and its journal when the shaft is rotating at 600 rpm. The coefficient of friction is 0.03.

15/29

Determine the torque on a shaft which transmits 600 kW when running at 200 rpm.

15/30

A guillotine relies for its operation on the kinetic energy stored in its flywheel. The flywheel has a mass of 400 kg, a radius of gyration of 1.3 metres, and is rotating at 300 rpm. During a one-second shearing operation, it slows to 270 rpm. Determine

  (i) the loss in kinetic energy of the flywheel during the cut;

  (ii) the power used during this cut, if the machine has an overall efficiency of 60%.

15/31
An electrically operated hoist drum of diameter
1.2 metres has a moment of inertia of 100 kg m^2.
The friction in the bearings of the hoist is equivalent
to a couple of 1 kN m. Determine the power of the
motor required to drive this hoist if it is to be
capable of lifting a maximum load of 2 tonnes
vertically with an acceleration of 1 m/s^2 for 4
seconds. Assume that motor efficiency is 87%.

15/32
A pressing machine has a flywheel of moment of
inertia 50 kg m^2 which provides part of the energy
for the pressing operation. This flywheel is driven
by a 3-kW motor and each pressing requires 5 kJ of
energy and takes 1 second to complete. Determine
the reduction in speed of the flywheel in rpm and
the maximum number of pressings possible per
minute, if the initial speed of the flywheel before
each pressing is 240 rpm.

16

Direct Stress and Strain

$$\sigma = \frac{P}{A}$$

$$\varepsilon = \frac{\Delta L}{L} = \frac{e}{L}$$

$$E = \frac{\sigma}{\varepsilon} = \frac{PL}{Ae}$$

$$\text{SE/unit volume} = \frac{\sigma^2}{2E}$$

Consider a structural member or component to which an external load or force is being applied. If the member remains in equilibrium and does not bend or buckle, a force equal in magnitude and opposite in direction to the external force or load must be present within the member itself, the internal force being a *reaction* to the external applied force (Newton's third law).

Since the internal force in the member is a reaction to the action of the externally applied force or load, it will be reduced when the external force is reduced and will increase when the external force is increased. Similarly, the internal force will disappear when the external force is removed, provided that the member or component has not suffered permanent deformation.

INTENSITY OF STRESS

Internal forces set up within externally loaded members are termed *stresses* and the *intensity of stress* is calculated as the amount of force per unit area of the member.

Consider the round bar shown in Figure 16.1, which is being subjected to the tensile load P and which has a cross-sectional area of A units.

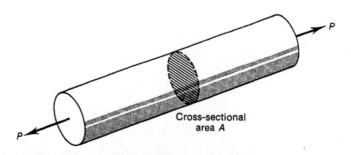

Figure 16.1 The stress due to the application of a tensile load P is at a maximum on planes at 90° to the axis of the bar; the cross-sectional area A is a typical such plane.

The intensity of stress, or simply the stress, within this bar can be calculated from the relationship:

$$\text{intensity of stress} = \frac{\text{internal reaction force}}{\text{cross-sectional area}}$$

or

$$\sigma = \frac{P}{A}$$

$$\therefore \text{unit of stress} = \frac{P \text{ (in newtons)}}{A \text{ (in square metres)}}$$

$$= N/m^2$$

Thus, the basic unit of stress is the newton per square metre (N/m^2) which is known as the *pascal* (Pa), common multiples being the kilopascal (kPa) and the megapascal (MPa). Since most areas of members used in structures are measured in square millimetres, it is useful to realise that

$$1 \text{ N/mm}^2 = 10^6 \text{ N/m}^2 = 1 \text{ MPa}$$

TYPES OF STRESS

There are three fundamental types of direct stresses and they are classified according to the type of external force system that produces them. They are:

(1) tensile stresses;
(2) compressive stresses;
(3) shear stresses.

Figure 16.2 illustrates the conditions under which these three types of stress are developed in members.

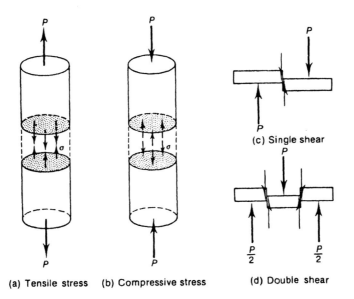

(a) Tensile stress   (b) Compressive stress   (c) Single shear   (d) Double shear

Figure 16.2 Types of stresses.

Tensile stresses are produced in bars subjected to tensile forces or loads as in Figure 16.2 (a), the stresses being set up normal to plane sections of the bar which are themselves at 90° to the direction of the external force.

Compressive stresses are produced by compressive forces acting on a member (Figure 16.2 (b)) and are also set up normal to plane sections of the member which are themselves perpendicular to the direction of the external force.

It is important to note that both the tensile and compressive stresses as illustrated are being produced by *axial loads*, thus no bending of the bars will occur. Stresses set up by the bending of members will be dealt with in Chapter 17.

Shear stresses, as shown in Figures 16.2 (c) and 16.2 (d), are produced by equal and opposite forces whose lines of action do not coincide. Shear stresses act along planes within the member which are parallel to the lines of action of the external forces. Thus, shear forces tend to make one section of a bar or member *slide* over an adjacent section. The bar shown in Figure 16.2 (d) is in double shear since it tends to fail along two separate parallel planes.

Many rivets, bolts, and locking pins are placed in double shear in their normal service conditions.

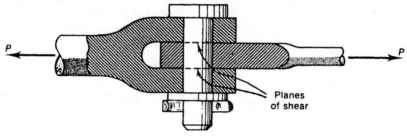

Figure 16.3 The pin shown as the coupling between two rods is placed in double shear by the force *P*. The two shear planes are illustrated by dashed lines.

SAMPLE PROBLEM 16/1

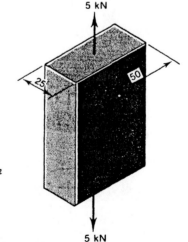

A 25-mm by 50-mm rectangular bar is subjected to a 5 kN axial tensile load. Determine the stress in the bar.

Calculation:

$$\sigma = \frac{P}{A}$$

where: $P = 5 \text{ kN}$
$A = 25 \times 50 \text{ mm}^2$

$$\therefore \sigma = \frac{5 \times 10^3}{25 \times 50} \text{ N/mm}^2$$
$$= 4 \text{ MPa}$$

Result:
The bar has a uniform tensile stress of 4 MPa.　　■　　Problem 16/1

SAMPLE PROBLEM 16/2

A circular pad is to have a machine of mass 4000 kg
placed upon it. Calculate the minimum diameter of
this pad if the compressive stress in it is not to
exceed 275 kPa.

Analysis:
Since the machine has a mass of 4000 kg, the load it places upon the circular pad is equal to
4000×9.8 newtons.

Calculation:

$$\sigma = \frac{P}{A}$$

$$\therefore A = \frac{P}{\sigma}$$

$$= \frac{4000 \times 9.8}{275 \times 10^3}$$

$$\therefore A = 0.143 \text{ m}^2$$

where: $\sigma = 275 \times 10^3$ Pa
$P = 4000 \times 9.8$ N

But

$$A = \frac{\pi d^2}{4}$$

$$\therefore d^2 = \frac{4A}{\pi}$$

$$= \frac{4 \times 0.143 \times 7}{22}$$

$$= 0.182 \text{ m}^2$$

$$\therefore d = 0.43 \text{ m}$$

$$= 430 \text{ mm}$$

where: d = diameter of pad

Result:
The pad diameter must be 430 mm if the stress in the pad is to not exceed 275 kPa. ■

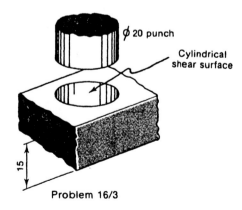

SAMPLE PROBLEM 16/3

A load of 20 kN is slowly applied to a piece of
15-mm thick metal plate by means of a 20-mm
diameter punch. Determine the shear stress set up
in the plate.

ϕ 20 punch

Cylindrical
shear surface

15

Problem 16/3

Analysis:
The shear stress is developed on a cylindrical plane of diameter 20 mm and height 15 mm, the area
of this plane being the area over which the shear load of 20 kN acts.

Calculation:

$$\text{Area of plane} = \pi dh$$

$$= \frac{22}{7} \times 20 \times 15$$

$$= 943 \text{ mm}^2$$

where: $d = 20$ mm
$h = 15$ mm

Now

$$\sigma = \frac{P}{A}$$

$$= \frac{20 \times 10^3}{943} \text{ (N/mm}^2)$$

$$= 21.2 \text{ MPa}$$

where: $P = 20$ kN

Result:
The shear stress developed in the plate is 21.2 MPa. ■

STRAIN

When a member is subjected to an external load, a change in shape occurs; that is, the member is *deformed* by the applied force or load. The actual amount of deformation depends upon several factors, including the magnitude of the applied force, the dimensions of the member and the *stiffness* of the material itself. (Stiffness is discussed later in this chapter.)

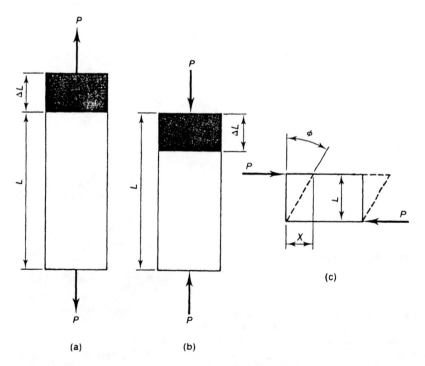

Figure 16.4 Three types of strain: (a) shows the tensile strain $\Delta L/L$: (b) shows the compressive strain $\Delta L/L$; and (c) shows the shear strain X/L which is equal to the angle ϕ expressed in radians.

The amount of *strain* produced in any member is calculated as *the ratio of the amount of deformation per unit length*, that is,

$$\text{strain} = \frac{\text{change in length}}{\text{original length}}$$

$$\therefore \varepsilon = \frac{\Delta L}{L}$$

Since strain is the *ratio* of two lengths, both of which have the same unit, *strain has no unit*.

Strains can be produced by
(a) the application of an external force or load; and,
(b) a change in temperature, which causes an expansion or contraction of the object.

SAMPLE PROBLEM 16/4

A rod in a machine compresses from 250 mm to 249 mm during the operation of the machine. Determine the strain in the rod.

Calculation:

$$\varepsilon = \frac{\Delta L}{L}$$

$$\therefore \varepsilon = \frac{-1}{250}$$

$$= -0.004$$

$$= -4 \times 10^{-3}$$

where: $\Delta L = -1$ mm
$L = 250$ mm

Result:
The strain in the rod is 4×10^{-3}, the negative sign being used in the calculation to indicate that the strain is compressive. ∎

HOOKE'S LAW

If the deformation in a member brought about by an applied load disappears when that load is removed, then the member is said to be behaving *elastically*. Most materials behave elastically over a limited range of stress which is known as the *elastic range* of the material.

In many engineering materials stressed within the elastic range the strain produced is directly proportional to the stress induced in the specimen. This proportionality between stress and strain was first expounded by Robert Hooke in 1678, and is known as *Hooke's law*.

Expressed simply, Hooke's law states that

$$\frac{\text{elastic stress}}{\text{elastic strain}} = \text{a constant}$$

that is

$$\frac{\sigma}{\varepsilon} = \text{a constant}$$

or

$$E = \frac{\sigma}{\varepsilon}$$

where the constant E is unique to the material involved and is, of course, the measure of the *stiffness* of that material.

The relationship between stress and strain for any given material is readily determined from the results of routine mechanical tests conducted on standard test specimens using standardised procedures in a mechanical testing laboratory, the most common mechanical test being the tension test.

THE TENSION TEST

This test consists of the application of a steadily increasing axial tensile load to a small specimen until failure occurs.

During the test, measures of applied load and corresponding extensions are recorded, the testing machine often graphing the readings automatically as the test proceeds. Figure 16.5 shows a typical load-extension graph for a low-carbon steel, the significant features being labelled.

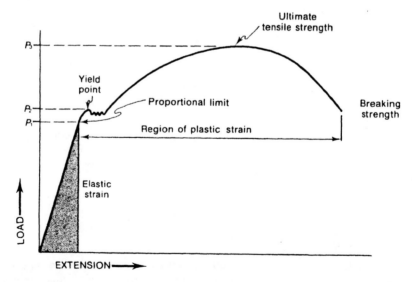

Figure 16.5 A typical tensile load-elongation graph for low carbon steel.

The following data are readily calculated from tensile tests:
(i) proportional limit stress;
(ii) yield stress (or proof stress);
(iii) ultimate stress (commonly called ultimate tensile strength, UTS);

(iv) Young's modulus (stiffness);
 (v) modulus of resilience.

The *proportional limit stress* is the greatest stress that can be developed in the specimen without losing straight line proportionality between stress and strain. In other words, up to the proportional limit the material obeys Hooke's law. Referring to Figure 16.5 it is obvious that the proportional limit stress is equal to the applied load at the proportional limit, P_1, divided by the cross-sectional area of the specimen. The proportional limit is often regarded as the elastic limit of the material but the true elastic limit stress is often lower than the stress at the proportional limit.

The *yield stress* is the stress at which some marked increase in strain occurs without a corresponding increase in stress. Some materials, notably low-carbon steels, exhibit a well-defined yield point; one such example being shown in Figure 16.5 for the indicated load P_2. Thus, the yield stress in this particular case is P_2 divided by the cross-sectional area of the specimen. Many metals and other materials do not exhibit a definite yield point; they yield progressively once the proportional limit is exceeded and another measure of yield stress, known as proof stress, must be used.

Proof stress is the stress necessary to produce a certain amount of strain within the specimen. Usually strains of 0.1% or 0.2% are used; and the stresses necessary to produce these strains can be determined by the "offset method" shown in Figure 16.6.

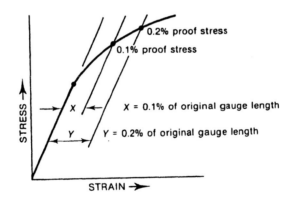

Figure 16.6 The offset method for determining 0.1% and 0.2% proof stress.

The *ultimate tensile strength* (UTS or ultimate stress) of a material is the maximum stress that can be produced in the material without failure occurring. In practice, the UTS is calculated by dividing the maximum test load (P_3 in Figure 16.5) by the *original cross-sectional area* of the specimen; thus the stress value is *nominal* only. (Nominal or engineering stresses are dealt with later in this chapter.) UTS values are often used in design work even though the material deforms plastically before the UTS is reached. However, larger *factors of safety* must be used if UTS values and not yield stresses are used in design work.

Table 16.1. Yield Stress, Proof Stress and UTS for Some Metals.

	Yield Stress	0.1 % Proof Stress (MPa)	UTS (MPa)
Copper, annealed	—	60	220
Copper, hard rolled	—	320	400
Aluminium, annealed	—	30	90
Aluminium, hard rolled	—	140	150
Mild steel	230-280	—	350-400
Structural steel	220-250	—	430-500
Stainless steel	230	—	600
Grey cast iron	—	—	280-340

Note:
(i) These values are approximate and should be regarded as average test figures only.
(ii) For a brittle metal like grey cast iron only the ultimate strength is measured in a tension test.
(iii) Yield stresses are generally measured only for some steels; for most other metals, 0.1 % or 0.2 % proof stress is measured. Stress at proportional limit is rarely calculated.

Young's modulus (modulus of elasticity or modulus of stiffness) is a measure of the slope of the straight-line portion of the test graph up to the proportional limit. *It is thus equal to the ratio of stress to strain and must be the constant E referred to previously in the Hooke's law equation.*

$$E = \frac{\sigma}{\varepsilon}$$

where σ is any stress less than that at the proportional limit and ε is the corresponding strain in the material.

$$\therefore E = \frac{P/A}{e/L} \qquad \text{where: } P = \text{load}$$
$$\therefore E = \frac{PL}{Ae} \qquad \qquad L = \text{gauge length}$$
$$A = \text{cross-sectional area}$$
$$e = \text{extension}$$

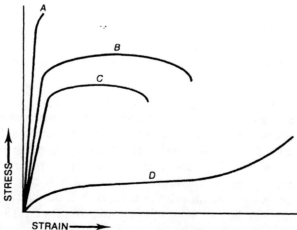

Figure 16.7 Young's modulus as an indication of stiffness. Materials *A*, *B* and *C* behave elastically up to their proportional limits, *A* being the stiffest and *C* being the least stiff material. Material *D* does not obey Hooke's law and thus a value for Young's modulus cannot be calculated for this material.

The steeper the slope of this straight-line portion of the test curve the higher the value of Young's modulus and the smaller the deformation that occurs in the specimen. Thus, Young's modulus measures the *resistance to elastic deformation* or the *stiffness* of the material.

Young's modulus has the same unit as that of stress:

$$E = \frac{\sigma}{\varepsilon}$$

$$\text{Unit} = \frac{\text{Pa}}{\text{dimensionless ratio}} = \text{Pa}$$

The most useful units are megapascals (MPa) and Table 16.2 lists the values of Young's modulus for some common engineering materials.

Table 16.2. Values of Young's Modulus for Some Materials

Material	E (MPa†)
Steel	206×10^3
Brass	84×10^3
Aluminium alloys	70×10^3
Copper	120×10^3
Spheroidal cast iron	175×10^3
Timber	10×10^3
*Grey cast iron	110×10^3
*Rubber	$0.4 - 1.6$

*Note that grey cast iron and rubber do not obey Hooke's law and thus these values are working approximations only.

† Although it would have been possible to express most of these values in gigapascals (GPa), megapascals are often used because of convenience in calculations.

SAMPLE PROBLEM 16/5

A tensile test specimen of a certain metal had a gauge length of 50 mm and an initial diameter of 10 mm. When tested, it produced the following results:

Load at proportional limit   = 24 500 N
Ultimate load   = 40 000 N
Extension at proportional limit = 0.1 mm

Calculate (i) the proportional limit stress, (ii) the ultimate tensile strength of the metal, and (iii) Young's modulus.

Calculation:

(i)
$$\sigma = \frac{P}{A}$$
$$= \frac{24\,500}{A}$$
$$\frac{24\,500 \times 7 \times 4}{22 \times 10^2} \text{ N/mm}^2$$
$$= 312 \text{ MPa}$$

where: $A = \frac{\pi d^2}{4}$

$d = 10$ mm

(ii)
$$\text{UTS} = \frac{40\,000 \times 7 \times 4}{22 \times 10^2} \text{ N/mm}^2$$
$$= 509 \text{ MPa}$$

(iii)
$$E = \frac{PL}{Ae}$$
$$= \frac{24\,500 \times 50}{\dfrac{\pi \times 10^2}{4} \times 0.1}$$
$$= 155\,909 \text{ MPa}$$
$$\approx 156 \times 10^3 \text{ MPa}$$
$$(\text{or} \approx 156 \text{ GPa})$$

where: $P = 24\,500$ N
$L = 50$ mm
$A = \dfrac{\pi \times 10^2}{4}$
$e = 0.1$ mm

Results:
The stress at proportional limit was 312 MPa; the ultimate tensile stress was 509 MPa; and the Young's modulus was 156×10^3 MPa. ∎

SAMPLE PROBLEM 16/6

A steel specimen of gauge length 56 mm and cross-sectional area 100 mm^2 has a tensile load of 1 kN applied to it. If $E = 206 \times 10^3$ MPa, determine the extension and the strain in the specimen.

Calculation:

$$E = \frac{PL}{Ae}$$
$$\therefore e = \frac{PL}{AE}$$
$$= \frac{1000 \times 56}{100 \times 206 \times 10^3} \text{ mm}$$
$$= 2.72 \times 10^{-3} \text{ mm}$$

where: $E = 206 \times 10^3$ MPa
$P = 1000$ N
$L = 56$ mm
$A = 100$ mm^2

$$\varepsilon = \frac{e}{L}$$
$$= \frac{2.72 \times 10^{-3}}{56}$$
$$\therefore \varepsilon = 48 \times 10^{-6}$$

where: $e = 2.72 \times 10^{-3}$ mm
$L = 56$ mm

Result:
The elongation of the specimen is about 2.72×10^{-3} mm and the strain is 48×10^{-6} ∎

Other Mechanical Tests Involving Direct Stress

Two other mechanical tests involving direct stresses are the *compression test* and the *shear test*. These tests are also conducted on standard specimens using standardised procedures, and the results of both tests may also be expressed in terms of load–deformation graphs.

The information given below is intended as a summary only and reference should be made to Chapter 5 of the *Introduction to Materials Science, SI Edition*, by B.R. Schlenker for details of these and other mechanical tests.

Compression test. The specimen is subjected to a uniaxial compressive load which is slowly increased in magnitude until failure occurs. The results are often given as a graph of load against resultant compression; however, for brittle materials like brick, stone and concrete only the ultimate load may be measured. Some measure of yield stress is calculated for ductile materials, the ultimate stress is calculated for brittle materials, and the elastic modulus (*E*) may be calculated for any type of material which obeys Hooke's law. The test is useful in evaluating brittle materials that are to be used in compression, e.g. concrete, stone, brick, cast iron, wood.

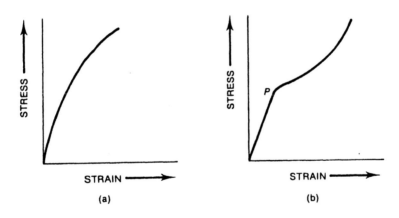

Figure 16.8 Typical load–elongation graphs for ductile and brittle materials: (a) shows the behaviour of grey cast iron, the graph revealing that only a small amount of plastic deformation occurred before failure; (b) shows that the very ductile mild steel obeyed Hooke's law until the point *P* (proportional limit) was exceeded, and then the specimen deformed plastically in compression with further increase in load.

Shear Test. Shear tests may be carried out using standardised specimens in *single shear* or *double shear*, or upon sheet or plate using a punching shear test. Usually the ultimate shear stress of the material is calculated by dividing the ultimate load by the area of shear. The numerical values obtained from shear tests are useful when calculating the thickness of material to be used in a given shear situation.

Table 16.3. Average Ultimate Strengths of
Non-metallic Materials (MPa).

	Compression	Tension
Bricks, best hard	83	2.8
Brickwork, common	10	0.35
Portland cement, (1 year old)	21	3.45
Concrete	7	1.4
Granite	131	4.8
Sandstone, dense	60	2
Limestone	60	2.1
Slate	96	3.45

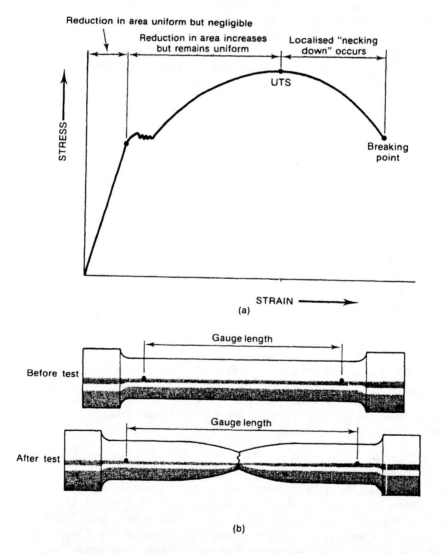

Figure 16.9 (a) Reduction in area of tensile test specimens related to the load–elongation curve. (b) The tensile test specimen before (above) and after (below) the tension test.

ENGINEERING STRESS

It is obvious that the cross-sectional area of a specimen undergoing a tensile test will decrease as the length of the specimen increases. Generally the amount of decrease in area is very small up to the proportional limit. However, once the proportional limit is exceeded, the decrease in area becomes greater and thus more significant. From the proportional limit to the point of ultimate tensile strength the decrease in area remains uniform along the parallel portion of the test specimen but, once the UTS is exceeded, localised reduction in area (known as "necking down") becomes very large in the region where failure will occur. Thus, while the load seems to reduce progressively from the UTS to the breaking point, the intensity of stress in the specimen at the zone of failure is actually increasing due to the rapid reduction in the cross-sectional area of the specimen in that region.

However, it is difficult to measure the changing diameter of a test specimen as the test progresses; thus stress calculations for the proportional limit, yield strength, proof stress, UTS and breaking strength are based upon the original cross-sectional area of the specimen. These stresses are termed *engineering stresses* or *nominal stresses* and are always less than the true stresses present.

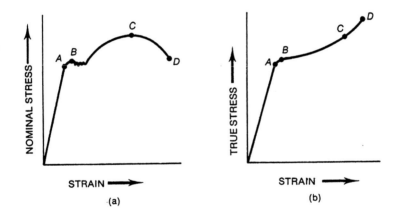

Figure 16.10 Graphical representation of (a) nominal stress and (b) true stress against strain in a tensile specimen showing a definite yield point, *B*. Note particularly the increase in true stress from the UTS (*C*) to the breaking point, *D*.

Since members or components in a structure are never stressed to anywhere near their proportional limits under conditions of normal loading, the use of nominal or engineering stress in design calculations is quite satisfactory.

SAMPLE PROBLEM 16/7

A metal bar of original diameter of 10 mm was
subjected to a tensile test and failed at a breaking
load of 40 kN. Calculate the nominal stress and the
true stress in the bar at failure if the diameter at
the section of failure was 8.5 mm.

Calculation:
(i) Nominal stress

$$\sigma = \frac{P}{A}$$

$$= \frac{40 \times 10^3 \times 7 \times 4}{22 \times 10^2} \text{ N/mm}^2$$

$$= 509 \text{ MPa}$$

where: $P = 40 \times 10^3$ N

$A = (\pi \times 10^2)/4$

(ii) True stress

$$\sigma = \frac{P}{A}$$

$$= \frac{40 \times 10^3 \times 7 \times 4}{22 \times (8.5)^2} \text{ N/mm}^2$$

$$= 705 \text{ MPa}$$

where: $P = 40 \times 10^3$ N

$A = (\pi \times (8.5)^2)/4$

Result:
The nominal stress at failure was 509 MPa but the true stress was 705 MPa. ■

RESILIENCE

Resilience is the amount of strain energy contained in a deformed body. When a piece
of material is deformed up to its proportional limit, the strain energy present is equal
to the *work done* during deformation and, since the deformation up to this point is
elastic, this strain energy is *recoverable* when the stress on the material is removed.

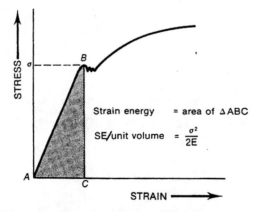

Figure 16.11 The strain energy present in a material stressed up to its
proportional limit, *B*. The strain energy per unit volume of material is
equal to area of $\triangle ABC = \frac{1}{2} \sigma\varepsilon = \frac{1}{2}\sigma\frac{\sigma}{E} = \frac{\sigma^2}{2E}$.

Consider the load–elongation curve in Figure 16.11. The strain energy per unit volume present in a piece of material stressed up to its proportional limit (point *B*) is known as its *modulus of resilience* and is a measure of the toughness of the material.

$$SE/\text{unit volume} = \frac{\sigma \times \varepsilon}{2}$$

$$= \frac{\sigma^2}{2E} \left(\text{since } \varepsilon = \frac{\sigma}{E}\right)$$

where: $\sigma = $ stress
$\varepsilon = $ strain
$E = $ Young's modulus

$$\text{Unit} = \text{force} \times \text{distance/unit volume}$$
$$= \text{N m/m}^3$$
$$= \text{J/m}^3$$

The unit for the modulus of resilience of a material is therefore *joule per cubic metre* (J/m^3). Thus, the more work that has to be done to deform unit volume of a material up to its proportional limit, the greater the amount of stored strain energy present in the deformed material at that point.

SAMPLE PROBLEM 16/8

A tensile load of 25 kN is applied to a metal bar of length 2 metres and cross-sectional area 500 mm². If Young's modulus for the metal is 200 GPa, determine
(a) the strain energy present in the bar;
(b) the modulus of resilience of the metal.

Analysis:
The strain energy stored in the bar is equal to the work done in deforming the bar, which is equal to the average force causing elongation multiplied by the resultant elongation.
The modulus of resilience is equal to the strain energy per unit volume.

Calculation:
(i) Extension of bar

$$E = \frac{PL}{Ae}$$

$$\therefore e = \frac{PL}{AE}$$

$$\therefore e = \frac{25 \times 10^3 \times 2}{500 \times 10^{-6} \times 200 \times 10^9}$$

$$= 5 \times 10^{-4} \text{ m}$$

where: $P = 25 \times 10^3$ N
$L = 2$ m
$A = 500 \times 10^{-6}$ m²
$E = 200 \times 10^9$ Pa

(ii) Strain energy in bar = work done

$$\therefore SE = P_{av} \times e$$

$$= \frac{25 \times 10^3 \times 5}{2 \times 10^4}$$

$$= 6.25 \text{ J}$$

where: $P_{av} = (25 \times 10^3)/2$ N
$e = 5 \times 10^{-4}$ m

(iii) Modulus of resilience $= \dfrac{\text{strain energy}}{\text{volume of rod}}$

Original volume of rod $= 2 \times 500 \times 10^{-6}$
$= 10^{-3} \text{ m}^3$

Modulus of resilience $= \dfrac{6.25}{10^{-3}} \text{ J/m}^3$

$= 6.25 \text{ kJ/m}^3$

(*Note.* An alternative method for part (iii) would be to calculate the stress in the bar and then apply the formula SE $= \sigma^2/2E$).

Result:
The strain energy stored in the bar is 6.25 J and the modulus of resilience is 6.25 kJ/m^3. ■

FACTOR OF SAFETY

Although materials loaded within their elastic limits do not suffer permanent deformation, the maximum stress that is allowed to develop in a given member or component is always considerably less than the yield stress (or proportional limit stress, or proof stress) for that material. Some reasons for this are as follows.

(a) The difficulty of determining accurately the elastic limits of materials, and the consequent need to use other measures of "elastic" strength such as proof stress.
(b) The need to limit deformation in members or components; if one member of a structure is allowed a relatively large deformation, this may cause unnecessarily large stresses to be developed in adjoining members of the same structure. Thus, all members of a structure must be *compatible.*
(c) The difficulty of estimating the actual service loads to which a member or component will be subjected during its working life. In some instances, load estimation to within 10% may be acceptable for design work.
(d) The need to consider the rate of load application. Allowable stresses must be severely reduced, often by as much as 50%, if the working loads are suddenly applied or are fluctuating.

Table 16.4. Allowable Working Stresses for Some Metals.*

| Metal | Type of Service Load | | | |
	Dead load (stress in MPa)	Load gradually applied (stress in MPa)	Fluctuating load (stress in MPa)	Impact load (stress in MPa)
Low-carbon steel	137.5	90	70	45
Copper	60	50	40	27.5
Aluminium	40	35	27.5	17
Brass	80	70	55	35
Grey cast iron	35	20	13.5	9.5
Stainless steel	137.5	113	85	60

*These are approximations and are intended to provide comparisons only. They will vary greatly with variations in composition.

Where mass reduction is important, such as in the design of aircraft or high-rise buildings, allowable stresses are kept high and design calculations must be extremely accurate to ensure safety. However, where mass is of lesser importance, allowable stresses are reduced and more economical design procedures are adopted.

The ratio between the stress determined from routine tests on standardised specimens and the maximum allowable working stress in any given member or component is known as the *factor of safety* for that member or component.

For ductile materials

$$\text{factor of safety} = \frac{\text{yield stress}}{\text{allowable working stress}}$$

where the yield stress can be the stress as measured at a definite yield point (as for mild steel), the stress at the proportional limit, or proof stress.

For brittle materials

$$\text{factor of safety} = \frac{\text{ultimate stress}}{\text{allowable working stress}}$$

where the ultimate stress is the stress caused by the maximum test load experienced by the material in a routine mechanical test.

Factors of safety can vary between 2 for low-carbon steel carrying a dead load to 20 for timber subjected to an impact load.

SAMPLE PROBLEM 16/9

20 kN

Root diameter

An anchor bolt has a 20 kN tensile load applied to it as shown. Determine the minimum root diameter of the bolt if the allowable working stress is 140 MPa.

Problem 16/9

Analysis:
The area over which tensile stresses will act is the root area of the thread

$$A = \frac{\pi d^2}{4}$$

where d = the root diameter of the bolt.

Calculation:

$$\text{Stress in bolt, } \sigma = \frac{P}{A}$$

$$\therefore A = \frac{P}{\sigma}$$

$$= \frac{20 \times 10^3}{140} \text{ mm}^2$$

$$\therefore A = 142.8 \text{ mm}^2$$

where: $P = 20 \times 10^3$ N
$A = (\pi d^2)/4$
$\sigma = 140$ MPa

$$\therefore \frac{\pi d^2}{4} = 142.8$$

$$\therefore d^2 = \frac{142.8 \times 4}{\pi}$$

$$= 181.7 \text{ mm}^2$$

$$\therefore d = 13.47 \text{ mm}$$

Result:

The root diameter of the bolt would need to be about 13.5 mm. In practice, a standard-sized bolt having a root diameter somewhat greater than 13.5 mm would be used (M16 or bigger). ∎

SAMPLE PROBLEM 16/10

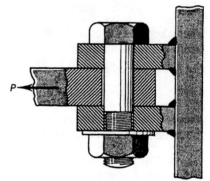

A 30-mm bolt holding three plates together in a machine is loaded in double shear as shown. Given that the factor of safety is 3, calculate the maximum value of the force P if the allowable shear stress in the bolt is 50 MPa.

Problem 16/10

Analysis:

Since the bolt is in double shear, the area of shear is twice the cross-sectional area of the bolt.

Calculation:

$$\text{Area of shear} = \frac{2\pi d^2}{4} \qquad \text{where: } d = 30 \text{ mm}$$

$$= \frac{2 \times 22 \times (30)^2}{7 \times 4}$$

$$= 1414 \text{ mm}^2$$

$$= 1414 \times 10^{-6} \text{ m}^2$$

Now,

$$\sigma = \frac{P}{A} \qquad \text{where: } \sigma = 50 \times 10^6 \text{ Pa}$$
$$\qquad\qquad\qquad\qquad A = 1414 \times 10^{-6} \text{ m}^2$$

$$P = \sigma A$$

$$= 50 \times 10^6 \times 1414 \times 10^{-6} \text{ N}$$

$$= 70\,700 \text{ N}$$

$$\therefore P = 70.7 \text{ kN}$$

Result:

The maximum value of the force P will be 70.7 kN. ∎

DIRECT STRESSES IN COMPOUND BARS

When two or more members are rigidly held together, they are said to be a *composite body*. For example, a concrete column reinforced with steel rods is a composite body in which the unique strength properties of each material in the composite are utilised to withstand the working loads imposed upon the composite member. Another name for this type of composite body is a *compound bar*.

The following points need to be considered when solving problems of compound bars.

(a) The members present *share* the working load, and the total load is equal to the sum of the loads taken by each member.

(b) Since they are rigidly fixed together, they must extend or compress *the same amount* when subjected to tensile or compressive loads.

(c) The load taken by each member of the compound bar is equal to the product of its stress and its area.

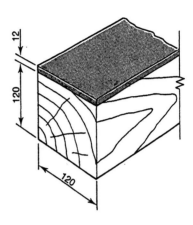

Problem 16/11

SAMPLE PROBLEM 16/11

A 120-mm square timber strut is reinforced by an aluminium plate of rectangular cross-section 120 mm × 12 mm. Calculate the stresses in the timber and the aluminium when the composite strut carries an axial load of 250 kN.

$$E_{timber} = 15 \times 10^3 \text{ MPa}$$
$$E_{aluminium} = 90 \times 10^3 \text{ MPa}$$

Analysis:
Let all knowns and unknowns for timber have the subscript t and those for aluminium, al.

Since the timber and aluminium plate form a compound bar, the total load is spread over the two components. Also, the strain in the timber will equal the strain in the aluminium, that is,

$$P_t + P_{al} = 250 \times 10^3 \text{ N}$$

and

$$\varepsilon_t = \varepsilon_{al}$$

Calculation: (loads are in newtons, areas in square millimetres and E in MPa)
(i) Calculation of loads

$$\varepsilon_t = \varepsilon_{al}$$

$$\frac{\sigma_t}{E_t} = \frac{\sigma_{al}}{E_{al}} \quad \text{(since } E = \frac{\sigma}{\varepsilon}\text{)}$$

$$\therefore \frac{P_t}{A_t E_t} = \frac{P_{al}}{A_{al} E_{al}} \quad \text{(since } \sigma = \frac{P}{A}\text{)}$$

However,

$$P_t + P_{al} = 250 \times 10^3 \text{ N}$$
$$\therefore P_{al} = (250 \times 10^3) - Pt$$

Substituting

$$\frac{P_t}{A_t E_t} = \frac{(250 \times 10^3) - P_t}{A_{al} E_{al}}$$

$$\therefore \frac{P_t}{120^2 \times 15 \times 10^3} = \frac{(250 \times 10^3) - P_t}{(120 \times 12) \times 90 \times 10^3}$$

where: $A_t = 120^2 \text{ mm}^2$
$E_t = 15 \times 10^3 \text{ MPa}$
$A_{al} = (120 \times 12) \text{ mm}^2$
$E_{al} = 90 \times 10^3 \text{ MPa}$

solving,

$$P_t = 156\ 250 \text{ N}$$

Substituting $P_t = 156\ 250$
in $P_t + P_{al} = 250 \times 10^3$
$$P_{al} = 250\ 000 - 156\ 250$$
$$\therefore P_{al} = 93\ 750 \text{ N}$$

(ii) Calculation of stresses

$$\sigma_t = \frac{P_t}{A_t}$$

$$= \frac{156\,250}{120^2} \text{ N/mm}^2$$

Result: stress in timber, $\sigma_t = 10.85$ MPa

$$\sigma_{al} = \frac{P_{al}}{A_{al}}$$

$$= \frac{93\,750}{120 \times 12} \text{ N/mm}^2$$

Result: stress in aluminium, $\sigma_{al} = 65.1$ MPa ∎

SAMPLE PROBLEM 16/12

A 500-mm × 300-mm concrete column, 1.5 metres long, is reinforced with two 30-mm diameter steel rods as shown. Determine the compressive stresses in the steel and the concrete and the contraction of the column if the column supports an axial compressive load of 60 tonnes.

$$E_{\text{steel}} = 200 \times 10^3 \text{ MPa}$$
$$E_{\text{concrete}} = 14 \times 10^3 \text{ MPa}$$

Analysis:

Total load $= 60 \times 10^3 \times 9.8$ N
$$= 588 \times 10^3 \text{ N}$$

Problem 16/12

Let the load supported by the steel rods be P_s and that supported by the concrete be P_c.
$$P_s + P_c = 588 \times 10^3 \text{ N}.$$
Since the column is a compound bar, the contraction of both steel and concrete is equal; therefore, their strains are equal, that is,

$$\varepsilon_s = \varepsilon_c$$

Area of steel, $A_s = 2 \times \dfrac{\pi d^2}{4}$ where: $d = 30$ mm

$$= \frac{2 \times 22 \times 30^2}{7 \times 4} \text{ mm}^2$$

$$= 1414 \text{ mm}^2$$

Area of concrete, $A_c = (500 \times 300) - 1414 \text{ mm}^2$
$$= (150\,000 - 1414) \text{ mm}^2$$
$$= 148\,586 \text{ mm}^2$$

Calculation:

(i) Proportions of loads on each member
$$P_s + P_c = 588 \times 10^3 \text{ N}$$
$$P_s = (588 \times 10^3) - P_c \text{ N}.$$

However,   strain in concrete = strain in steel
$$\varepsilon_c = \varepsilon_s$$
$$\therefore \frac{\sigma_c}{E_c} = \frac{\sigma_s}{E_s} \quad \left(\text{since } E = \frac{\sigma}{\varepsilon}\right)$$
$$\therefore \frac{P_c}{A_c E_c} = \frac{P_s}{A_s E_s} \quad \left(\text{since } \sigma = \frac{P}{A}\right)$$

Substituting $P_s = (588 \times 10^3) - P_c$ in the above equation

$$\frac{P_c}{A_c E_c} = \frac{(588 \times 10^3) - P_c}{A_s E_s}$$

$$\therefore \frac{P_c}{148\,586 \times 14 \times 10^3} = \frac{588 \times 10^3 - P_c}{1414 \times 200 \times 10^3}$$

$$\therefore P_c = 517.6 \times 10^3 \text{ N}$$

Substituting $P_c = 517.6 \times 10^3$ N   in $P_s = (588 \times 10^3) - P_c$

$$P_s = [(588 \times 10^3) - (517.6 \times 10^3)] \text{ N}$$

$$\therefore P_s = 60.4 \times 10^3 \text{ N}$$

(ii) Stresses in each member

$$\sigma_s = \frac{P_s}{A_s}$$

$$= \frac{60.4 \times 10^3}{1414} \text{ N/mm}^2$$

$$= 0.0427 \times 10^3 \text{ MPa}$$

$$\therefore \text{ stress in steel, } \sigma_s = 42.7 \text{ MPa}$$

$$\sigma_c = \frac{P_c}{A_c}$$

$$= \frac{517.6 \times 10^3}{148\,586} \text{ N/mm}^2$$

$$= 0.003\,48 \times 10^3 \text{ MPa}$$

$$\therefore \text{ stress in concrete, } \sigma_c = 3.48 \text{ MPa}$$

(iii) Calculation of the contraction

$$E = \frac{\sigma}{\varepsilon}$$

$$\therefore \varepsilon = \frac{\sigma}{E}$$

$$\frac{\Delta L}{L} = \frac{\sigma}{E} \quad (\text{since } \varepsilon = \frac{\Delta L}{L}) \qquad \text{where: } \Delta L = \text{contraction}$$

$$\therefore \Delta L = \frac{\sigma L}{E}$$

$$\Delta L_s = \frac{\sigma_s \times L_s}{E_s} \qquad\qquad \text{where: } \Delta L_s = \text{contraction}$$
$$\text{of steel}$$

$$= \frac{42.7 \times 10^6 \times 1.5}{200 \times 10^9} \text{ m}$$

$$= 32 \times 10^{-5} \text{ m}$$

$$\therefore \Delta L_s = 0.003\,2 \text{ mm}$$

$$\text{contraction of column} = 0.003\,2 \text{ mm}$$

Result:

$$\text{Compressive stresses} \begin{cases} \text{steel} & = 42.7 \text{ MPa} \\ \text{concrete} = 3.48 \text{ MPa} \end{cases}$$

$$\text{Contraction of column} = 0.003\,2 \text{ mm}$$

■

REVIEW PROBLEMS

(Use $g = 9.8$ m/s^2 unless otherwise specified.)

16/13

A press tool is required to punch out discs 42 mm diameter from steel sheet 3 mm thick. If the ultimate shearing stress of the sheet steel is 320 megapascals, calculate:

(i) the minimum force required to punch out a disc,

(ii) the average compressive stress set up in the punch.

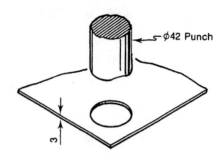

Problem 16/13

16/14

A 20 mm diameter bar has a tensile load of 36kN steadily applied to it. It is observed that a gauge length of 200 mm on the bar increased to 200.3 mm under the full load. If the elastic limit was not exceeded, determine

(i) the stress in the bar;

(ii) the strain in the bar;

(iii) the modulus of elasticity of the material.

16/15

A steel tie has a cross-section as shown in the diagram. Determine the stress in the section *A-A* if the tie is subjected to a tensile load of 20 kN.

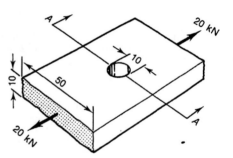

Problem 16/15

16/16

A press can exert a maximum force of 3.5 MN. What maximum thickness of sheet metal can be sheared by a cylindrical punch of diameter 100 m, if the ultimate shear strength of the sheet material is 400 MPa? What maximum compressive stress will be developed in the punch during this operation?

16/17

A hydraulic master cylinder is shown in the diagram.

(i) Determine the compressive force in the push-rod when the foot-pressure on the hydraulic brake pedal is 300 N.

(ii) What is the compressive stress in the piston if the master cylinder has a cross-sectional area of 600 mm^2?

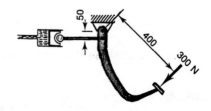

Problem 16/17

16/18
A tie rod in a machine is 3.140 m long and 40 mm in diameter. What is the extension under a load of 100 kN? The elastic modulus of the material is 200 GPa.

16/19
Some years ago airlines banned the wearing of stiletto heels by women because of damage to the aluminium alloy skins and floors of aircraft. Determine the approximate mass of a woman passenger whose heel just perforates the aluminium floor, given that the floor is 0.35 mm thick and its ultimate shear stress is 45 MPa. Assume that her full weight-force acted on one heel. The size of her heel is as shown in the diagram.

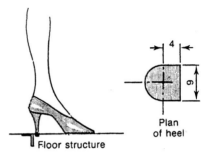

Plan of heel

Floor structure

Problem 16/19

16/20
Calculate the length of a steel tie rod of cross-sectional area 300 mm² which, when suspended vertically, contains a maximum stress of 100 kPa due to its own weight-force. The weight density of steel is 60 kN/m³.

16/21
Given that the shear stress in the bolt is not to exceed 150 MPa and that the maximum axial load to be applied to the rod coupling is 25 kN, calculate the minimum diameter bolt that should be used.
What size bolt should be used if a factor of safety of 2 is introduced in the design calculations?

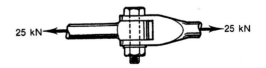

25 kN ⟵   ⟶ 25 kN

Problem 16/21

16/22

A wire 2.5 m long and of uniform circular cross-section is stretched 5 mm by the application of a load of 2 kN. Calculate the wire diameter, if the modulus of elasticity of the wire material is known to be 200 GPa.

16/23

A 5-metre length of 310 UBP 79 (steel universal bearing pile, mass 79 kg/m) has a cross-sectional shape as shown in the diagram. Determine the stress in the column (i) near the top and (ii) near the bottom, when it is supporting an axial compressive load of 800 kN. The cross-section of 310 UBP 79 has an area of 10 000 mm².

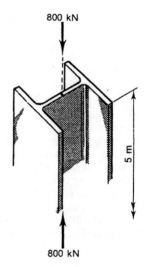

Problem 16/23

16/24

A compound bar is shown in the diagram. Determine:

(i) the tensile stress in the steel section of the bar and the tensile stress in the copper section of the bar;

(ii) the elongation of the bar if the modulus of elasticity of the steel is 200 GPa and that of the copper is 100 GPa.

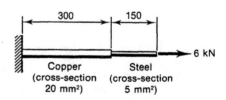

Problem 16/24

16/25

When a bolt is in tension, the load on the nut is transmitted through the root area of the bolt. An ISO metric bolt 24 mm in diameter (root area = 353 mm²) carries a tensile load. Find the percentage error in the calculated value of the stress if the shank area is used instead of the root area.

16/26

A quick-acting vice is operated by the air cylinder as shown in the diagram. The air pressure used to clamp the small object A in this vice is 75 kPa. Determine:

(i) the clamping force exerted on the object A;

(ii) the compressive stress in the air-cylinder piston, given that its cross-sectional area is 2000 mm^2;

(iii) the stress in the activating rod at B, given that it does not bend significantly and that its cross-sectional area is 300 mm^2 at this section;

(iv) the compressive stress in the toggle links, given that they are 15 mm square; and

(v) the shear stress in the pin at C, given that its diameter is 8 mm. (Each side of the pin is in single shear.)

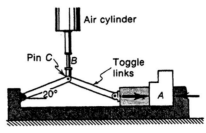

Problem 16/26

16/27

The permissible extension of a steel tie is 100 mm and the fractional strain in it is not to exceed 0.002. Calculate the maximum length of the tie.

16/28

The tractor shown in the diagram has a mass of 5 tonnes. Determine

(i) the maximum drawbar pull of the tractor if the coefficient of friction between wheels and the ground is 0.6;

(ii) the diameter of the drawbar hitch pin, given that the ultimate shear stress of the material of manufacture is 180 MPa and a factor of safety of 3 is to be used in its design.

Consider that the total weight force of the tractor acts on the two rear wheels when maximum drawbar pull is applied.

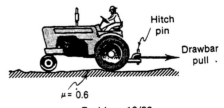

Problem 16/28

16/29

A pull of 1.25 kN is applied to a wire which operates a railway signal. The wire is 60 metres long and has a cross-section of 20 mm². If the movement of the end of the wire in the signal box is 250 mm, determine the length of movement of the end of the wire at the signal. The modulus of elasticity of the wire material is 200 GPa.

16/30

One side of the mechanism for a hydraulically operated loader is shown in the diagram. In removing a large rock from the face of the excavation, the loader bucket exerts a vertically upward force of 30 kN. Assuming that this force acts equally on both sides of the loader mechanism, determine

(i) the compressive stress in the lift-cylinder rod, given that its effective diameter is 50 mm;

(ii) the shear force on the bucket swivel pin *A* and the stress present within it, given that its diameter is 60 mm.

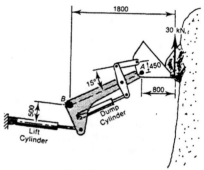

Problem 16/30

16/31

A tensile load of 5 kN is applied to an aluminium alloy bar of length 1 metre and of cross-sectional area 500 mm². Given that *E* for this material is 70 GPa, calculate

(i) the extension of the bar;

(ii) the total strain energy present in the deformed bar;

(iii) the modulus of resilience of the alloy;

provided that its proportional limit was not exceeded by the application of this load.

16/32
A 400-mm long steel bar of diameter 20 mm is placed concentrically inside a gunmetal tube of 22-mm inside diameter and 4-mm wall thickness. The length of the gunmetal tube exceeds the length of the steel bar by 0.1 mm. A compressive load is then applied to the ends of the gunmetal tube through the rigid plates as shown. Find
(i) the load needed to compress the tube to the same length as the steel bar;
(ii) the stress in the gunmetal and in the steel when an axial compressive load of 50 kN is applied.

$E_{steel} = 200$ GPa $\qquad E_{gunmetal} = 100$ GPa

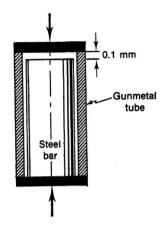

Problem 16/32

16/33
A concrete column reinforced by 4 steel rods is designed to carry a total load of 30 tonnes. If the combined cross-sectional area of the steel rods is 2000 mm², calculate the stress in the steel and in the concrete.

$E_{steel} = 200$ GPa $\qquad E_{concrete} = 14$ GPa

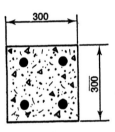

Problem 16/33

16/34
A large tipping trailer is shown. Its 15-tonne load of soil is stable but on the verge of tipping. Given that G, the combined centre of mass of the trailer plus load, is 500 mm above the pin B, determine the forces in the hydraulic ram AC and in the two ties AB (note that one lies directly behind the other). The mass of the empty trailer is 5 tonnes.

The ties are made from thick-walled pipe, outside diameter 75 mm, inside diameter 55 mm. Determine the tensile stress in these ties when in the position shown. Given that the elastic modulus for the tie material is 200 GPa, determine their strains.

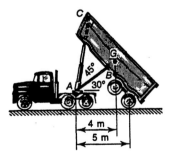

Problem 16/34

17

Beams and Bending

The ore bridge at the BHP, Newcastle, is used to unload iron ore and limestone at more than 1000 tonnes per hour. The cantilevered section of these huge crane beams can be raised to clear the ship's superstructure.

Beams are simple structural members used in building and other engineering applications; they are comparatively long and slender and are usually used as the horizontal members of structures. Besides providing some of the basic shape of the structure in which they are used, the primary function of any beam is to *support* and *resist* any loads which may be impressed upon it. These imposed loads include the weight-force of the beam itself and the weight-force of any other structural component or object supported on or by the beam. As well as these *dead loads*, variable loads such as those created by winds acting upon the structure or by people moving around inside a structure such as a building must be considered. These variable loads are termed *live loads* and often are of the greatest importance in design calculations for beams and other structural components.

As an example of a structure using a simple horizontal beam, consider the cross-section of the typical timber floor shown in Figure 17.1 (a). In this case the dead load acting upon the main structural beam, known as the *bearer*, is the weight-force of the rest of the floor; that is, the combined weight-forces of the floor joists and flooring timber. Once these weight-forces are determined, a force diagram such as that shown in Figure 17.1 (b) can be drawn and the design calculations needed to determine the cross-section of this bearer can begin.

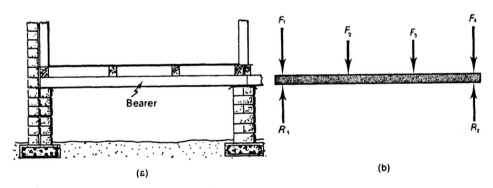

Figure 17.1 (a) Section through floor showing timber bearer in position. (b) The bearer as a free body. The reactions R_1 and R_2 are provided by the wall and pier, and the forces F_1, F_2, F_3 and F_4 are weight-forces acting on the bearer.

All beams bend or *deflect* in service under the influence of applied loads, including their own weight-force. Thus, when designing a beam such as the floor bearer shown in Figure 17.1 it is important to consider possible *deflection* under load as well as the ability of the beam to support the applied load. For example, a bearer may support the load of the floor adequately but may deflect too much in doing so and we would say that the floor structure *lacks rigidity*. Obviously a much larger bearer must be used to reduce this deflection to an acceptable amount. Thus, while both bearers could be said to be strong enough to support the floor, the first one lacks the necessary rigidity to reduce deflection to an acceptable amount. This type of rigidity in beams can be increased by altering the sectional size and shape of the beam.

However, as this chapter is concerned with bending moments, shear stresses and direct stresses induced in beams by bending moments, deflections of beams are not considered here and reference to more advanced texts is necessary for information on this aspect of beam design.

TYPES OF BEAMS

Horizontal beams can be grouped into four different types depending upon the way in which a beam is supported in service.

Simple beams are supported at their ends only; the bearer shown in Figure 17.1 (a) and the painter's plank shown in Figure 17.2 (a) being examples of simple (or simply-supported) beams. A variation of the simple beam is the beam that is supported at a number of places along its length. For example, the steel beam shown in Figure 17.2 (b) is centrally supported by a steel post as well as being supported at its ends by the brick walls. The use of this type of central support means that a beam of smaller cross-section (that is, of less rigidity) can be used and cost thus may be reduced.

Cantilevers are beams that are supported at one end only, often by that end being built into a wall or other rigid structure capable of providing a *reaction moment* (refer back to Chapter 7, "Reactions at Supports"). The diving board shown in Figure 17.2 (c) and the light support shown in Figure 17.2 (d) are both examples of cantilevered beams.

Some other beams are termed *overhanging* because they have at least one support some distance in from one end of the beam. This is illustrated by the different type of diving board shown in Figure 17.2 (e) and the decking joist shown in Figure 17.2 (f).

While beam calculations* differ slightly depending upon which one of the above types of beams is being analysed, a common starting point is to consider

(i) the pattern of external loading;
(ii) the reactions at the supports of the beams;
(iii) the internal forces that the beam must provide at various points along its length in order to resist the external force system set up by (i) and (ii) above.

---

*Only statically determinate beams are analysed in this chapter. Statically determinate beams and structures are those in which reactions and internal forces in members can be completely determined by the force analysis alone. For the more complicated analysis of statically indeterminate beams and structures reference to a more advanced text is necessary.

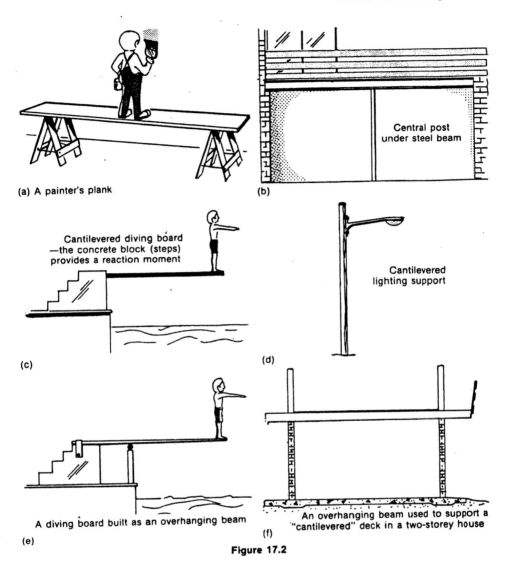

(a) A painter's plank

(b)

Central post
under steel beam

Cantilevered diving board
—the concrete block (steps)
provides a reaction moment

(c)

Cantilevered
lighting support

(d)

A diving board built as an overhanging beam

(e)

An overhanging beam used to support a
"cantilevered" deck in a two-storey house

(f)

Figure 17.2

SHEAR FORCE AND BENDING MOMENT

Suppose that the painter's plank already shown in Figure 17.2 (a) is 3 metres long and is supporting a painter of mass approximately 90 kilograms standing 1 metre from the left-hand trestle (refer to Figure 17.3 (a)). If the weight-force exerted by the painter is taken to be 900 N and the weight-force of the plank is considered to be negligible, then, *by inspection*, the reactions R_1 and R_2 at the two trestles are seen to be 600 N and 300 N respectively. Thus, the complete force diagram for this plank is as shown in Figure 17.3 (b).

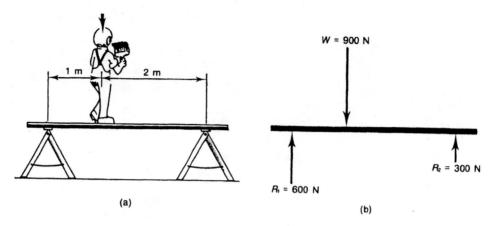

Figure 17.3   A painter standing on a plank. The reactions R_1 and R_2 can be seen by inspection or, alternatively, they can be calculated by taking moments about R_1 and R_2 respectively.

This painter's plank, since it is supporting the painter, must obviously be providing *internal resistance* to this external force system. In order to determine the internal resistance at a certain point along this plank, let us suppose that the plank is sectioned as shown at its mid-point and the two halves of the plank are moved slightly apart (refer to Figures 17.4 (a) and (b)).

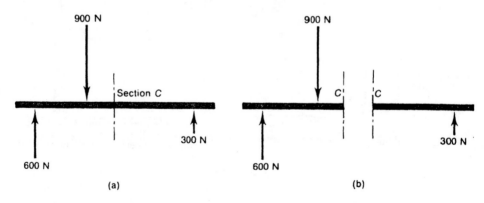

Figure 17.4   (a) The painter's plank is sectioned at C, its mid-point. (b) The two halves are then moved slightly apart.

Now, the left- and right-hand portions of this plank can be considered as *free-bodies* in their own right and thus the forces necessary to maintain equilibrium at each section can be readily determined.

Equilibrium of the Left-hand Portion of the Plank

It is obvious that, to maintain *vertical equilibrium* of the left-hand portion, a vertical force, F_V, as shown in Figure 17.5 (a) must be introduced at the section itself. Thus, since $\Sigma F_V = 0$ in all instances of vertical equilibrium, and considering upward forces to be positive,

$$\Sigma F_V = 0 \;{}_{+}\uparrow$$
$$600 - 900 + F_V = 0$$
$$\therefore F_V = 300 \text{ N}$$

Thus, vertical equilibrium of the left-hand portion of the plank is maintained if a vertically upward force of 300 N acts along the mid-section itself.

However, this left-hand portion of the plank is not in *moment equilibrium* since the sum of the moments taken about any point on the left-hand portion is not zero. Thus a moment, M, can be introduced at the mid-section and if it has the necessary magnitude and sense it will provide for the moment (rotational) equilibrium of the left-hand section. This is shown in Figure 17.5 (b) where M is arbitrarily shown as an anti-clockwise moment. Thus, since $\Sigma M = 0$ in all instances of moment equilibrium of free-bodies, and considering anti-clockwise moments to be positive,

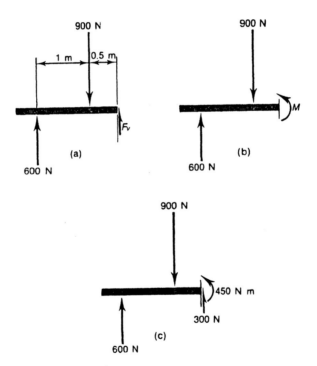

Figure 17.5 The left-hand half of the painter's plank is considered as a free-body: (a) shows that a vertical force, F_V, must be introduced at the section in order to provide vertical equilibrium; (b) shows the moment M necessary for moment equilibrium; and (c) shows F_V and M in terms of their calculated values.

$$\Sigma M = 0 \;\overset{+}{\curvearrowleft}$$

Moments about mid-section
$$(-600 \times 1.5) + (900 \times 0.5) + M = 0$$
$$\therefore M = 450 \text{ N m} \;\curvearrowright$$

Thus, for moment equilibrium the moment, M, placed at the section on the left-hand portion of the plank must have a magnitude of 450 N m and must act anti-clockwise.

Therefore, for the left-hand portion of the painter's plank to be in equilibrium a vertically upward force of 300 N and an anti-clockwise moment of 450 N m must be introduced at the cut section, that is, at the mid-point of the plank. This is shown in Figure 17.5 (c).

Equilibrium of the Right-hand Portion of the Plank

The right-hand portion shown in Figure 17.6 (a) must also be unbalanced both vertically and rotationally. Thus a suitable force, F_V, and a suitable moment, M, must be introduced at the section itself if equilibrium is to be attained.

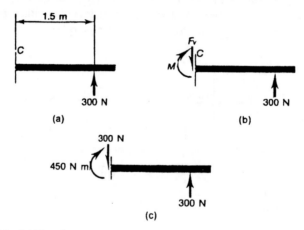

Figure 17.6 (a) The right-hand portion of the painter's plank. (b) Diagram showing how a vertical force F_V and a moment M need to be introduced at the section C in order to maintain equilibrium. (c) The complete free-body diagram showing all forces and moments.

Suppose that F_V and M are introduced at this section as shown in Figure 17.6 (b). Considering the right-hand portion as a free-body:

$$\Sigma F_V = 0 \quad +\uparrow$$
$$- F_V + 300 = 0$$
$$\therefore F_V = 300 \text{ N acting } vertically \ down$$
$$\Sigma M = 0 \ \circlearrowright$$
$$+ (300 \times 1.5) - M = 0$$
$$\therefore M = 450 \text{ N m acting } clockwise$$

Thus, for the right-hand portion of the painters plank to be in equilibrium, a vertically downward force of 300 N and a clockwise moment of 450 N m must be introduced at the cut section. This is shown in Figure 17.6 (c).

If the two portions of the painter's plank are now placed together again so that the plank is once again *continuous*, it is obvious that internal opposing vertical forces of 300 N and internal opposing moments of 450 N m are acting on this mid-section already considered.

The internal opposing vertical forces are a *reaction* to the shearing action of the external force system as it affects the mid-section of the plank, and are known as the *shear force* present at that section.

The internal moment is a *reaction* to the bending effect of the external force system as it affects the mid-section of the plank, and is known as the *bending moment* present at that section.*

SAMPLE PROBLEM 17/1

Determine the shear force and bending moment at points 1 metre and 4 metres from the left-hand support of the simple beam shown. Assume that the beam has negligible mass.

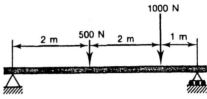

Problem 17/1

Analysis:
The shear force and bending moment at any given section of a beam are the internal reactions at that section and are found by considering one portion of the beam as a free-body.

In this problem the first step is to determine the reaction at the supports and then the shear force and bending moment at each section can readily be found by considering in each case the left-hand portion of the sectioned beam.

Calculation of Reactions:
Let reactions be R_1 and R_2 as shown.

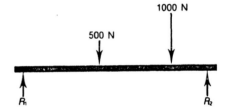

Consider moments about R_2
$\Sigma M = 0$ Assume $\overset{+}{\curvearrowright}$
$(R_2 \times 0) - (1000 \times 1) - (500 \times 3) + (R_1 \times 5) = 0$
$$\therefore R_1 = 500 \text{ N} \uparrow$$

Consider moments about R_1
$\Sigma M = 0$ Assume $\overset{+}{\curvearrowright}$
$(R_1 \times 0) + (500 \times 2) + (4 \times 1000) - 5R_2 = 0$
$$\therefore R_2 = 1000 \text{ N} \uparrow$$

---

*In this chapter the method employed to calculate the shear force and the bending moment at a section of a rigid bar assumes that they are the *equilibrants* necessary to balance the external force system as it affects that section of the bar. An often used but no more valid procedure is to calculate shear force and bending moments as the *resultants* of the external force system at any given section. The numerical figure for shear force and bending moment obtained by either method is, of course, the same, since any given section of a rigid bar is in equilibrium unless the bar collapses under the applied loads. The directions of the external resultants (shear and bending) oppose those of the internal equilibrants (shear and bending); this is of no consequence, provided a consistent procedure together with an adequate sign convention for positive and negative shear forces and bending moments are used in calculations.

Check $$\Sigma F_V = 0; \qquad R_1 + R_2 = 1500 \text{ N}$$
$$\therefore 1000 + 500 = 1500 \text{ N}$$
$\therefore R_1 + R_2$ are correctly determined.

Calculations of Shear Forces and Bending Moments:
(a) Consider the section 1 metre from the left-hand support.

Introduce a vertical force, F_V, and a moment, M, at this section to provide for equilibrium of this left-hand section.

$\Sigma F_V = 0$ Assume F_V acts vertically upward and that vertically upward forces are positive.
$$500 + F_V = 0$$
$$\therefore F_V = -500 \text{ N}$$

$\Sigma M = 0$ Assume M acts anti-clockwise and that anti-clockwise moments are positive.
Take moments about the cut section.
$$-(500 \times 1) + M = 0$$
$$\therefore M = 500 \text{ N m}$$

Thus the shear force 1 metre along the bar is 500 N and the bending moment is 500 N m. (Note that the original direction assumed for F_V was incorrect, thus the answer was negative at the end of the calculation. However, the correct sense was assumed for the moment M).

(b) Consider the section 4 metres from the left-hand support.

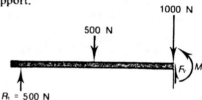

Introduce a vertical force and a moment M at the section as shown to provide equilibrium of the left-hand section.
$\Sigma F_V = 0$ Assume F_V acts vertically upward and that vertically upward forces are positive.
$$+500 - 500 - 1000 + F_V = 0$$
$$\therefore F_V = 1000 \text{ N}$$

$\Sigma M = 0$ Assume M acts anti-clockwise and that anti-clockwise moments are positive.
Take moments about the section
$$-(500 \times 4) + (500 \times 2) + (1000 \times 0) + M = 0$$
$$\therefore M = +1000 \text{ N m}$$

Thus the shear force at the point of application of the 1000 N force is 1000 N, while the bending moment is 1000 N m. (Note that the assumed direction of the bending moment was correct since the answer is positive in our calculation.)

Answers:
(a) At section 1 m from left-hand support:
$$\text{shear force} = 500 \text{ N}$$
$$\text{bending moment} = 500 \text{ N m}$$
(b) At section 4 m from left-hand support:
$$\text{shear force} = 1000 \text{ N}$$
$$\text{bending moment} = 1000 \text{ N m}$$

■

Sample Problem 17/1 illustrates the fact that values for shear force and bending moment can vary along the length of any given beam. As will become apparent, this variation will depend upon

(i) the type of beam and its supports, and

(ii) the type of external force system present.

Thus it is obvious that the *distribution* of shear forces and bending moments along the length of a beam need to be plotted before design calculations for that particular beam can be undertaken. Obviously the *maximum values* of shear force and bending moment are of considerable importance in such calculations and these maxima can be readily identified from shear force and bending moment diagrams.

SIGN CONVENTION FOR SHEAR FORCE AND BENDING MOMENTS

In all calculations of shear force and bending moments in beams it is essential that a consistent system of sign conventions be adopted for forces and moments. This is particularly true if three-dimensional analysis of a beam or a framework structure is to be carried out on a computer. Even though three-dimensional force systems and computer analysis are outside the scope of this book, the following sign convention is consistent with the latest techniques and is particularly suited to computer programming.*

The first step in assigning either positive or negative signs to internal forces and moments in a beam is to draw an x-axis through the beam. Since we are dealing only with horizontal beams in all cases, this x-axis is to be drawn horizontally from left to right (see Figure 17.7 (a)). Any axial forces present (tension or compression) will be along this x-axis, tension being positive and compression being negative.

Now, if the beam is sectioned at C and only *the left-hand piece is considered* (see Figure 17.7 (b)):

(i) positive y is vertically upwards and shear forces acting in this direction are always considered to be positive;

(ii) positive bending moments act anti-clockwise about the section as shown.

Similarly, *if the right-hand piece is considered* (see Figure 17.7 (c)):

(i) positive shear forces are always considered to act downwards; and,

(ii) positive bending moments are clockwise about the section.

Figure 17.7 (d) thus presents the simplified convention that this book adopts for positive shear forces and bending moments at a section in a rigid bar. *The reader should note that this convention is not widely adopted in elementary texts of engineering mechanics* but is consistent with current engineering practice.

---

*The authors wish to make special acknowledgement to Professor A. S. Hall, Professor of Civil Engineering, University of New South Wales, who features this type of convention in his book, *An Introduction to the Mechanics of Solids*, John Wiley, Sydney, 1st ed. 1969, SI ed. 1973.

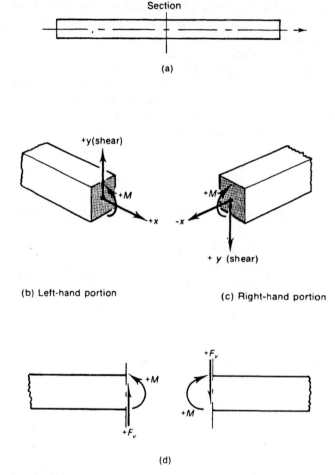

(a)

(b) Left-hand portion

(c) Right-hand portion

(d)

Figure 17.7 Sign convention for positive values of axial force, shear force and bending moment in a beam. (d) The simple two-dimensional correction adopted in this book for positive bending moment and shear force in any beam.

SHEAR FORCE DIAGRAMS

A shear force (SF) diagram is a graphical method of representing the variation of shear force along the length of a loaded beam. To construct such a diagram, *remember* that the shear force at a section of a loaded beam is the vertical equilibrant force that must be provided at that section in order to maintain vertical equilibrium if one part of the sectioned beam is considered as a *free-body*.

Consider a cantilever loaded at its free end with the force P (as in Figure 17.8 (a)). The shear force at any section C along its length is equal to $+ P$ units, this being readily seen when the left-hand portion of the cantilever is considered as a free-body, as in Figure 17.8 (b). Thus, the shear force diagram for this cantilever, loaded as shown, illustrates uniform positive shear force along its entire length. This is shown in Figure 17.8 (c).

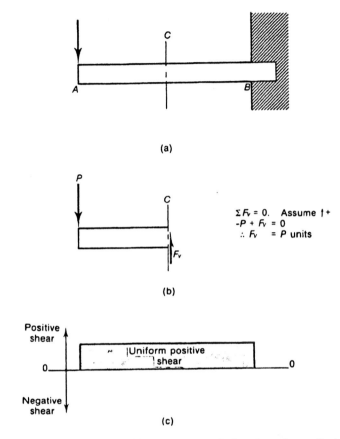

(a)

$$\Sigma F_v = 0. \quad \text{Assume } \uparrow +$$
$$-P + F_v = 0$$
$$\therefore F_v = P \text{ units}$$

(b)

(c)

Figure 17.8 The cantilever *AB* is sectioned at *C* and the shear force F_v is introduced at the section as in (b). From $\Sigma F_v = 0$, the shear force F_v must equal +*P* units and, since *C* can lie anywhere between ends *A* and *B*, the shear force along the loaded cantilever must be uniform and equal to +*P* newtons.

General Method for Constructing Shear Force Diagrams
(Refer to Sample Problem 17/2.)
(1) Draw the beam showing all forces, including reactions. If necessary, calculate the reactions using the equations of equilibrium, or determine them graphically.
(2) Draw the line of zero shear about which the shear force diagram will be constructed.
(3) Commence at the left-hand end of the beam and work to the right, keeping in mind that positive shear forces act vertically upward for the left-hand section. Always assume that the shear force does act vertically upward for the left-hand

section; thus, a negative sign in your answer will indicate a negative shear force at that section.

(4) Calculate the shear forces at all necessary sections, including at reactions and between all adjacent forces. (You will find it useful to know that, at the point of application of a concentrated load, the shear force changes by an amount equal to the magnitude of that load.)

(5) Draw the shear force diagram on the previously drawn zero line, placing it in projection underneath the diagram of the loaded beam.

SAMPLE PROBLEM 17/2

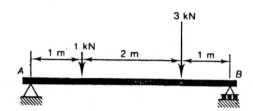

Draw the shear force diagram for the beam AB loaded as shown.

Problem 17/2

Analysis:

(i) The reactions at each end of the beam, R_A and R_B, must be determined.

(ii) The shear forces at the points of application of the external forces and at sections between R_A and the 1-kN force; between the 1-kN and the 3-kN forces; and between the 3-kN force and the reaction R_B must be calculated from $\Sigma F_V = 0$.

Calculations:

(i) Reactions

For reaction R_A, take moments about R_B and assume that anti-clockwise moments are positive.
$$(R_B \times 0) + (3 \times 1) + (1 \times 3) - (R_A \times 4) = 0$$
$$4R_A = 6$$
$$R_A = 1.5 \text{ kN}$$

For reaction R_B, take moments about R_A and assume that anti-clockwise moments are positive.
$$(R_A \times 0) - (1 \times 1) - (3 \times 3) + (R_B \times 4) = 0$$
$$4R_B = 10$$
$$\therefore R_B = 2.5 \text{ kN}$$

(ii) Shear Forces

Considering the left-hand section in all instances, and assuming that the shear force F_s is acting upwards in all instances and that vertically upward forces are positive.

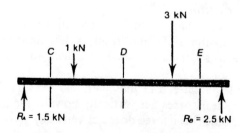

Shear force at section C:
$$1.5 + F_s = 0$$
$$F_s = -1.5 \text{ kN}$$

Shear force at 1-kN force:
$$1.5 - 1 + F_s = 0$$
$$F_s = -0.5 \text{ kN}$$

Shear force at section D:
$$1.5 - 1 + F_s = 0$$
$$F_s = -0.5 \text{ kN}$$

Shear force at 3-kN force:
$$1.5 - 1 - 3 + F_s = 0$$
$$F_s = +2.5 \text{ kN}$$

Shear force at section E:
$$1.5 - 1 - 3 + F_s = 0$$
$$F_s = +2.5 \text{ kN}$$

Result:
The shear force diagram is shown below, and several points deserve special attention.

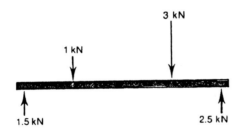

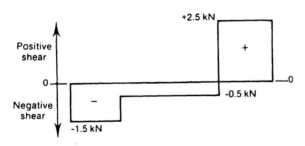

Problem 17/2

(i) Since section C can be anywhere between R_A and the 1-kN force, it follows that the shear force is constant along the bar from R_A to the 1-kN force.

Similar conclusions can be drawn for the shear force between the 1-kN and 3-kN forces and between the 3-kN force and the reaction R_B.

(ii) At the point of application of the 1-kN force the shear force changes by 1 kN, and at the point of application of the 3-kN force the shear force changes by 3 kN. Thus *it was really unnecessary to calculate the shear forces present at the points of application of these concentrated loads.* ∎

BENDING MOMENT DIAGRAMS

A bending moment (BM) diagram is a graphical method of representing the variation of bending moment along the length of a loaded beam. To construct this diagram, *remember* that the *bending moment* acting at a section of a loaded beam is the equilibrant moment that must be applied at that section in order to maintain rotational equilibrium if one part of the sectioned beam is considered as a *free-body*.

Sign convention: For the left-hand section, the BM is positive if it acts anti-clockwise at the section itself (refer back to Figure 17.7).

According to the adopted sign convention, a horizontal beam that has a positive bending moment will *sag* or suffer a downward deflection. This is the normal type of deflection that occurs in a horizontal beam which supports vertical loads.

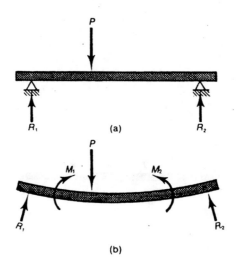

Figure 17.9 The beam shown in the diagram has a positive bending moment along its length. The effect of the external force system which consists of the forces R_1, P and R_2 causes this beam to deflect downwards or sag. In fact, the effect of the external force system is similar to that of the opposing external moments M_1 and M_2. However, note that, according to the system used in this book, the bending moment is the internal reaction moment within the beam.

General Method for Constructing Bending Moment Diagrams

(1) Draw the beam (usually to scale) showing all forces, *including reactions*. If necessary, calculate the beam reactions using the equations of equilibrium, or determine them graphically.

(2) It is usually most convenient to commence at the left-hand end of the beam, keeping in mind that positive bending moment is anti-clockwise for the left-hand section. Always assume that the bending moment acts anti-clockwise for the left-hand section; thus a negative answer indicates a negative bending moment at that section.

(3) Calculate the bending moments at the points of application of all concentrated forces and remember that on unloaded sections the bending moment is represented by a straight line. (Refer to Figure 17.12 and Table 17.1 to see how uniformly distributed loads are treated.)

(4) Draw the bending moment diagram about a horizontal base or zero line, in projection underneath the (scaled) drawing of the beam or underneath the shear force diagram if one has been constructed. Positive bending moments are to be plotted vertically upwards from this base or zero line.

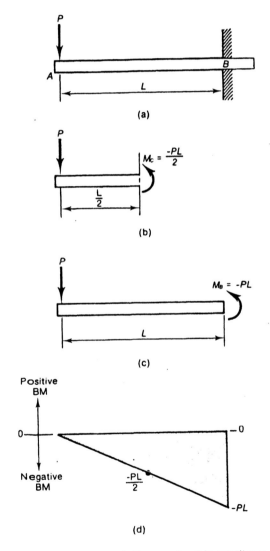

Figure 17.10 The bending moment diagram for this cantilever shows that the bending moment is always zero at a free end and that it reaches its maximum at the point of support.

As a simple example, refer to the cantilever of length L units illustrated in Figure 17.10 (a) which is loaded at its free end with the force P.

Working from the left-hand end and always considering the left-hand section:

Let the BM at the free end A be equal to M_A.

Take moments about A and assume that anti-clockwise moments are positive.

$$\Sigma M = 0$$
$$\therefore M_A + (P \times 0) = 0$$
$$\therefore M_A = 0$$

At section C, halfway along the cantilever:

Let the BM be equal to M_C as shown in Figure 17.10 (b).

Take moments about C and assume that anti-clockwise moments are positive.

$$\Sigma M = 0$$
$$M_C + \left(P \times \frac{L}{2} \right) = 0$$
$$M_C = - \frac{PL}{2}$$

At the supported end B:

Let the BM be equal to M_B as shown in Figure 17.10 (c).

Take moments about B and assume that anti-clockwise moments are positive.

$$\Sigma M = 0$$
$$\therefore M_B + (P \times L) = 0$$
$$\therefore M_B = - PL$$

The BM diagram for the cantilever loaded at its free end appears in Figure 17.10 (d). It should be noted that bending moment diagrams are normally drawn underneath the shear force diagram if one has been drawn. However, they have been separated up to now in an attempt to clarify the concepts involved in each.

SAMPLE PROBLEM 17/3

Draw the bending moment diagram for the beam in Sample Problem 17/2.

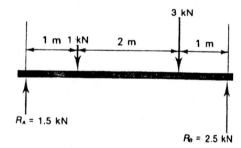

Problem 17/3

Analysis:

Since the reactions R_A and R_B have already been calculated, they are simply shown as (reaction) forces acting on the beam.

Now, bending moments need to be calculated at the point of application of each load, including reactions.

Use the left-hand sections in all calculations and assume that anti-clockwise moments are positive.

Calculations:
(i) At R_A $\Sigma M = 0$

$$\therefore M_A + (R_A \times 0) = 0$$
$$\therefore M_A = 0$$

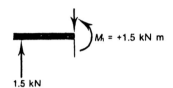

$M_A = 0$

(ii) At 1-kN force $\Sigma M = 0$

$$\therefore (1 \times 0) + M_1 - (1.5 \times 1) = 0$$
$$\therefore M_1 = + 1.5 \text{ kN m}$$

$M_1 = +1.5$ kN m

1.5 kN

(iii) At 3-kN force $\Sigma M = 0$

$$\therefore (3 \times 0) + M_3 + (1 \times 2) - (1.5 \times 3) = 0$$
$$\therefore M_3 = + 2.5 \text{ kN m}$$

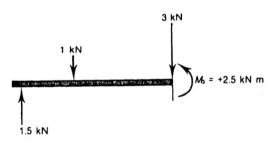

3 kN

1 kN

$M_3 = +2.5$ kN m

1.5 kN

(iv) At R_B $\Sigma M = 0$

$$\therefore (2.5 \times 0) + M_B + (3 \times 1) + (3 \times 1) - (1.5 \times 4) = 0$$
$$\therefore M_B = 0$$

$M_B = 0$

Answer:

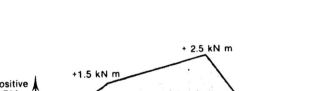

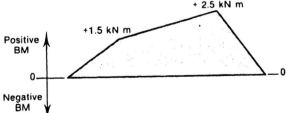

2.5 kN

+ 2.5 kN m

+1.5 kN m

Positive
BM

0

0

Negative
BM

Notes:

(i) It would obviously have been simpler to calculate the bending moment at the end *B* by considering the right-hand section.

(ii) The positive bending moment means that the beam *sags* under the applied loads. Do not become confused between this fact and the technique of plotting positive bending moment vertically *up* (that is, in the *positive* sense along the *y*-axis of the graph).   ■

SHEAR FORCE AND BENDING MOMENTS FOR UNIFORMLY DISTRIBUTED LOADS

Consider the beam *AB* shown in Figure 17.11 which carries a uniformly distributed load of *P* kilonewtons per metre and is *L* metres in length.

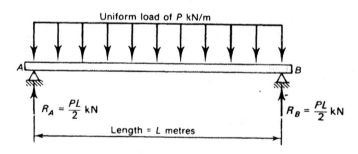

Figure 17.11

To determine the reactions R_A and R_B the total load of *PL* kN can be assumed to be concentrated at the midpoint of the beam; thus $R_A = R_B = \dfrac{PL}{2}$ kN.

Shear Force: The shear force will be at a maximum at each end of the beam and will be zero at the midpoint where it changes from negative to positive.

Bending Moment: The BM will be zero at the ends and will be at a maximum at the midpoint; it will also be *positive*, that is, the beam *sags* under the applied load.

The shear force and bending moment diagrams for this beam are given in Figure 17.12 and the necessary calculations are shown in Table 17.1.

Table 17.1.

UDL of P kN/m over length L

A $\frac{PL}{2}$ B $\frac{PL}{2}$

Shear Force Diagram
For all calculations assume $\uparrow+$ and shear force is $\uparrow+$

Bending Moment Diagram
For all calculations assume $\smile +$ and BM $= +$

At end A:

$R_A + F_s = 0$

$\therefore F_s = -\dfrac{PL}{2}$ N

At end A:

$M + (R_A \times 0) = 0$

$\therefore M = 0$

At section C:

$\dfrac{PL}{2} - \dfrac{PL}{4} + F_s = 0$

$\therefore F_s = -\dfrac{PL}{4}$ N

At section C:

$M + \left(\dfrac{PL}{4} \times \dfrac{L}{8}\right) - \left(\dfrac{PL}{2} \times \dfrac{L}{4}\right) = 0$

$\therefore M = \dfrac{3PL^2}{32}$ kN m

At section D:

$\dfrac{PL}{2} - \dfrac{PL}{4} + F_s = 0$

$\therefore F_s = 0$

At section D:

$M + \left(\dfrac{PL}{2} \times \dfrac{L}{4}\right) - \left(\dfrac{PL}{2} \times \dfrac{L}{2}\right) = 0$

$\therefore M = \dfrac{PL^2}{8}$ kN m

At section E:

$\dfrac{PL}{2} - \dfrac{3PL}{4} + F_s = 0$

$\therefore F_s = \dfrac{PL}{4}$ N

At section E:

$M + \left(\dfrac{3PL}{4} \times \dfrac{3L}{8}\right) - \left(\dfrac{PL}{2} \times \dfrac{3L}{4}\right) = 0$

$\therefore M = \dfrac{3PL^2}{32}$ kN m

At end B:

$\dfrac{PL}{2} - PL + F_s = 0$

$\therefore F_s = \dfrac{PL}{2}$ N

At end B:

$M + \left(\dfrac{PL}{2} \times 0\right) + \left(PL \times \dfrac{L}{L}\right) - \left(\dfrac{PL}{2} \times L\right) = 0$

$\therefore M = 0$

Note: The right-hand section could have been used to simplify calculations for shear and BM at section E and end B.

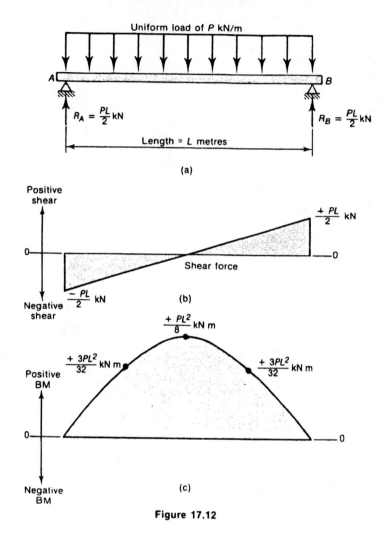

Figure 17.12

Any beam sags under its own weight-force and the weight-force of a beam is, in fact, another example of a *uniformly distributed load* producing a bending moment and shear force within the beam itself.

Generally the weight-force of a beam is negligible compared with the loads supported by the beam; however, in some situations the weight-force must be considered when determining total shear and bending at any given section within that beam.

Table 17.2 shows some typical beams together with their shear force and bending moment diagrams. The position of the maximum bending moment, together with its value, is clearly seen and the formula used to calculate the maximum deflection of the beam is also given.

Table 17.2. Shear Force and Bending Moment Diagrams
and Maximum Deflection for Some Beams.

	CANTILEVER		SIMPLY SUPPORTED		FIXED END SUPPORTS	
	point load	UDL	point load	UDL	point load	UDL
CASE						
SHEAR FORCE						
BENDING MOMENT						
MAXIMUM DEFLECTION	$\dfrac{1}{3}\dfrac{W\ell^3}{EI}$	$\dfrac{1}{8}\dfrac{w\ell^4}{EI}$	$\dfrac{1}{48}\dfrac{W\ell^3}{EI}$	$\dfrac{5}{384}\dfrac{w\ell^4}{EI}$	$\dfrac{1}{192}\dfrac{w\ell^3}{EI}$	$\dfrac{1}{384}\dfrac{w\ell^4}{EI}$

Notes:

(1) In all instances, the weight-force of the beam is assumed to be negligible compared with the supported loads.

(2) SF and BM diagrams are drawn according to the conventions adopted in this chapter.

(3) No proof is offered for the formulas for maximum deflections; however, engineers are usually concerned with maximum bending moments together with the associated beam deflections and thus these formulas are offered for comparison value only.

(4) Most beams are designed on the basis of bending and the resulting cross-section so obtained is then checked to see if it will be strong enough in shear.

(5) The significance of values for I, the second moment of area, are discussed in the remainder of this chapter.

STRESSES DUE TO BENDING

It has already been shown that shear forces and bending moments are set up in beams which are subjected to an external force system and that the variation in intensity of shear force and bending moment for any given beam can be plotted using shear force and bending moment diagrams.

The presence of bending moments in a beam causes *internal stresses* to be set up within various parts of the beam; these stresses are known as *bending stresses* and their presence can be readily demonstrated by taking a piece of flexible rubber at least 15 mm thick by about 10 mm wide and gluing strips of tissue paper along its edges. If, when the glue has set, this rubber beam is bent as shown in Figure 17.13, you will observe that the tissue paper is torn on the outer edge since it is placed in tension, while on the inner edge the paper is wrinkled since it is placed in compression. Thus, tensile stresses were obviously set up in parts of this beam due to bending, while compressive stresses were set up in other parts which were on the opposite side of the central axis.

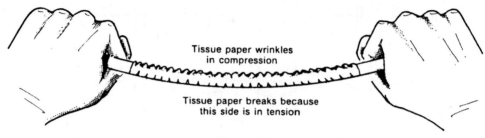

Tissue paper wrinkles
in compression

Tissue paper breaks because
this side is in tension

Figure 17.13

Consider the centrally loaded beam shown in Figure 17.14 (a). Obviously the underside of this beam has tensile stresses set up in it while compressive stresses are set up in the uppermost portion. Thus, there must be some portion of the beam that is under neither tension nor compression; this portion is known as the *neutral axis* of the beam and remains unchanged and unstressed after bending. This is shown diagrammatically in Figure 17.14 (b).

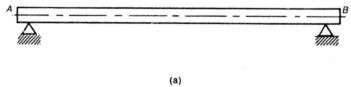

(a)

Upper surface is the area
of maximum compressive stress

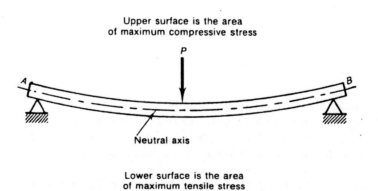

Neutral axis

Lower surface is the area
of maximum tensile stress

(b)

Figure 17.14 (a) Beam without load applied. (b) Beam after load application the neutral axis *AB* remains unchanged in length.

While the location of the neutral axis of a beam is of importance when considering the stresses due to bending in beams, it is equally important to consider the *effects of the distribution of the mass of the beam about the neutral axis.*

To demonstrate these effects, two pieces of flat mild steel of 25 mm × 6 mm cross-section are cut to a suitable length, say 1 metre, and supported as shown in Figures 17.15 (a) and (b) so that one piece lies on its flat face and the other lies on its edge. If equal masses are now centrally placed on each of these simply supported steel beams, the beam lying flat (that is, on the 25-mm face) will deflect or sag to a much greater extent that the other piece which was placed on edge. This is shown in Figures 17.15 (d) and (e). Obviously tensile forces are set up in the underside of each piece of steel. However, the greater the deflection the larger the magnitude of these tensile forces since the *stretching* (or tensile strain) present in the lower surface or "skin" of the metal increases with increasing deflection.

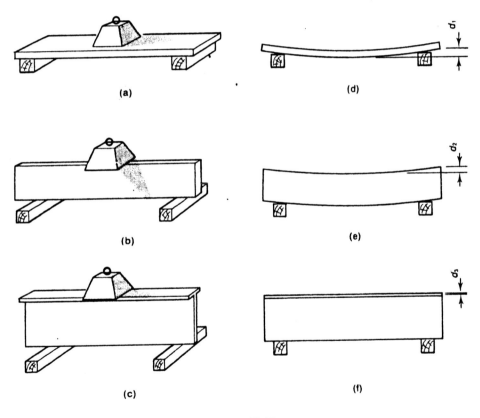

Figure 17.15

Thus, *beams of equal masses per unit length will not necessarily deflect equally when subjected to identical external force systems.* To demonstrate further the importance of mass distribution in a beam, a steel beam of T-section about 12 mm × 38 mm × 3 mm thick (shown in cross-section in Figure 17.16) has about the same mass per unit length as the previously discussed mild steel strip, but, if placed with T upper-most, deflects even less than the 25 mm × 6 mm section placed on edge (see Figures 17.15 (c) and (f)).

Obviously there is some property related to the distribution of the mass of a beam in relation to its neutral axis which is important in determining the degree of deflection that occurs under any given system of external forces (or loads and reactions).

This property is known as *the second moment of area* of the section and, as shown in Figure 17.16, is *related to the way in which the cross-sectional shape or area of a beam is distributed in relation to its neutral axis.*

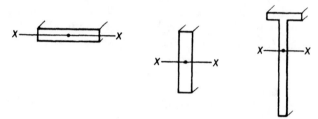

Figure 17.16 Distribution of mass in relation to the neutral axis *XX*. These three beams have approximately same cross-sectional area, but because of the distribution of the cross-sectional areas, the T-section is much more rigid than the other two sections.

CENTROIDS AND SECOND MOMENTS OF AREA

The centroid of any solid object, such as a cube or sphere, is the same as its centre of gravity or *centre of mass*.

However, in relation to plane figures, such as the cross-sections of beams, *the centroid is the centre or mean position of all the elements of that area.* While the centroids of regular plane areas such as rectangles, squares and circles are easily located as shown in Figure 17.17 (b), the centroids of irregular plane areas are not as readily determined and need to be calculated mathematically.

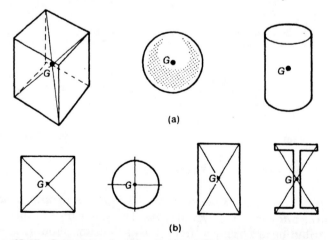

Figure 17.17 Centroids of some simple solids and plane figures (areas). (a) Centroids of solids (or volumes) are their geometrical centres. (b) Centroids of symmetrical plane figures are also readily located geometrically.

Consider the irregular plane figure shown in Figure 17.18 (a). Suppose that its centroid is located at position G and lies $\bar{x}$ units from the axis YY and $\bar{y}$ units from the axis XX.

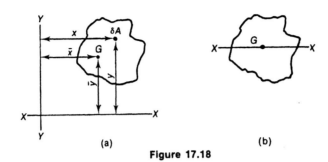

(a)   (b)

Figure 17.18

If the total area of this figure is A units and this total area is thought of as being concentrated at G, the centroid of the area, then

$$\text{moment of whole area } A \text{ about axis } XX = A\bar{y}$$

(*Note:* This is analogous to the taking of *a moment of a force* where the force is concentrated through some point, as, for instance, in the case of the weight-force of a man being concentrated through his centre of gravity.)

Now, consider that δA is any very small element of area of this irregular figure and is located x units from the axis YY and y units from the axis XX. Then

$$\text{moment of area } \delta A \text{ about the axis } XX = y \times \delta A$$

However, δA is very small and many such elements of area can exist within the given plane figure. Each element δA will possess a moment of area equal to $y \times \delta A$ with respect to axis XX, provided that the element is y units from that axis. Thus, the total moment of area of this section must be equal to the sum of all of the individual moments $y \times \delta A$, that is

$$\text{moment of whole area about axis } XX = \Sigma y \times \delta A$$

But
$$\text{moment of whole area} = A \times \bar{y}$$

$$\therefore A \times \bar{y} = \Sigma y \times \delta A$$
$$\therefore \bar{y} = \frac{\Sigma y \times \delta A}{A}$$

When the chosen axis XX passes through the centroid of the area as shown in Figure 17.18 (b), then the distance $\bar{y}$ will be zero, and thus

$$\frac{\Sigma y \times \delta A}{A} = 0$$

The expression $\dfrac{\Sigma y \times \delta A}{A}$ is really the first moment of area of a given section about the X (or horizontal) axis.

However, if this value is again multiplied by y, the distance of this section from the X axis, the second moment of area is obtained for this section. This is denoted by the symbol I to which suffixes are attached to indicate which axis is being considered. Thus, the second moment of area, I, about the axis XX is equal to

$$I_{XX} = \Sigma y^2 \times \delta A$$

Similar reasoning will bring forward the conclusion that the second moment of area about the axis YY is equal to

$$I_{YY} = \Sigma x^2 \times \delta A$$

The calculation of second moments of area of plane figures invariably involves the use of elementary calculus and, *as an example*, the second moment of area of a rectangle about a horizontal axis passing through its centroid will be calculated.

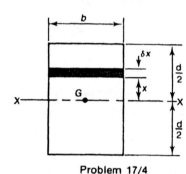

SAMPLE PROBLEM 17/4

Determine the *second moment of area of a rectangle*. Consider the rectangle of width b and depth d as shown in the diagram. The centroid lies at the geometrical centre of this rectangle and the second moment of area I_{XX} is to be calculated about the axis XX which passes through this centroid.

Problem 17/4

Calculation:

Consider the thin slice of thickness δx which lies x units from the axis XX. The area of this slice is $b\delta x$. Thus, the total second moment of area of the rectangle is expressed as

$$I_{XX} = \Sigma b x^2 \times \delta x$$

Applying simple calculus to this expression, the second moment of area for this rectangle can be written as

$$I_{XX} = \int_x^x b x^2 \times dx$$

$$\therefore I_{XX} = \int_{-\frac{d}{2}}^{+\frac{d}{2}} b x^2 \times dx, \quad \text{since } x \text{ varies from } +\frac{d}{2} \text{ to } -\frac{d}{2}$$

Thus,

$$I_{XX} = \int_{-\frac{d}{2}}^{+\frac{d}{2}} b x^2 \times dx$$

$$= \left[\frac{bx^3}{3} \right]_{-\frac{d}{2}}^{+\frac{d}{2}}$$

$$= \frac{1}{3} \left[b\left(\frac{d}{2}\right)^3 - b\left(\frac{-d}{2}\right)^3 \right]$$

$$= \frac{1}{3} \left[\frac{bd^3}{8} + \frac{bd^3}{8} \right]$$

$$= \frac{1}{12} bd^3$$

That is, the second moment of area of any rectangle about a horizontal axis passing through its centroid is $\frac{1}{12}bd^3$, where b is the breadth and d the depth of that rectangle. ∎

Figure 17.19 shows some common cross-sectional shapes and their second moments of area taken about horizontal or XX axes passing through their centroids.

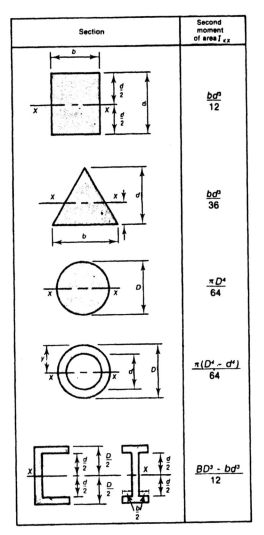

Section	Second moment of area I_{xx}
	$\dfrac{bd^3}{12}$
	$\dfrac{bd^3}{36}$
	$\dfrac{\pi D^4}{64}$
	$\dfrac{\pi(D^4 - d^4)}{64}$
	$\dfrac{BD^3 - bd^3}{12}$

Figure 17.19 Formulas to calculate the second moments of areas of common symmetrical sections about an XX axis passing through their centroids.

In practice, it is not necessary to calculate second moments of area for standard beam sections since these are provided in manufacturers' tables. In fact, it would be difficult to calculate the second moments of area for many commercial sections with their fillets and non-parallel surfaces. For this reason, the *second moments of area* of sections used in problems are always specified in this book unless they are those of simple rectangles, squares or circles.

Units of Second Moment of Area

The basic unit of length used for the cross-sectional shapes of beams are millimetres and therefore the corresponding units for I, the second moment of area, will be mm^4. However, it is possible and often desirable to express I in terms of m^4; thus, consistent units must be selected and *adhered to* throughout any given calculation.

There is some confusion between the terms *second moment of area* and the *moment of inertia* of a rotating mass (refer back to Chapter 13). A plane section having no mass cannot have any inertia and thus cannot possess a moment of inertia. Therefore confusion in usage of these two terms must be avoided; in fact, they involve different concepts and units.

CALCULATION OF STRESSES DUE TO BENDING

The portion of a beam shown in Figure 17.20 (b) is bent so that its underside is in tension and its upper surface in compression. The neutral axis of this beam is shown as *EF* and of course remains in the same condition, that is, neither stretched nor compressed. If the rectangle *ABCD* was marked on the side of the unbent beam (Figure 17.20 (a)), its shape after bending becomes the area *A'B'C'D'*. Thus, the length *CD* has increased to *C'D'* while the length *AB* has decreased to *A'B'*. However, the length of the neutral axis *EF* remains the same, so $EF = E'F'$.

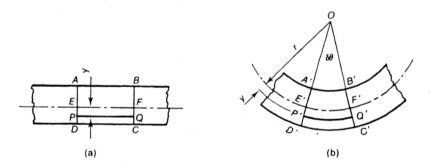

(a)   (b)

Figure 17.20 Relationship between stress and the distance of the section from the neutral axis.

If the distance between the lines *AD* and *BC* is very small, then *A' B'* becomes the arc of a circle which subtends a very small angle $\delta\theta$ at its centre *O* as shown, and *E' F'* is an arc of a concentric circle of radius r units.

Suppose that PQ is any general section on this beam: it is y units distance from the neutral axis and, after bending, it becomes the line $P'Q'$. The length of $P'Q'$ is found from

$$\frac{EF'}{OE'} = \frac{P'Q'}{OP'} = \delta\theta \quad \text{(in radians)}$$

$$\therefore P'Q' = EF' \times \frac{OP'}{OE'}$$

$$= EF' \frac{(r+y)}{r}$$

$$= EF \frac{(r+y)}{r}, \quad \text{since } EF' = EF$$

$$= \frac{PQ\,(r+y)}{r}, \quad \text{since } EF = PQ$$

$$\therefore \frac{P'Q'}{PQ} = \frac{r+y}{r}$$

$$= 1 + \frac{y}{r}$$

However, the *strain* on the section PQ caused by bending is equal to

$$\frac{\text{the change in length of } PQ}{PQ} = \frac{P'Q' - PQ}{PQ}$$

$$= \frac{P'Q'}{PQ} - 1$$

$$= 1 + \frac{y}{r} - 1, \quad \text{since } \frac{P'Q'}{PQ} = 1 + \frac{y}{r}$$

$$\therefore \text{strain on } PQ = \frac{y}{r}$$

This means that the strain on any section of the beam is directly proportional to the distance of that section from the neutral axis, a fact which appeared obvious from the previous experiment with the rubber beam.

Let the stress on the section PQ be σ. If Young's modulus for the beam is E, then

$$E = \frac{\text{stress}}{\text{strain}}$$

$$\therefore E = \frac{\sigma}{y/r}$$

$$\therefore \frac{\sigma}{y} = \frac{E}{r}$$

That is, the ratio between the stress and the strain in any section of the bent beam is equal to the ratio between Young's modulus for the material and the radius of curvature, r, at that section. From

$$\frac{\sigma}{y} = \frac{E}{r}$$

$$\therefore \sigma = \frac{Ey}{r}$$

Thus, the stress increases with increasing distance, y, from the neutral axis.

Figure 17.21 shows the variation of stress across a simple rectangular beam in bending.

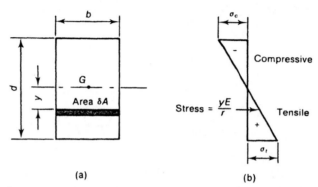

(a)   (b)

Figure 17.21 Stress variation across the section of a rectangular beam subjected to pure bending. In this figure, the top surface has the maximum compressive stress. (a) The cross-section of the beam and a small area δA which is y units from the neutral axis passing through the centroid G. (b) The diagram showing how tensile and compressive stresses increase from zero on the neutral axis to maxima on the lower and upper surfaces of the beam respectively. The stress on the small area δA is shown as yE/r, where r is the radius of curvature of the beam.

Now consider again a beam subjected to simple bending. If this beam is imagined to be sectioned and separated as shown in Figure 17.22, the small area δA (shaded black) is under a tensile stress of $\sigma = Ey/r$ since it is y units from the neutral axis. It is possible to consider that such a tensile stress could also be produced by the tensile force F acting on the area δA; the magnitude of this force F then is found from

$$\text{force} = \text{stress} \times \text{area}$$

$$\therefore F = \frac{Ey}{r} \times \delta A$$

This force F has a *moment* about the neutral axis of the beam, where

$$M = Fy$$

$$\therefore M = \frac{Ey}{r} \delta A \, y$$

$$= \frac{E}{r} y^2 \, \delta A$$

However, many similar such strips δA exist across the beam and each has a similar force F acting on it. Also every one of these forces will have a moment $M = \frac{E}{r} y^2 \, \delta A$ about the neutral axis of the beam.

The sum of all these individual moments is known as the *moment of resistance*, M_R, of the beam and

$$M_R = \frac{E}{r} \Sigma y^2 \delta A$$

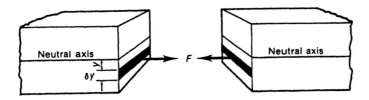

Figure 17.22 A beam subjected to simple bending, and sectioned and separated. The force F can be considered to be producing the tensile stress in the small area δ A.

However, it has already been shown that $\Sigma y^2 \delta A$ is equal to I_{XX}, the second moment of area of a beam about the axis XX passing through the centroid of the beam.

$$\therefore M_R = \frac{E}{r} \times I$$

or

$$\frac{M_R}{I} = \frac{E}{r}$$

In the earlier work in this chapter, bending moment diagrams were used to show the variation of bending moment in loaded beams. It is obvious that the bending moment induced in any section of a loaded beam must be equal to the moment of resistance, M_R, in all cases where the elastic limit of the material is not exceeded. If the elastic limit is exceeded, the beam is incapable of providing the necessary moment and failure will occur. Thus, for design considerations, the position and value of the maximum bending moment must be known.

The fundamental formula for considering the relationships between stresses and the bending of beams is most conveniently written as:

$$\frac{M}{I} = \frac{E}{r} = \frac{\sigma}{y}$$

where:

M = bending moment (or moment of resistance);
I = second moment of area of the beam section;
E = Young's modulus for the material;
r = radius of curvature at the point of application of the moment M_R;
σ = stress in the beam;
y = distance from neutral axis.

This formula can be used to solve many problems of pure bending in beams.

Given the same *area of section* (or weight per unit length), the value of I, the second moment of area, varies markedly with variations in cross-sectional shape. This was illustrated in a comparative manner on page 397, but is illustrated quite clearly in Figure 17.23 which shows four typical beam sections each having a cross-sectional area of about 8750 mm². The accompanying table shows the value of I_{XX} for these sections, and the bending moment which will produce a stress of about 75 MPa in each section. Clearly the universal beam is by far the strongest of these four sections, *indicating the advantage gained by placing the greatest amount of material as far away from the neutral axis of the beam as possible.*

	ϕ 105	$\mapsto$ 50 $\mapsto$	150	150
I (mm⁴)	6 × 10⁶	22.3 × 10⁶	44 × 10⁶	240 × 10⁶
M_{max} (kN m)	8.6	19.2	22.2	90

Figure 17.23 A comparison of second moment of area I and maximum BM for four beams of about the same cross-sectional area.

SAMPLE PROBLEM 17/5

A symmetrical I-beam has a section approximating to that shown in the diagram. Using the method shown in Figure 17.19, calculate the second moment of area of this section about a horizontal axis passing through its centroid.

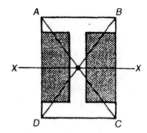

Problem 17/5

Analysis:

Since the I-section is symmetrical, the centroid is the geometrical centre found by joining the diagonals AC and BD as shown. The second moment of area can be found by taking I_{xx} for the whole rectangle $ABCD$ and subtracting I_{xx} for the two shaded areas as shown.

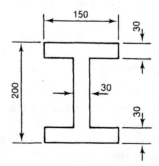

Problem 17/5

Calculation:

$$I_{xx} \text{ for rectangle } ABCD = \frac{bd^3}{12}$$

$$= \frac{150 \times 200^3}{12}$$

$$= 10^8 \text{ mm}^4$$

where: $b = 150$ mm
$d = 200$ mm

$$I_{xx} \text{ for } each \text{ shaded area} = \frac{bd^3}{12}$$

$$= \frac{60 \times 140^3}{12}$$

$$= 0.137 \times 10^8 \text{ mm}^4$$

where: $b = 60$ mm
$d = 140$ mm

Therefore, I_{xx} for I-section $= 10^8 - (2 \times 0.137 \times 10^8) \text{ mm}^4$
$$= 72.6 \times 10^6 \text{ mm}^4$$

Result:
The second moment of area of the I-section about a horizontal axis passing through its centroid is $72.6 \times 10^6 \text{ mm}^4$. ∎

SAMPLE PROBLEM 17/6

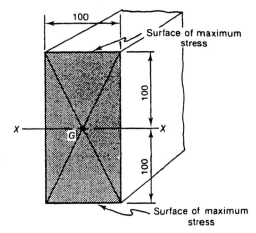

100

Surface of maximum stress

100

100

100

X —— G —— X

Surface of maximum stress

Problem 17/6

A rectangular beam 100 mm wide by 200 mm deep is subjected to a maximum bending moment of 240 kN m.
(a) Determine the maximum stress in the beam.
(b) Determine the radius of curvature for that portion of the beam where the BM is 240 kN m, if $E = 200$ GPa.

Analysis:
The maximum stress in the beam occurs in the surfaces furthest from the centroid of the section. In this case, one side of the beam is in tension and the other in compression, and these stresses will be equal since the section is symmetrical about the XX axis passing through its centroid.

The use of consistent units is essential in this kind of problem; m^4 for I, Pa (N/m²) for stress and m for radius of curvature are to be used.

Calculation:

$$I_{xx} \text{ for beam} = \frac{bd^3}{12}$$

$$= \frac{100 \times 200^3}{12}$$

$$= \tfrac{2}{3} \times 10^8 \text{ mm}^4$$

$$= \tfrac{2}{3} \times 10^{-4} \text{ m}^4 \text{ (since 1 m}^4 = 10^{12} \text{ mm}^4)$$

where: $b = 100$ mm
$d = 200$ mm

Maximum stress in beam

$$\frac{M}{I} = \frac{\sigma}{y}$$

$$\therefore \sigma = \frac{My}{I}$$

However, M = maximum bending moment

y = distance from centroid to upper and lower surfaces of beam

$$\therefore \sigma_{max} = \frac{My}{I}$$ where M = 240 kN m = 24×10^4 N m

$$y = 100\,mm = 0.1\,m$$

$$\therefore \sigma_{max} = \frac{24 \times 10^4 \times 0.1}{\frac{2}{3} \times 10^{-4}}$$ (Units, $\frac{Nm \times m}{m^4} = N/m^2$)

$$= 36 \times 10^7 N/m^2$$

$$= 360\,MPa$$

Radius of curvature

$$\frac{M}{I} = \frac{E}{r}$$ where: E = 200 GPa = 200×10^9 Pa (N/m^2)

$$I = \tfrac{2}{3} \times 10^{-4}\,m^4$$

$$\therefore r = \frac{EI}{M}$$ $M = 24 \times 10^4$ N m

$$\therefore r = \frac{200 \times 10^9 \times \frac{2}{3} \times 10^{-4}}{24 \times 10^4}$$ (Units: $\frac{N/m^2 \times m^4}{Nm} = m$)

$$= \frac{200 \times 2}{24 \times 3} \times 10\,m$$

$$\therefore r = 55.5\,m$$

Result:

At the point where the BM was 240 kN m, the maximum stress in the beam was 360 MPa and the radius of curvature was 55.5 m.   ∎

REVIEW PROBLEMS
(Use $g = 10\,\text{m/s}^2$ unless otherwise specified.)

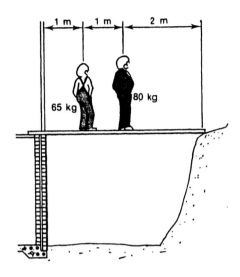

17/7
A plank is used to provide temporary access over a small excavation to the door of a house. Consider this plank as a simply supported beam and determine the distribution of shear force and bending moments within it when two people of masses 65 kg and 80 kg respectively stand on the plank as shown. Ignore the mass of the plank.

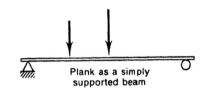

Plank as a simply supported beam

Problem 17/7

17/8
A workman lifting the edge of a dimension stone stands on one end of his crowbar. If the man, by exerting a force of 400 N vertically down, just raises the edge of the stone, determine
  (i) the vertical force F_v acting between bar and stone;
  (ii) the vertical component R_v of the reaction at the fulcrum;
  (iii) the distribution of shear force and bending moments along the length of the crowbar.
Neglect the mass of the crowbar since it is small compared to the mass of the stone being lifted.

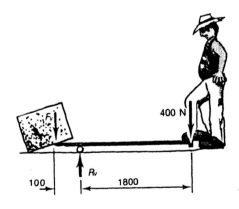

Problem 17/8

17/9
A see-saw of effective length 2.2 metres is horizontal and supporting two children as shown. If
each child has a mass of 35 kg, determine the
distribution of shear force and bending moments
along the length of the see-saw, if the mass of the
plank is negligible.

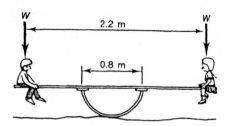

Problem 17/9

17/10
A diving board is 3 metres long and is cantilevered
as shown in the diagram. By jumping on its free
end, a diver exerts a force of 1.5 kN vertically
down. Determine the maximum bending moment
induced in the diving board due to this load and
specify its position from the strap at *A*.
A partial free-body diagram of the board is drawn
to assist you.

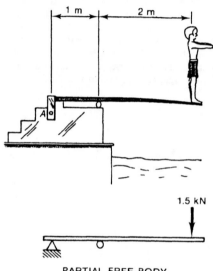

PARTIAL FREE-BODY
DIAGRAM

Problem 17/10

17/11
A section th ough a timber floor giving the loads
transmitted to the bearer by the floor joists is
shown in the diagram. Neglecting the mass of the
bearer, determine the distribution of shear force
and bending moments along this bearer.
Joist centres are at 450 mm.

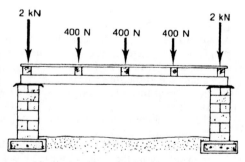

Problem 17/11

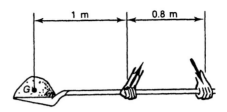

17/12

A long-handled shovel is held horizontally as shown. Given that its 5-kg load of soil has its centre of mass at G, determine the distribution of shear force and bending moments present in the handle due to the load.

Problem 17/12

17/13

Overhanging floor joists are used to cantilever a timber deck at first floor level in a house structure. The weight of a module of the floor, as supported by the one joist shown in the diagram, is known to be 80 N/m. Given that the walls exert vertical loads of 2 kN and that three 80-kg people stand centrally on this joist, determine the distributions of shear force and bending moments along the length of the joist. (Ignore the mass of the deck railing.)

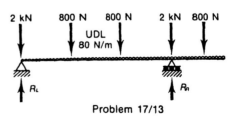

Problem 17/13

17/14
An adjustable hanger is positioned as shown. Determine the shear force and bending moment at the points A and B when this hanger supports a load of 250 N. Since the mass of the hanger is very small, it may be neglected in your calculation.

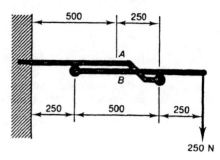

Problem 17/14

17/15
A metal plate is sheared by a blade which exerts a force of 1 kN on it at C. The plate rests on fixed supports at A and B and may be treated as a simply supported beam. Sketch the shear force and bending moment diagrams for the plate.

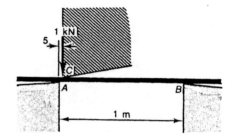

Problem 17/15

17/16
The steel channel shown forms one side of a short flight of stairs. If the channel has a mass of 18 kg per metre, determine the axial force, shear force and bending moment at its centre section A-A due to its own weight-force. Assume that the beam has its centre of mass on the section A-A, the joint at B is a roller and the joint at C is a hinge.

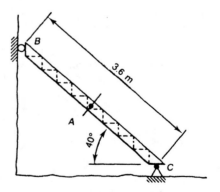

Problem 17/16

17/17

Calculate the second moments of area about the
axis X-X of the sections shown in the diagram.

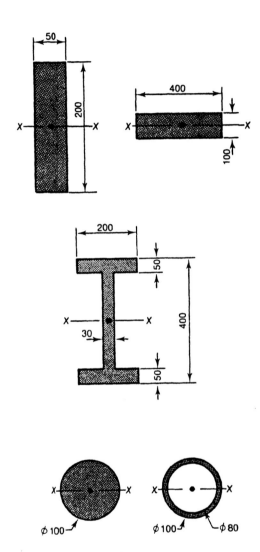

Problem 17/17

17/18

A length of 530 UB 94 is stacked horizontally across two other beams which are 4 metres apart. Considering the 4-metre length as a simply supported beam, determine

(i) the bending moment diagram for the beam;

(ii) the maximum bending moment present and its position;

(iii) the maximum stress in the upper and lower flanges, if I_{xx} for this section is 554×10^6 mm^4;

(iv) the radius of curvature at the midpoint of the span.

530 UB 94 has a mass of 94 kg per metre and a cross-section as shown in the diagram.

UDL 94 kg/m ($\approx$ 940 N/m)

PARTIAL FREE-BODY DIAGRAM

Problem 17/18

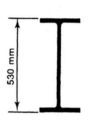

530 mm

17/19

A weightlifter who has just completed his lift holds the bar and weights directly above his head. Neglecting the mass of the bar, determine the distribution of shear force and bending moment along the bar, given that the 120-kg load at each end of the bar has its centre of mass at G. What is the radius of curvature of the bar at the weightlifter's hands if the bar has a diameter of 36 mm and E for the bar material is 200 GPa?

Problem 17/19

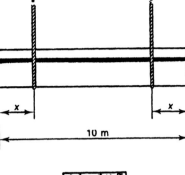

17/20
A prestressed concrete deck beam for a bridge is being lifted into position as shown. It has not been designed to accept a negative bending moment greater than 2 kN m. If the beam has a mass of 150 kg/m, at what maximum distance x from each end of the beam can the slings be placed?

Cross-section
of beam

Problem 17/20

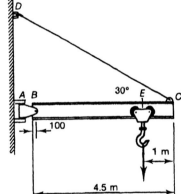

17/21
A small jib crane in a foundry is designed so that its beam can swing in a horizontal arc about the pivot at A. For the position shown, determine
  (i) the reaction at the pin B, in terms of its horizontal and vertical components, and the tension in the cable CD, given that the total load on the crane is 1 tonne and the beam has a mass of 82 kg/m.
  (ii) the shear force and bending moment present in the beam at the section E (midway between the trolley wheels of the monorail hoist); and
  (iii) the maximum stress in the upper and lower flanges at the section E.
The beam is a 460 UB 82; its cross-section is as shown in the diagram and I_{xx} for this section is 371×10^6 mm^4.

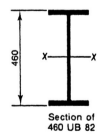

Section of
460 UB 82

Problem 17/21

17/22

During its operation the 10-mm wide bandsaw blade is wrapped around the 400-mm diameter driving wheel as shown. Given that the blade thickness is 1 mm, determine the maximum stress induced in the blade by this bending operation. $E_{blade} = 200$ GPa.

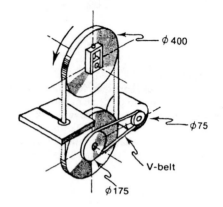

Problem 17/22

17/23

A rocker arm used in a steam engine has a I-section whose dimensions at the section *A-A* approximate to those given in the diagram. Determine the load F at the end of this arm which causes a maximum tensile stress of 10 MPa to occur at this section.

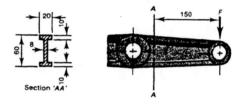

Problem 17/23

17/24

The trailer of a car transporter carries three identical cars whose centres of mass are shown at G_1, G_2 and G_3. Each car has a mass of 1.4 tonnes. The mass of each deck is a uniform 100 kg/m. The overall length of the upper deck is 10 metres and the lower deck 6 metres.

Construct diagrams showing the distribution of shear force and bending moment in both the upper and the lower decks of the trailer. Ignore the mass of the pillars.

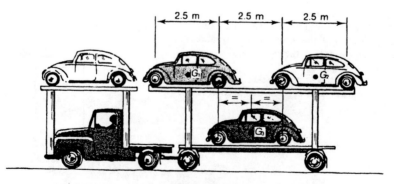

Problem 17/24

18

Principles of Machines

"When I get the bugs outa this ..."

A *machine* is a device for doing work. It assists man to do work with less apparent effort, or at greater speed, or more conveniently than would otherwise be possible. In this chapter we will consider some *simple machines* and compare the work done to the effort expended in using these machines.

A machine is said to have a *mechanical advantage* if a large load can be moved with a smaller effort. The mechanical advantage is expressed as the ratio of the forces involved, and is the *ratio of the load moved to the effort required.*

$$\text{mechanical advantage} = \frac{\text{load}}{\text{effort}}$$

LOAD
1000 N

EFFORT
100 N

100

1000

Figure 18.1

Consider the lever shown in Figure 18.1. If we take moments about the fulcrum, we discover that the effort needed to balance the load of 1000 N is only 100 N. When we compare the magnitudes of these two forces, it becomes apparent that the load is ten times as big as the effort required to balance it. *Thus, this simple machine is said to have a mechanical advantage of 10 to 1 (10 ∴1).*

It would seem that we have made a profit and got something for nothing, but, of course, this is not possible. When we compare the distances moved by the load and the effort we discover that their relationship is the reciprocal of the mechanical advantage, that is 1 to 10.

It is now obvious that, if the lever is rotated about its fulcrum, *the total work done on one end of the lever is the same as the total work done on the other end.*

$$\text{Work} = \text{force} \times \text{distance moved}$$
$$U = Fs \text{ (See Chapter 14.)}$$

The time taken to complete this movement is the same for each end of the lever but, as one end moves ten times as far as the other it must move ten times as fast. A comparison of the velocity of the effort with the velocity of the load enables us to determine the *velocity ratio* of the machine.

$$\text{velocity ratio} = \frac{\text{distance moved by EFFORT}}{\text{distance moved by LOAD}}$$

If the mechanical advantage and the velocity ratio of a machine are now compared, then the *efficiency* of the machine can be determined. The efficiency is usually expressed as a percentage.

$$\% \text{ efficiency, } \eta = \frac{MA}{VR} \times \frac{100}{1}$$

A machine with 100% efficiency would be known as the "ideal machine" but as in practice there are always some losses, such machine cannot exist. Even the simple lever shown in Figure 18.1 cannot quite reach 100% efficiency due to the energy used to bend the lever itself and to deform the surfaces in contact. In most other machines, *friction* significantly reduces the practical efficiency of the machine, sometimes by more than half.

Most machines are only combinations of several simpler machines, and if the principles of *the lever, the wheel, the inclined plane* and *pulleys* are understood, then more complex machines can be easily analysed.

LEVERS

The most common machine is the lever. It was certainly the first machine to be used by man, as our bodies consist of hundreds of levers (our bones). Levers fall into three basic types, first order, second order and third order, depending on the relative positions of the load, the fulcrum, and the effort. The examples of these types are shown in Figures 18.2, 18.3 and 18.4.

First Order Levers

In all first order levers the fulcrum lies between the load and the effort.

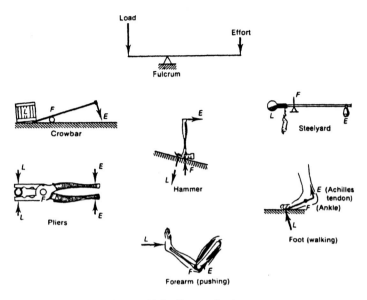

Figure 18.2 First order levers.

Second Order Levers

In all second order levers the load is between the fulcrum and the effort.

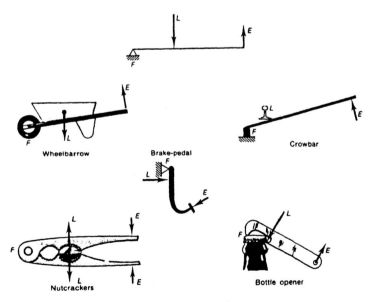

Figure 18.3 Second order levers.

Third Order Levers

In all third order levers the effort is between the fulcrum and the load.

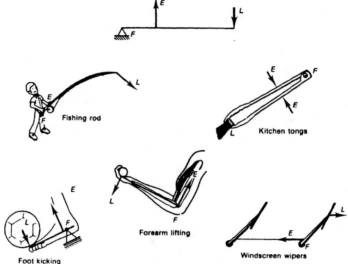

Figure 18.4   Third order levers.

MECHANICAL ADVANTAGE

First order levers may have a mechanical advantage or a mechanical disadvantage depending on the relative lengths of their lever arms. This is shown in Figure 18.5.

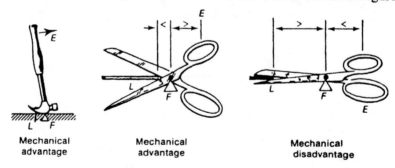

Mechanical advantage   Mechanical advantage   Mechanical disadvantage

Figure 18.5   The hammer always provides a mechanical advantage when extracting a nail. However, the scissors can provide either a mechanical advantage or disadvantage depending upon where the material being cut is located between the scissor blades.

Second order levers *always* provide a mechanical advantage, irrespective of the lengths of the arms. Thus, the effort is always smaller than the load. The wheelbarrow shown in Figure 18.6 illustrates the positive MA of second order levers.

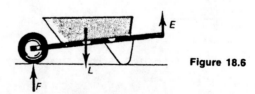

Figure 18.6

Third order levers always provide a *mechanical advantage less than 1* (that is, a mechanical disadvantage). The effort must always be greater than the load. However, this apparent disadvantage is offset by the greater movement of the load; for example, a small powerful effort on the fishing-rod by the forearm produces a large sweeping movement at the other end of the lever. Thus, third order levers enable small loads to be moved along relatively large distances fairly quickly.

Figure 18.7

Some levers produce *a couple*. These cannot be simply categorised as first, second, or third order levers. For instance, the spanner shown in Figure 18.8 is obviously a lever, but is it a first order lever or a second order lever?

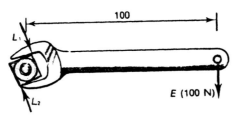

Figure 18.8

The effort produces a couple on the nut which rotates (consider the nut as a small wheel in this instance). An effort of 100 N on a spanner of 100 mm effective length will produce a couple (turning moment) of 10 000 N mm or 10 N m irrespective of the locations of the loads L_1 and L_2 on the nut, shown in Figure 18.8.

Equilibrium of Levers

When any lever is in equilibrium then the basic equilibrium conditions must be met; these are:

$$\Sigma F_x = 0, \quad \Sigma F_y = 0 \quad \text{and} \quad \Sigma M = 0.$$

When calculations are necessary, it will usually be convenient to take moments about the fulcrum, as this procedure ignores the fulcrum reaction. Then, only the moments need to be considered and the usual moment equation, $\Sigma M = 0$, becomes:

$$\text{effort} \times \text{effort arm} = \text{load} \times \text{load arm}$$
$$E \times E \, Arm = L \times L \, Arm$$

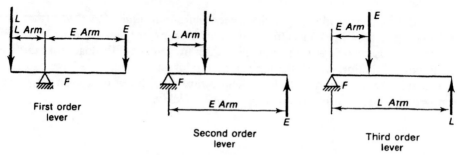

Figure 18.9 Equilibrium of levers.

TYPICAL LEVERS

(1) *The common balance.* The balance has equal lever arms, and compares the weight-forces exerted by the masses on the scale pans. This method of "weighing" actually compares masses, and would give the same result anywhere in the world (and on the moon).

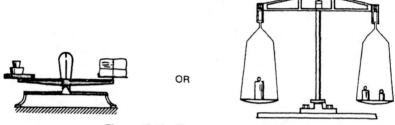

OR

Figure 18.10 The common balance.

(2) *The steelyard.* This is also a balance which compares masses, but whereas the unknown mass remains at a fixed distance from the fulcrum, the "comparison mass" has its moment varied by adjusting its position along a graduated lever arm. This lever arm is graduated in mass units, enabling a direct reading of the mass of the load. (This is shown in Figure 18.11.)

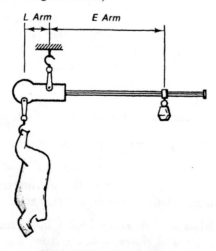

Figure 18.11 The steelyard.

(3) In many situations the simple lever is well disguised. For instance, *the screwdriver is obviously a lever* when used to open a paint tin, but is not quite as obviously a lever when used to turn a screw, until it is realised that the differences in diameters of the handle and the screw constitute a form of leverage. (This is shown in Figure 18.12.)

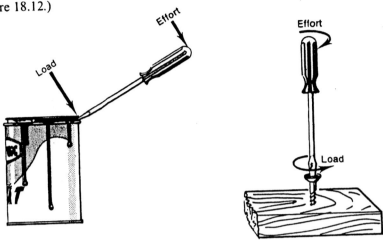

Figure 18.12 The screwdriver as a lever.

Often the mass of the lever itself is ignored, either because equal masses of lever on either side of the fulcrum balance each other (as in the common balance), or because the different masses on each side of the fulcrum are counterbalanced by having their centres of mass at suitable distances from the fulcrum (as in the steelyard).

Sometimes the mass of the lever is ignored to simplify the calculations, but in practice the mass of the lever can often be quite significant. It is important to recognise when it should be included in calculations, and when it can be safely ignored. For example, it is obviously easier to push down on a heavy crowbar than to lift the same load plus the crowbar, provided the same mechanical advantage is maintained. (This is shown in Figure 18.13.)

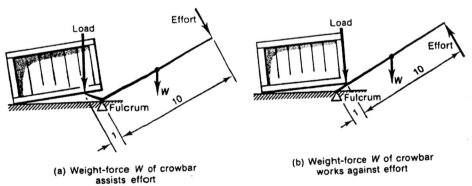

(a) Weight-force *W* of crowbar assists effort

(b) Weight-force *W* of crowbar works against effort

Figure 18.13

REVIEW PROBLEMS

(Use $g = 10$ m/s^2 unless otherwise specified.)

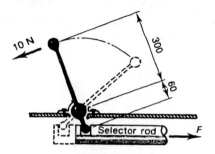

18/1

Determine the force, F, in the selector rod when an effort of 10 N is applied to the gear lever shown. What is the mechanical advantage?

Problem 18/1

18/2

What is the tension, T, in the handbrake cable when a force of 50 N is applied to the handle? What is the mechanical advantage?

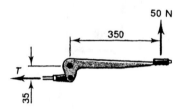

Problem 18/2

18/3

Determine the effort, E, on the clutch pedal shown needed to obtain a force of 1000 N in the linkage. What is the velocity ratio?

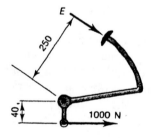

Problem 18/3

18/4

The pendant brake pedal shown has a force of 150 N applied at an angle of 30°. What is the load in the push-rod? What is the MA of this lever?

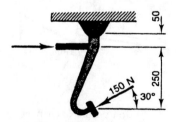

Problem 18/4

18/5
Determine the gripping force on the rod if an effort of 100 N is applied to the handles of the pliers.

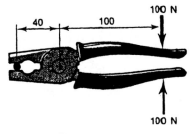

Problem 18/5

18/6
What effort, *E*, will be needed to support the barrow shown?

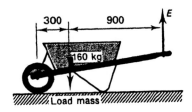

Problem 18/6

18/7
A crowbar has a mass of 10 kg. What effort will be required to just support one end of it?

Problem 18/7

In each of the above situations the simplest solution is derived by taking moments about the fulcrum, being careful to use the correct length of moment arm.

THE WHEEL

The second machine to be used by man was probably the *wheel*. In its simplest form it was a roller which had no axle and was simply placed under heavy loads. It is much easier to roll a rock than slide it, rolling friction being so much less than sliding friction (see Chapter 8).

Originally the roller was probably used to move building stones. This application was considerably refined and extended from two or three saplings under a rock, to a series of lateral logs under a sled rolling on a path of longitudinal logs. This was the forerunner of the road and the railway, and was improved by the advent of the axle, which enabled one or two rollers to be kept permanently under the load. Modern roller bearings and ball bearings are just a further refinement of the roller principle in the continuing effort to reduce friction. Rollers and wheels on axles are shown in Figure 18.14.

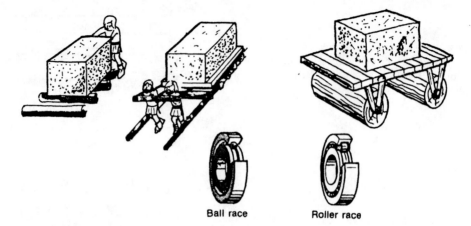

Ball race   **Roller race**

Figure 18.14

The roller has a velocity ratio of 1 and its main advantage is that it reduces friction which allows the work to be done more conveniently.

There are many combinations of these two simple machines, the lever and the wheel. These include *the wheel-barrow* (Figure 18.15 (a)) which has the mechanical advantage of the lever, and the convenience of the wheel; and *the hand winch* that was first used in the form of the *windlass* (Figure 18.15 (b)) for winding water out of the well and as a capstan (Figure 18.15 (c)) for hauling heavy tackle on windjammers. It is still used today for tensioning sheets and halyards on racing yachts (Figure 18.15 (d)), and winding the family boat on to its trailer (Figure 18.15 (e)).

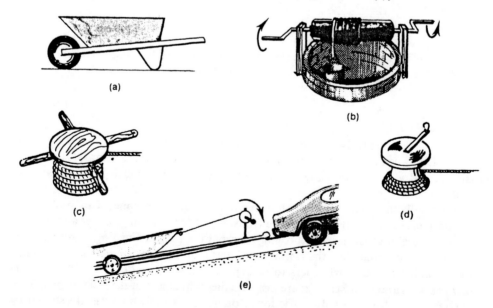

Figure 18.15

The winch is essentially a simple lever capable of continuous operation. The effort acts on a lever arm moving in a circle of radius R and this winds a load rope around a drum of radius r.

The velocity ratio of a simple winch (or wheel and axle) is

$$\frac{\text{distance moved by EFFORT}}{\text{distance moved by LOAD}} = \frac{2\pi R}{2\pi r}$$

$$\therefore VR = \frac{R}{r}$$

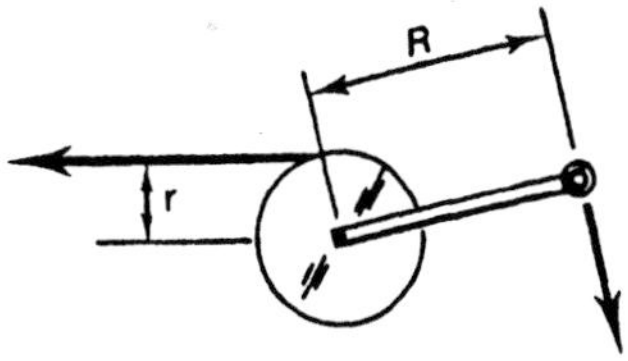

Figure 18.16

The windmill and the water wheel shown in Figure 18.17 are also early combinations of the lever and the wheel, using the kinetic energy of the wind, or the potential energy of the water behind a weir, to provide useful power for the pumping of water and grinding of grain. Hydroelectric turbines are only modern refinements of these principles. No doubt many other applications will easily come to mind.

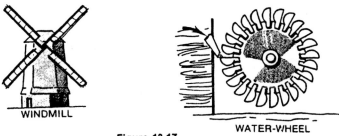

WINDMILL

WATER-WHEEL

Figure 18.17

Assignment: *Identify and analyse as many simple combinations of the lever and wheel as possible. Discuss their relative advantages and why they are, or were, used to benefit man.*

The *theoretical mechanical advantage* of each of these combinations is most easily established from their velocity ratio, which *is directly proportional to the distances moved by the effort and the load.*

$$\text{velocity ratio (VR)} = \frac{\text{distance moved by EFFORT}}{\text{distance moved by LOAD}}$$

$$\text{VR (relating circumferences)} = \frac{2\pi R}{2\pi r}$$

$$\text{VR (relating diameters)} = \frac{D}{d}$$

$$\text{VR (relating radii)} = \frac{R}{r} \quad \text{(the lever arms)}$$

Provided the efficiency of the machine is known, the actual mechanical advantage can easily be determined, since

$$\text{efficiency} = \frac{\text{MA}}{\text{VR}}$$

$$\text{MA} = \text{VR} \times \text{efficiency}$$

REVIEW PROBLEMS

(Use $g = 10\,\text{m/s}^2$ unless otherwise specified.)

18/8

The barrow and its load have their combined centre of mass located as shown. Determine the mechanical advantage of the wheelbarrow as a lever, and discuss the convenience of the wheel.

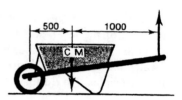

Problem 18/8

18/9

The windlass has a 200-mm winding drum and crank handles operating in a 600-mm diameter path. Determine the velocity ratio of this machine.

Approximately half of the input effort is used up overcoming friction, bending the rope around the drum, and raising the mass of the bucket. Considering these losses, suggest a suitable bucket size, if the effort on the handle is not to exceed 160 N.

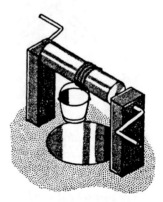

Problem 18/9

18/10

Design a suitable length, *L*, of the crank arm for the boat trailer winch if the winding drum plus the cable averages 60 mm diameter and the tension in the cable does not exceed 1000 N. (Estimate that frictional and other losses will use about 10% of the input effort of 250 N).

Problem 18/10

EFFICIENCY

We have already established in the previous problems that even the simple machines used as examples were not 100% efficient due to internal losses. It is impossible, and in many machines undesirable, to overcome friction entirely.

The windlass loses about half of its power and could therefore be termed only 50% efficient. The boat winch loses only about 10% of the power (better bearings) and is therefore about 90% efficient. The efficiency of any particular machine can be really discovered only by experiment (though, with experience, some fairly accurate predictions can be made).

The efficiency (usually expressed as a percentage) *is the relationship between the useful work done by the machine on the load to the total work done on the machine by the effort.*

$$\text{efficiency, } \eta = \frac{\text{useful work output}}{\text{total work input}}$$

Bearing in mind that work = force × distance, the above relationship for efficiency can be expressed as:

$$\% \text{ efficiency, } \eta = \frac{\text{experimentally determined MA}}{\text{theoretically determined MA} (= \text{VR})}$$

$$= \frac{\text{MA}}{\text{VR}} \times 100$$

Only theoretical machines are 100% efficient and in problems dealing with real machines it is wise to consider the efficiency of the machine, as it will have a significant effect on the solution to the problem.

REVIEW PROBLEMS

(Use $g = 10$ m/s² unless otherwise specified.)

18/11

The efficiency of the boat trailer winch (shown in Problem 18/10) is found by experiment to be only 40%. What length of crank arm is now needed? Is this length practicable? What other design modifications could be made to have a similar effect?

18/12

The windlass shown in Problem 18/9 is updated by addition of a light nylon cord and ball-race bearings and achieves an efficiency of 90%. The crank now feels far too easy to wind and the effort seems wasted. What new combination of drum diameter and bucket size can you suggest so that the previous effort on the crank is required?

PULLEYS

A pulley is a wheel whose sectional shape is designed to accommodate a rope, belt or chain. It is usually a part of a larger machine.

Grooved rope pulleys are known as *sheaves* and are used for changing the direction of a hauling rope.

The *flat-belt* and *V-belt pulleys* provide friction drives for industrial machines; for example, the lathes and drill presses in a workshop.

Chain pulleys provide a positive, non-slip drive where the relationship between drive pulley and driven pulley must remain constant; for example, from crankshaft to camshaft in a car engine.

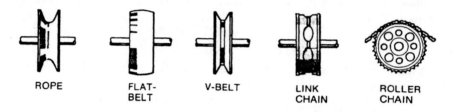

ROPE   FLAT-BELT   V-BELT   LINK CHAIN   ROLLER CHAIN

Figure 18.18 Various types of pulleys.

The single fixed pulley

This type of pulley is used when a force is required in a certain place and a suitable effort can be more conveniently applied elsewhere, that is, a single fixed pulley changes the direction of an applied force (or effort); for example, raising a venetian blind, opening and closing curtains, etc.

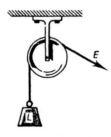

Figure 18.19 Single fixed pulley.

Suppose that the load L is to be raised by the effort E using the single fixed pulley shown in Figure 18.19. If there is no friction in the pulley bearing, and the cord is perfectly flexible (that is, it offers no resistance to bending around the pulley) then the mechanical advantage will be 1 : 1. If the cord is also inextensible (that is, does not stretch) then the velocity ratio will also be 1 : 1. In practice, of course, we have to pay for the convenience of being able to change the direction of the applied force (the effort E), and the velocity ratio will be slightly less than 1 : 1, whereas the mechanical advantage will be considerably less than 1 : 1.

The single movable pulley

This is the simplest form of *block and tackle*. In operation, the movable pulley is hooked to the load, one end of the rope is anchored, and the effort is applied to the other end of this rope. The result is a machine with a theoretical mechanical advantage of 2: 1.

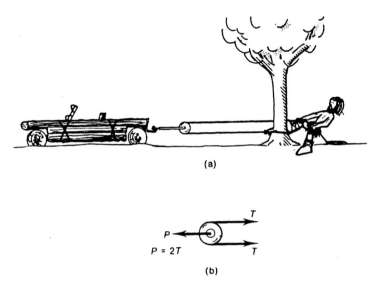

(a)

$P = 2T$

(b)

Figure 18.20

This is illustrated in Figure 18.20 (a) where the car is shown being hauled out of a bog by the driver. For every metre of rope he pulls, the car moves out only half a metre, *but the force pulling on the car is nearly twice the driver's effort.* The explanation for the apparent doubling of his effort is simple. If friction is ignored, then the pull of the hook on the car is twice the tension in the rope, because two ropes are pulling. This is shown in the free-body diagram (Figure 18.20 (b)). The driver has hearly doubled his pulling force on the car (MA $\approx$ 2 : 1) but he has to pull twice as far (VR = 2 : 1). In practice, of course, the mechanical advantage would be a little less than 2 : 1, due to such things as friction in the bearings, the weight of the pulley and the rope, the flexibility and the extension of the rope.

The principle of the movable pulley can be extended many times by using multiple sheaves, and the *velocity ratio of the resulting combination of pulleys is determined by the number of ropes supporting the load.*

A typical *block and tackle* is shown in Figure 18.21 (a). In this particular configuration there are four ropes supporting the load, and the velocity ratio is therefore 4 : 1. Thus, if the load rises by one metre, then each supporting rope must shorten by one metre. As there are four supporting ropes, the total length of rope must shorten by four metres, that is, VR = 4 : 1.

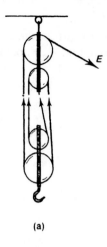

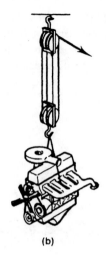

(a)

(b)

Figure 18.21

REVIEW PROBLEMS

(Use $g = 10 \, \text{m/s}^2$ unless otherwise specified.)

18/13

From a strong-point in the classroom ceiling, a group is experimenting with a block and tackle. The boys are sitting in a "bosun's chair" and being hauled up and down, when Tom discovers it is easier to pull himself up, than to pull up Dick, who has the same mass. Why is it easier?

18/14

Tom has a mass of 50 kg. By pulling the rope with a spring balance, Harry discovers that it needs a force of approximately 60 N to just support Tom (that is, prevent "overhaul") and a force of about 170 N to slowly raise him further. If the system has a velocity ratio of 5 : 1, what is its mechanical advantage? What is its efficiency? How could this be improved?

Problems 18/13 and 18/14

18/15

Tom and Dick have the same mass, and hold onto a block each, with the rope stretched out along the floor. Harry pulls the free end of the rope. Which boy will slide along the floor? Why? How much rope will Harry have pulled when that boy has slid one metre if there is a single and a double sheave rigged as shown?

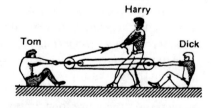

Problem 18/15

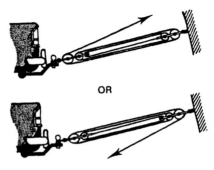

18/16

Your car will not start, and you need to haul it uphill into the garage, to be repaired. You have a suitable anchor point in the garage and a block and tackle with five sheaves. Which block will you hook onto the car, the three-sheave block, or the two-sheave block? Why?

Problem 18/16

18/17

The car engine has a mass of 150 kg and needs to be lifted out for repair. Your block and tackle has five sheaves, and is slung from the ridge of the garage roof. It is about 60% efficient. Will your 60-kg mass be heavy enough to lift it out alone, or will it "overhaul" (that is, lift you)?

Problem 18/17

Sometimes the block and tackle system is used *in reverse* to magnify movement rather than force. A large, slow effort can then give a big movement to a small, light load. This was the method of operating many ancient elevators, and is still in use today, raising people over 300 m in the Eiffel Tower (see Figure 18.22).

A wire rope block-and-tackle system with many sheaves is strung between a strong frame and huge floating weights. The *free ends* of the cables run to the lift cages. Small changes in the heights of the weights are made by changing the water level, the water flowing in from a reservoir and out to the Seine. The resultant small movements of the floating weights are magnified many times by the pulley system, thus allowing the lifts to rise and fall as the control valves are opened and closed.

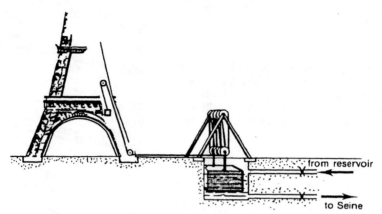

Figure 18.22

Other systems of slight interest involve further combinations of the principles already discussed and will be given brief mention only. The student may research and experiment further.

Multiple movable pulleys

For simplification of the following explanation we will assume that all ropes and blocks are weightless and that no friction is present in the pulley bearings. Start considering the situation shown in Figure 18.23 at the original load, L. Suppose that the load L produces a tension T in its supporting rope. This tension T now becomes the load and is supported equally by two ropes. Thus the tension in each of these ropes becomes $T/2$, and $T/2$ becomes the load on the rest of the pulley system.

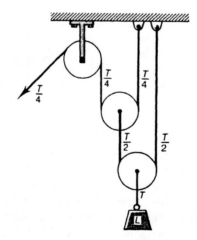

Figure 18.23

Now, this load ($T/2$) is anchored at one end and supported at the other by two ropes, the tension in each of which will therefore be $T/4$. This kind of analysis can be carried further for any given number of movable pulleys. *Note that the single fixed pulley only changes the direction of the last force, T/4.*

If we try to make this situation real, and consider the mass of each "snatch block" (that is, the movable pulley assembly) the amount of cable each assembly supports, friction and the resistance to bending in each cable, the problems become quite beyond the intended scope of this book. This type of block rigging is used on large cranes, where the load is supported on multiple cables, because one cable would be too inflexible.

The Weston Differential Pulley Block

When the Weston Differential system is used as a rope block, several turns are made around each pulley because friction between rope and pulley is necessary to prevent slipping. Therefore a "chain block", using special pulleys which engage the links in the chain, is often used instead of the rope system.

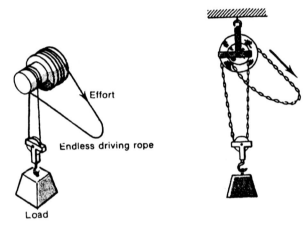

(a) Weston differential rope block

(b) Weston differential chain block

Figure 18.24

Principle of operation. The pulley wheels on the top block are cast together, and revolve as one, but they have different diameters. Those shown in Figure 18.24 (b) have radii, diameters, circumferences, and link slots all in the proportion of 5 : 3.

When the pull P on the slack loop pulls five links of the chain over the big pulley, three links of the chain are lowered from the small pulley. The tight loop, carrying the snatch block and load, has now been shortened by two links, and the load has been raised by a distance equal to one link. This means that the machine has a velocity ratio of 5 : 1. *In practice, the efficiency of these machines is less than 50% and thus the machine does not "overhaul", or run backwards, when the effort (the pull P) is removed.* This is a very useful feature.

The Weston Differential pulley block is particularly suitable for raising heavy loads over short distances—such as raising heavy castings into position in a machine shop.

REVIEW PROBLEM

(Use $g = 10 \, \text{m/s}^2$.)

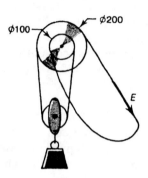

18/18
Determine the effort required to (a) support; (b) raise the load of 100 kg if the Weston Differential rope block shown is only 50% efficient.

Problem 18/18

PULLEYS AND BELTS

The most efficient speed of many electric motors and internal combustion engines is between 1000 and 2000 rpm. This spindle speed is not always suitable for the needs of the machine being driven, and thus direct drive is not often possible. For example, a large water pump may need to operate more slowly, while a small circular saw or drill may need to revolve much faster than the spindle speed of the motor being used.

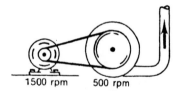
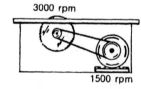

Figure 18.25

A common method of driving such machines is with a belt and pulleys. The belts originally were made of leather, working on "flat" pulleys (which had a small camber to keep the belt on). The modern V-belt is made from rubber reinforced with cotton, and it grips the pulley by wedging itself in a specially shaped groove. V-belts can transmit much more power for their size than flat belts, and have almost entirely superseded them.

When two pulleys (as shown in Figure 18.26) are connected by an endless belt, the smaller pulley will revolve faster than the larger one *if no slip occurs* between pulleys and the belt. The difference in speeds of the two pulleys is *inversely proportional* to their radii, their diameters, and their circumferences (whichever is more conveniently applicable from the information given).

$$\text{VR (A : B)} = \frac{\text{radius } b}{\text{radius } a} = \frac{\text{diameter } b}{\text{diameter } a} = \frac{\text{circumference } b}{\text{circumference } a}$$

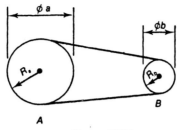

Figure 18.26

If we do not use a belt but allow the two pulleys to rub together, we have a *friction drive*. This is a very old method which has only recently been revived, one reason being that we now have the knowledge to manufacture more suitable friction materials for the mating surfaces. Two methods of direct friction drives are shown in Figure 18.27.

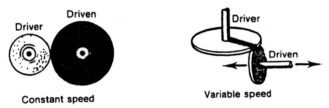

Figure 18.27

GEAR DRIVES

The inefficiencies of the friction drive were first overcome by casting matching teeth in the mating wheels. These are known as "spur-tooth gears", and their design has evolved over many years through various basic tooth shapes, such as "involute", "cycloidal", and "helical", in an endeavour to improve the smoothness of operation, and reduce tooth wear and noise. The modern car gear-boxes are a typical result of this continuing design process.

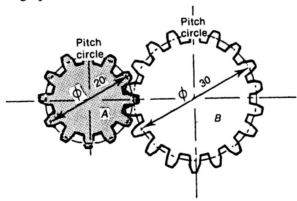

Figure 18.28

The relationship between the speeds of the mating gears is inversely proportional to the effective diameter of the gears, known as the *pitch circle diameter* (PCD), and also to the number of teeth on each gear. (The smaller gear will revolve faster than the larger gear, in that ratio.) Thus, in the instance of the two gears shown in Figure 18.28,

$$\text{VR (A : B)} = \frac{\text{number of teeth on B (18)}}{\text{number of teeth on A (12)}} = \frac{\text{PCD of B (30)}}{\text{PCD of A (20)}}$$

Due to its positive nature, this type of connection between shafts is very suitable for maintaining *synchronisation* between them. A clock has many shafts, the final three of which carry gear wheels with a velocity ratio of 1 : 60. Maintaining synchronisation between these is important—you may not be pleased if your watch suggests there are seventy minutes in the hour or five days in the week! The camshaft and distributor of a petrol engine require accurate synchronisation with the crankshaft, and gears are often used for this purpose. See Figure 18.29.

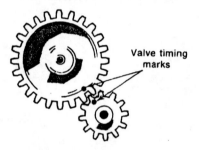

Figure 18.29 Simple timing gears.

The screw-cutting lathe is probably one of the most accessible examples of accurate synchronisation achieved between revolving shafts through use of gears. In order to cut a screw-thread, a constant relationship must be maintained between the revolving work-piece and the movement of the screw-cutting tool. The lead-screw (which feeds the tool) is connected to the spindle by gears known as "change-wheels"

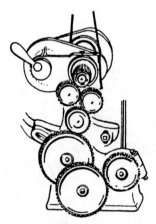

Figure 18.30 Hercus lathe

and these can be varied to suit the particular thread being cut. The type of gearing used is shown in Figure 18.30; however, the student is encouraged to examine a small lathe for himself.

When synchronisation between gears is not important, such as in the car gear-box, the washing machine gear-box, and the back gears on the lathe, then it is carefully avoided by adding one extra tooth to the larger gear. This helps the teeth wear more evenly, and reduces vibration.

CHAIN DRIVES

Another method of maintaining a certain orientation between revolving shafts is by the use of a *roller chain* and *chain wheels*. The chain sprocket on the crankshaft of a petrol engine has only half as many teeth as the one on the camshaft (refer to Figure 18.31). This ensures that the camshaft revolves at exactly half crankshaft speed, an essential feature of the 4-stroke internal combustion engine.

Figure 18.31

A roller chain is also used where synchronisation may not be necessary but where it is important that no slip occurs. The bicycle and motorcycle use roller chains to provide the final drive. (Suggest reasons why this method is chosen instead of belts or gears.) The motorcycle uses a small drive sprocket and a large wheel sprocket because the engine needs to revolve faster than the back wheel. The bicycle, on the other hand, uses a large drive sprocket, and a small wheel sprocket because the rider's legs are more suitable for a slow, powerful effort (refer to Figure 18.32).

(a)   (b)

Figure 18.32

REVIEW PROBLEMS
(Use $g = 10$ m/s^2 unless otherwise specified.)

18/19
Examine the various types of bicycles available, and count the teeth on the drive sprockets and driven sprockets of each type. Measure the crank lengths and the various diameters of the drive wheels. Calculate the velocity ratio for each machine overall, and tabulate the results.

On the basis of this information and practical tests, recommend the most useful combinations of crank lengths, wheel diameters, sprocket sizes, and overall velocity ratios. Determine which bicycle is
(a) easiest to push;
(b) goes furthest for the effort expended;
(c) is most comfortable over long distances;
(d) any other criteria which you may feel relevant to the sensible, informed choice of a bicycle.

18/20
Make a similar investigation of mini-bikes, trail bikes, or go-carts. Tabulate all your findings and model your reports on the reports to be found in consumer publications such as *Choice* or *Which?*. Do not hesitate to criticise any or all of the current models which may be more "market-oriented" than well designed. Allow your imagination to run free when considering viable alternatives to the models available.

THE INCLINED PLANE

One of the most frequently used simple machines is *the inclined plane* and its use as a *ramp* for raising or lowering loads is well known. Some common examples are shown in Figure 18.33.

Another inclined plane is the wedge. Wedges can be driven into rock or timber to split it open; between two mating parts to separate them; or under a heavy machine to adjust its height. Wedges are often driven under heavy loads to lift them enough to enable the crane sling to be slipped under them. When used to adjust the height of a prop or tom during building construction, a pair of folding wedges is usually used. (See Figure 18.34.)

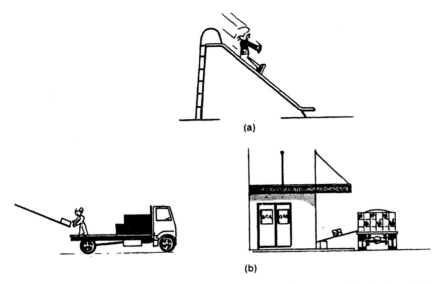

(a)

(b)

Boxes, mailbags and drums slide down planks to speed the loading and unloading of lorries.

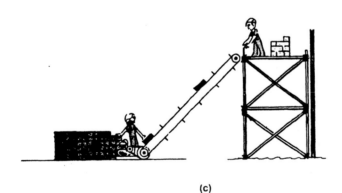

(c)

The builder's brick elevator chugs away all day, pushing bricks up an inclined plane to the bricklayers.

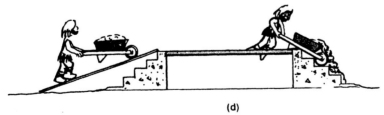

(d)

It is easier to wheel a barrow up and down a plank than up and down steps. The smaller the slope the easier it becomes.

Figure 18.33

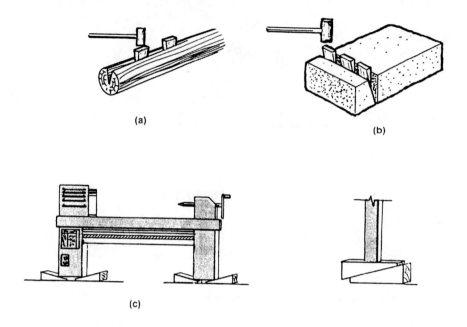

(a)

(b)

(c)

Figure 18.34

The wedge can also be used to hold two parts tightly together. In this case it can hardly be called a machine. For example, the mortise and tenon joint is often wedged; pulleys are wedged onto shafts with keys, and the head is wedged on to the handle of a hammer (refer to Figure 18.35).

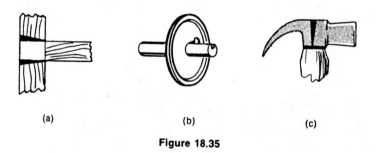

(a)   (b)   (c)

Figure 18.35

To the engineer, roads, railways and drains are all inclined planes (see Figure 18.36).

(a) Roads are cambered (that is, given a slope to drain the water off the surface).

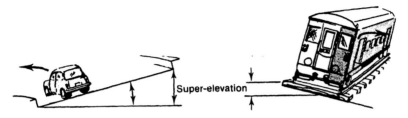

(b) Roads and railways are always super-elevated on curves to assist the vehicle to change direction.

(c) Roads and railways have carefully designed, even gradients up and down hills.

(d) Rivers and drains have a constant fall to the sea (or to the pumping station).

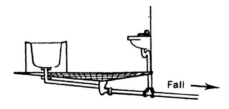

(e) Floors which get wet in service are given a fall to a drain. The bathroom and laundry floors have a gentle slope enabling splashed water to run into the house drains.

Figure 18.36

There are apparently many specialised terms to describe the inclination of a surface, but fortunately there are only a few methods of measuring this inclination. The angle is expressed in degrees when it is of suitable magnitude, but in the case of very small angles such as the gradients of roads, railways and drains, the slope is quoted as the sine or tangent of the angle of inclination. For example, a gradient of 1 in 40 means a rise or fall of 1 metre for every 40 metres travelled. (For $\theta > 5°$, $\sin \theta \approx \tan \theta$).

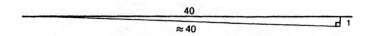

Figure 18.37

When calculating the forces involved for any inclined plane, a free-body diagram should always be drawn. This will show the forces acting on the object of interest only. Problems involving inclined planes, particularly wedges, can be complex due to the frictional forces involved and graphical solutions are often useful. Refer to Chapters 8, 9 and 10 for details, particularly to Sample Problems 8/5, 8/8, 9/10, 10/14 and 10/15.

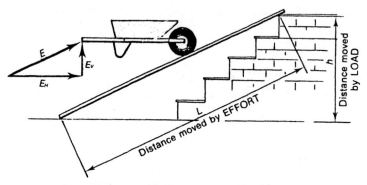

Figure 18.38 The vertical component E_v does useful work. Thus the velocity ratio is equal to L/h; the mechanical advantage will be less due to friction.

The velocity ratio for an inclined plane is found from the usual formula

$$VR = \frac{\text{distance moved by EFFORT (along the inclined plane)}}{\text{distance moved by LOAD (vertically up)}}$$

$$MA = VR \times \text{efficiency}$$

The efficiencies of inclined planes are often very low due to frictional forces. However, the convenience and practicability offered by inclined planes more than offset these low efficiencies; for example, wheeling the barrow up the plank is really much easier than lifting or dragging it up the steps.

During the building of the Great Pyramids, an inclined plane was built around each pyramid to make a gradual slope up which the huge stones could be hauled. This ramp was extended as necessary, and removed at the completion of the structures. Our roads traverse mountains in a similar fashion (see Figure 18.39 (a)).

If a length of wire is wrapped around a cylinder, a coil spring is formed. The name of the three-dimensional curve so produced is a *helix*, and the spring is known as a *helical coil spring*. The distance from one turn of wire to the next is known as the *pitch* (Figure 18.39 (b)).

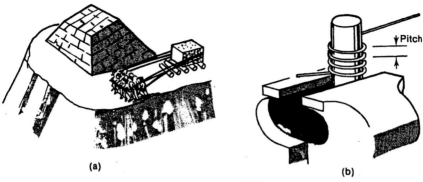

(a)

(b)

Figure 18.39

A helix is only a simple inclined plane on the surface of a cylinder, instead of one on a flat surface. The angle of this inclined plane is known as the *helix angle*, and is found by relating the distance travelled axially along the cylinder, to the distance moved around the circumference of the cylinder (πD). Thus,

$$\frac{\text{pitch}}{\text{circumference}} = \text{gradient of inclined plane}$$

$$= \text{tangent of helix angle}$$

$$\frac{P}{\pi D} = \tan \theta$$

This is easily seen when we unroll one turn of wire. (Roll a spring on a stamp-pad and then along a sheet of paper—the gradient can be calculated from any convenient length, L. (Refer to Figure 18.40.)

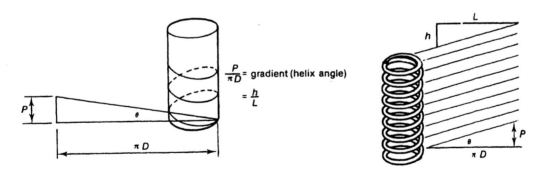

$\frac{P}{\pi D}$ = gradient (helix angle)

$= \frac{h}{L}$

Figure 18.40

THE ARCHIMEDEAN SCREW

One of the earliest uses of the helix was the Archimedean screw. It has been lifting water out of the Nile for thousands of years. A flat blade is coiled around the inside of a pipe, forming a ramp rather like a spiral staircase, up which water can be made to climb. The pipe is tilted at an angle greater than the helix angle, and rotated. The water *runs down* the part of the helix which slopes downhill, *but rotation continually transfers the water along the helix*, and therefore upwards (and out the top into an irrigation ditch).

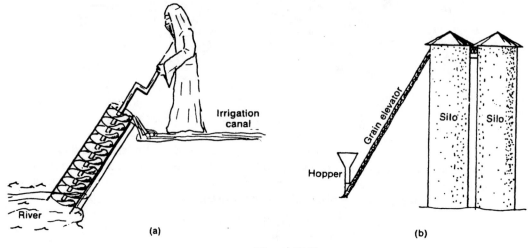

Figure 18.41

Modern versions of the Archimedean screw are used in grain elevators, and in the ready-mixed concrete trucks. When the agitator drum of the concrete truck is revolved in one direction the contents are evenly blended and agitated (this extends the setting time of the concrete). Rotation in the opposite direction discharges the concrete. A two-start helix is usually employed, as shown in Figure 18.42.

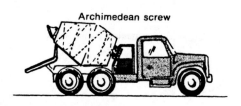

Figure 18.42

THE WORM AND WORM WHEEL

When a helix is meshed with a gear wheel, the helix is called a *worm* and the gear a *worm wheel*. The worm and worm wheel are very suitable for connecting two shafts at right angles to each other, they provide smooth, silent gearing. However, due to the sliding contact surfaces involved good lubrication is a problem.

Figure 18.43

A very wide range of velocity ratios is possible. When used in the differential in a car or bus, the VR used is about 5 : 1 but when used in reduction gear-boxes for speedometer drives and hoists and winches, the VR may be as high as 500 : 1. The velocity ratio depends on the helix angle of the worm and the effective diameter of the worm wheel. Very steep helix angles are used in the mechanism that must overhaul, such as the car's differential gear, and small helix angles are used if the worm drives a winch, such as the chain-block, which must not overhaul.

SCREW THREADS

The version of the helix which probably has proved most useful to man is the common *screw-thread*. The bolt, nut and spanner form the most magnificent combination of the inclined plane, the wheel and the lever yet known (all of them are basic machines). The pure simplicity of design must make the nut and bolt the greatest invention of all time, and yet we take them for granted.

Research assignment: *Find out who invented or first used the idea.*

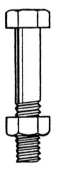

Figure 18.44

There are many types of screw-threads, but they all fall into two basic categories: screw-threads for *transmitting power or motion* and screw-threads used for *fastening*. The fastening threads have a V-section and introduce friction, bursting and wedging forces beyond the scope of this chapter, so we will consider only those used for transmitting power and motion. These have a square-type thread (square, acme or buttress) and so are able to deliver all their thrust in the intended direction. (No extraneous nut-bursting forces are present to complicate the calculations.) These types of threads are shown in Figure 18.45.

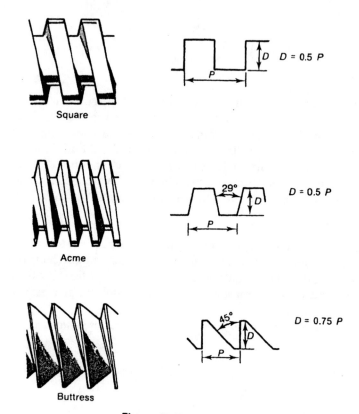

Figure 18.45

They are used in vices, car jacks, and universal testing machines (such as the Hounsfield Tensometer) and on lead screws and feed screws of lathes. When a screw-thread is used to do work, such as raising a car, the velocity ratio can be calculated from

$$VR = \frac{\text{distance moved by EFFORT}}{\text{distance moved by LOAD}}$$

The effort is usually applied at the end of a spanner, tommy-bar, or other lever (such as a vice handle) and the circumference of the path along which the effort travels will be the distance the effort moves. The load will rise a distance equal to the lead for every full turn of the thread (this is equal to the pitch for a single-start thread).

Pitch

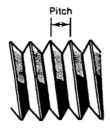

Figure 18.46

SAMPLE PROBLEM 18/21

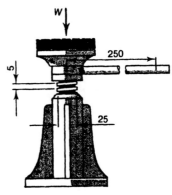

Determine the velocity ratio of the car jack if it has a single-start square thread of 25-mm effective diameter and 5-mm pitch. The effort is applied to a tommy bar at a point 250 mm from the axis of the thread. What is its probable mechanical advantage?

Problem 18/21

Solution:

$$VR = \frac{\text{distance moved by EFFORT}}{\text{distance moved by LOAD}}$$

During one revolution of the screw, the effort will move πD mm in a circle around the jack, and the load will move a distance of 5 mm upwards.

$$VR = \frac{\pi \times 500}{5}$$
$$= \frac{\pi \times 100}{1}$$
$$= \frac{314.2}{1}$$

Analysis:
The mechanical advantage is always less than the velocity ratio due to friction and other losses. In a machine such as this, friction must be high enough to prevent the machine from running backwards when the effort is removed (in this case, the helix angle is designed to be less than the angle of friction) and therefore the efficiency is usually less than 50%. The actual mechanical advantage would be found by experiment to be no greater than 150 : 1.

Note:
The given diameter of the screw-thread is an important part of the design of the jack, affecting the strength, stability, and life of the machine, but is quite irrelevant to this problem. ∎

THE HYDRAULIC RAM

A much more convenient device for raising a motor car is the *hydraulic jack*. The *effort* on a small piston pushes it down into a fixed volume of incompressible fluid, such as oil. The pressure in the fluid then forces a larger piston, the *ram*, to be pushed up (see Figures 18.47 and 18.48). The volume of the effort piston which is pushed into the oil must equal the volume of the load piston, the ram, pushed out of the oil ($\pi r^2 h = \pi R^2 H$).

$$VR = \frac{\text{distance moved by EFFORT}}{\text{distance moved by LOAD}} = \frac{h}{H}$$

Since

$$\pi r^2 h = \pi R^2 H$$

therefore

$$\frac{h}{H} = \frac{\pi R^2}{\pi r^2} \left(= \frac{\text{area of ram}}{\text{area of piston}} \right)$$

$$\therefore VR = \frac{R^2}{r^2} \left(= \frac{\text{radius of ram}^2}{\text{radius of piston}^2} \right)$$

$$= \frac{D^2}{d^2} \left(= \frac{\text{diameter of ram}^2}{\text{diameter of piston}^2} \right)$$

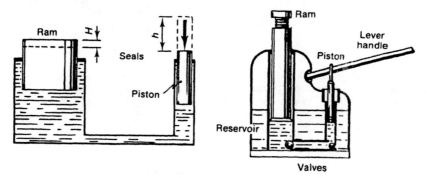

Figure 18.47   **Figure 18.48**

In practice, valves allow many strokes of the effort piston to draw fluid from a reservoir, and so raise the car by a useful amount.

The hydraulic brakes of a motor car operate on a similar principle (see Figure 18.49). The fact that when fluids are under pressure, the pressure is the same throughout the fluid, enables the brakes on all wheels to have exactly the same pressure applied to them.

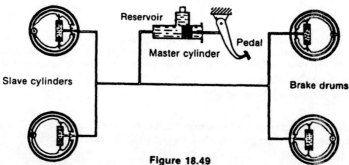

Figure 18.49

The hydraulic cylinders which operate tractor attachments, bulldozer and grader blades, front-end loader and back hoe buckets, etc., also use oil under high pressure to force a piston up or down in the cylinder. Oil is pumped continuously from a reservoir past a pressure relief valve to the control valves. When the operator moves a control valve lever, this fluid under very high pressure flows into one end of the cylinder, and at the same time the fluid in the other end of the cylinder is allowed to return to the reservoir (Figure 18.50 (b)).

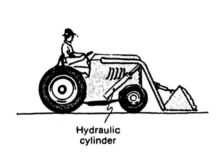

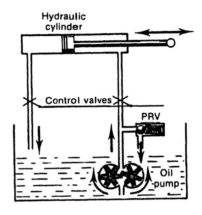

Figure 18.50

Hydraulic systems are very flexible, allowing great power to be transmitted easily to the place where it is to be used. They have very few moving parts to wear out, and as most of these parts work in oil, hydraulic systems are smooth, silent and nearly friction-free. They have a high degree of efficiency and hence possess mechanical advantages almost equal to their velocity ratios.

REVIEW PROBLEMS
(Use $g = 10 \text{ m/s}^2$ unless otherwise specified.)

18/22
Fred is carefully reeling in his catch on his 50-mm diameter reel and the fish continues to pull at 250-N force. If the length of the handle is 100 mm, what force must delighted Fred exert to secure his catch?

Problem 18/22

18/23

In a test on a lifting machine, it was found that an effort of 120 N was required to lift a load of 300 kg. The effort moved through a distance of 10 m while the load moved 300 mm. Determine

(i) the distance moved by the effort while the load is moving 1 m;

(ii) the work done (output) by the machine while the load is moving 1 m;

(iii) the efficiency per cent of the machine when the load is 300 kg;

(iv) the effort required to lift the 300 kg load, if the machine were perfect.

18/24

Fred's D4 Caterpillar tractor (bulldozer) has a maximum drawbar pull of 150 kN near stalling point, under full engine power. Fred is about to use it to remove a stump, and is undecided where he should attach his cable to the stump in order to get the maximum bending moment at the base of the stump. Explain to Fred where he should attach it, and why.

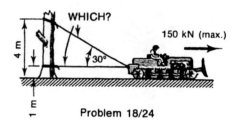

Problem 18/24

18/25

A remote lever-actuated electrical circuit breaker at the top of a pole is shown. Given that the normal force between the blade and the switch contacts is 100 N, determine the magnitude of the force F_H that must be applied to the handle when operating the switch, if the coefficient of static friction present between the switch contacts and the blade is 0.5.

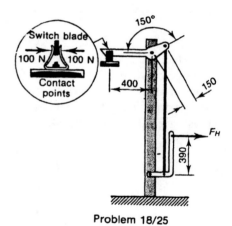

Problem 18/25

18/26

If a 14-mm steel bar requires a force of 36 kN to cut it with the bolt-cutters shown, what force does the operator have to exert on the handles?

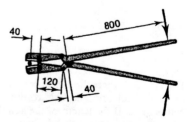

Problem 18/26

18/27

A set of pulley blocks is used to lift a load of 200 kg.
(i) If the effort required is 300 N and the velocity ratio of the pulley blocks is 8 to 1, what is the efficiency of the equipment?
(ii) Explain briefly why the efficiency is less than 100 per cent.

18/28

A socket spanner is designed for use by a crewman of a spacecraft where he has no platform against which to push.

The pin *A* fits into a hole in the spacecraft near the bolt to be turned. By squeezing the handles of the tool, the bolt turns. One side of the tool is used for tightening and the opposite side for loosening a bolt. The reaction against the pin *A* provides the "anti-torque" characteristic of the tool.

For a gripping force of $P = 30$ N, determine the torque transmitted to the bolt.

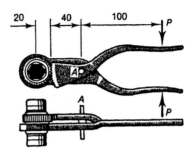

Problem 18/28

18/29

Determine the theoretical torque available to drive the drill bit with a force of 30 N on the handle. The pitch circle diameter of the hand wheel is 90 mm and that of the pinion is 20 mm. The crank is 90 mm long.

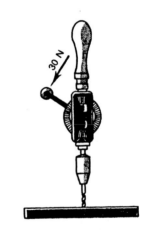

Problem 18/29

18/30

A garden roller of mass 50 kg is to be moved up the step. Is it easier to lift it vertically, or to drag it up the step using the handle inclined as shown?

18/31

For a given lifting machine the effort moves 10 times as fast as the load. If an effort of 200 N lifts a load of 160 kg, find the mechanical advantage and the efficiency per cent.

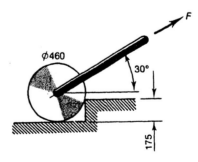

Problem 18/30

18/32

A power operated loading platform for the back of a truck is shown in the diagram. The position of the platform is controlled by the hydraulic cylinder which pulls at *C*. The links are pivoted to the truck frame at *A, B*.

Determine the force *P* supplied by the cylinder in order to support the platform in the position shown. The mass of the platform and links may be neglected compared with that of the 200 kg washing machine on it.

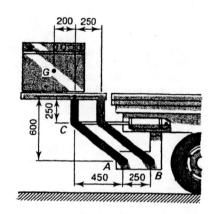

Problem 18/32

18/33

What load may be lifted by an effort of 40 N applied to a lifting machine, if the velocity ratio is 80, and the efficiency is 60 per cent?

18/34

An electric motor drives a steel shaft by means of a belt drive. The pulley fitted to the motor shaft is 100 mm diameter, while the one fitted to the driven shaft is 200 mm diameter. If it is assumed that there is no belt slip and the speed of the electric motor is 1450 revolutions per minute, determine
(i) the linear speed of the belt;
(ii) the angular speed of the driven shaft.

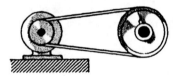

Problem 18/34

18/35

For the compound gear train shown in the diagram determine the speed of the driven gear and the overall velocity ratio.

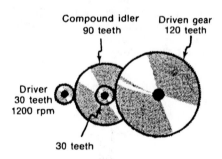

Problem 18/35

18/36

A rope pulley block and tackle has a velocity ratio of 5 : 1. When lifting a load of 200 kg its efficiency is 80 per cent. What effort would be necessary to lift the 200-kg load?

18/37

A crawler tractor with a dozing blade (a bulldozer) is pushing a mound of earth with a force of 100 kN.
(i) Determine the force in each of the pair of hydraulic piston rods which hold the blade in position.
(ii) Is it a tensile force or a compressive force?
(iii) What is the pressure in the hydraulic lines, if each cylinder has a cross-sectional area of 8000 mm² ?

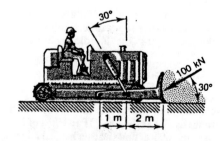

Problem 18/37

18/38

The trailer plus load has a mass of 1 tonne and is equipped with two jacking cams. By lowering the cams and reversing the trailer the load can be removed from the wheels. What minimum horizontal force F must be exerted on the drawbar in order to just lift the loaded trailer off its wheels?

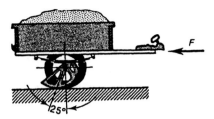

Problem 18/38

18/39

A crawler tractor with a single ripper attached is excavating a building site. When removing a large rock, the tine point of the ripper is acted upon by a force having vertical and horizontal components 20 kN and 50 kN. Determine the forces present in the hydraulic cylinder and in the top and bottom links for this system of loading.

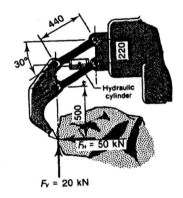

Problem 18/39

18/40

A screwjack has a screw with a pitch of 5 mm and is operated by a tommy bar of 350 mm in effective radius.

  (i) What is the velocity ratio?
 (ii) If the efficiency is 50 per cent, what is the effort required to lift 1 tonne?
(iii) What is the mechanical advantage?

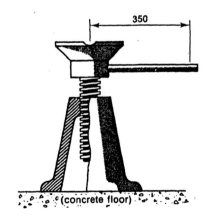

Problem 18/40

18/41

A hand winch has the dimensions shown in the diagram. It is 60 % efficient. Determine the tension, T, in the cable when there is a force of 100 N on the handle.

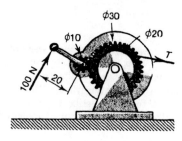

Problem 18/41

18/42

Fred has his first car accident and has to have his car towed away. The tow-truck driver's hands on the winch handle travel 40 m while he is winding up the crane cable and raising the car 500 mm. Of the work done, 35 % is wasted due to friction. The crane cable is supporting a load of 0.7 tonne. What is the force exerted by the tow-truck driver's hand?

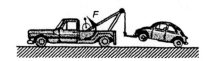

Problem 18/42

18/43

A row of wedges is used to split a hardwood log. If the coefficient of friction between wood and wedge is 0.15, calculate the maximum included angle θ for which the wedges are self-locking in the log.

Problem 18/43

18/44

In the car-jack shown in the diagram, the bevel gears have a velocity ratio of 3 : 1 and the screw thread has a pitch of 7 mm. Its efficiency is found to be 20 %. Find the approximate effort required to raise a 500-kg load.

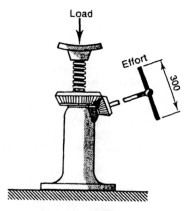

Problem 18/44

18/45

The main features of a travelling luffing crane are shown in the diagram. The distances *AC*, *AD* and *CD* are each 5 metres. The position of the boom is controlled by the winch at *C* winding a greasy wire rope which operates through a series of pulleys, providing a velocity ratio of 12 : 1. The winding system is 80% efficient. The centre of mass of the 100-tonne turntable and winchhouse is near the winding drum below the point *C* and the gantry is symmetrical.

Determine the tension in the cable at the winding drum and the force on the wheels near the edge of the wharf when the 20-tonne load is being raised at uniform velocity.

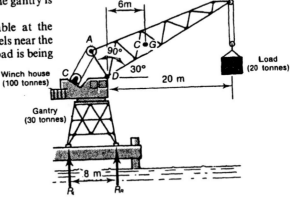

Problem 18/45

18/46

An automobile front-wheel assembly supports 250 kg. Determine the force exerted by the spring and the shear force in the swivel pins at *A* and *B*.

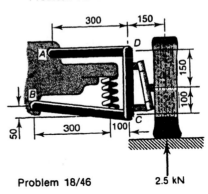

Problem 18/46

18/47

The device shown is commonly used on yachts, and grips a rope (a line) under tension because of the very large friction forces developed. Given that the rope has a 5-kN force in it and the coefficient of friction present is 0.4, determine the total shear force acting on the swivel pin at *A* for the position shown.

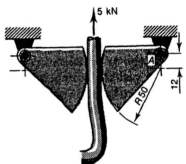

Problem 18/47

19

An Approach to Design Analysis

Engineering design is an activity directed towards the goal of satisfying those human needs which can be met by the technology of our society. Whenever man has a design problem, he finds a solution. It is not always a good solution, but with each successive attempt he learns a little more about the real nature of the problem and the new knowledge helps with the solution to the next problem of a similar nature. There are two types of design: *design by evolution* and *design by innovation*.

Design by evolution occurs when devices or systems change gradually as time goes on, each change making a small improvement to the previous model. This type of approach reduces the possibility of major errors.

Design by innovation often follows a scientific discovery when a new body of technical knowledge develops rapidly. The proper use of this new technology may dictate a break with the past, and so a new design based on untried or unproven ideas is projected. The risk of technical error in innovation is immense, and great care must be taken at each stage of the design to prevent costly and possibly disastrous mistakes.

Very little need be said about design by evolution, as we all regularly modify and improve our surroundings and possessions. However, design by innovation is a much more involved process which can be very exciting and rewarding when the results of calculated risks prove to be worthwhile.

It is proposed in this chapter to discuss very broadly some of the common engineering approaches to the analysis of design problems. The procedures adopted by a design engineer are usually simple, logical, step-by-step processes which can perhaps be summarised as in the Figure 19.1.

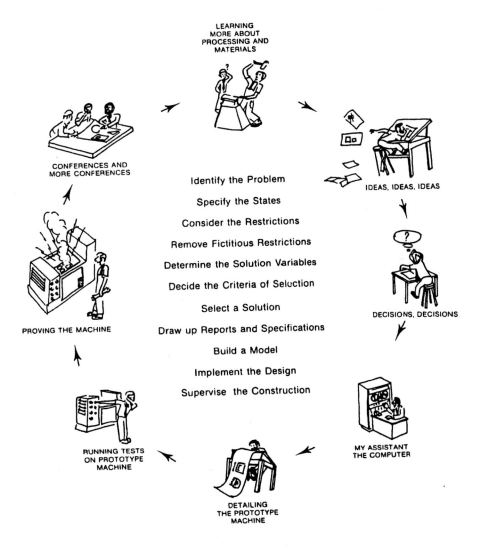

LEARNING MORE ABOUT PROCESSING AND MATERIALS

CONFERENCES AND MORE CONFERENCES

IDEAS, IDEAS, IDEAS

PROVING THE MACHINE

DECISIONS, DECISIONS

RUNNING TESTS ON PROTOTYPE MACHINE

MY ASSISTANT THE COMPUTER

DETAILING THE PROTOTYPE MACHINE

Identify the Problem

Specify the States

Consider the Restrictions

Remove Fictitious Restrictions

Determine the Solution Variables

Decide the Criteria of Selection

Select a Solution

Draw up Reports and Specifications

Build a Model

Implement the Design

Supervise the Construction

Figure 19.1 The design cycle.

IDENTIFYING THE PROBLEM

The first step in design is always to identify the *real* problem, as the problem posed by the engineer's employer is often only the apparent problem. For example, consider the following situation.

Many years ago, Fred's wife asked him to design a more efficient tool for splitting wood for the cooking fire. Fred promptly invented the axe. His grand-children found that coal would also burn and so they invented the pick and shovel. Fred identified the problem as one of "chopping wood"; his grand-children saw it as one of "digging coal", but actually these were only the *apparent* problems. The *real* problem was to simplify the process of cooking.

Later solutions to the same problem have included the use of oil, gas, electricity and microwaves. It is only recently that the obvious source of all energy in the solar system is being tapped directly by man. There are problems associated with the efficient and economical collection and storage of solar energy yet to be solved, but these sub-problems are a lot closer to the real problem than Fred's axe.

SPECIFY THE STATES

To help to identify the *real* problem it is often useful to specify the states before and after the proposed design exercise. These are sometimes known as "input" and "output" or State A and State B.

In Fred's case these were *apparently*

Before (input or State A)	big pieces of wood
After (output or State B)	small pieces of wood

Apparent problem

but they were *really*

Before (input or State A)	uncooked food
After (output or State B)	cooked food

Real problem

If Fred had realised the nature of his real problem, he would have immediately lined his umbrella with aluminium foil—thereby inventing the solar barbecue, and depriving *his* world of the axe!

The true nature of a problem is often obscured by irrelevant information, mis-leading opinions, and also by the solutions currently in use. This difficulty is com-pounded by the fact that, while learning, trainee engineers are usually presented with "textbook" problems which have an unrealistically pure form. Thus they have little experience in defining and formulating the real problem for themselves. A common tendency is to try to improve the existing procedure or the existing solution; that is, to use *design by evolution*. This can usually be done but rarely provides the most satis-factory solution.

For example, it is desired to reduce the traffic congestion in major cities. The chosen solutions have usually involved wider roads, more miles of freeways and bigger parking stations. These solutions only make it easier to get more cars into the city and thus provide a greater potential for even bigger traffic jams! Only the more enlightened designers have considered the development of cheap, comfortable mass transport systems so that cars *would not need to come into the city*! An even broader solution to this problem is decentralisation which effectively reduces the *need* for so many people in one central area.

Usually a knowledge of the historical background to a situation will justify the existence of the current design. Often an examination of the ways in which a particular problem has been approached in the past proves to be worth while, even if it only helps the engineer to guard against repeating the same mistakes. However, in many cases it becomes apparent that any improvement on an existing design would be quite unsuitable for the present conditions, and an entirely new solution must be found. For instance, the horse and carriage reached a peak of development with springs, rubber tyres, and many horses to pull it. Further improvements were impossible without a radical change in the design philosophy—thus the horseless carriage! Even this was really only a relatively minor change in design since it was the type of motive power, rather than the type of vehicle, that was modified.

RESTRICTIONS

Once the real problem has been identified, the next step is to list and consider the restrictions placed on the designer. These restrictions can be imposed by his employer, the forces of nature, the need to comply with particular laws, and other limiting factors with which the designer must work. Some of the restrictions imposed on the designer cannot always be completely met. Compromises are often necessary, and indeed often desirable. When restrictions will be incompatible with each other, or economically unsound, the engineer's duty to his client or employer is to acquaint him with the facts, and then present him with viable alternatives.

FICTITIOUS RESTRICTIONS

Consider the following problem:
You are required to connect the nine dots in Figure 19.2 with four straight lines, without removing your pencil from the paper.

Figure 19.2

Many people are unable to solve this problem, because they automatically restrict the solution to the confines of the square formed by the dots, even though no such restriction was mentioned in the original statement of the problem.

Everyone has the tendency to impose these unnecessary boundaries on his thinking, and to accept *what* IS as *what* SHOULD BE. Realisation that this is not necessarily the case is the first step towards opening up a problem to worth-while solutions.

SOLUTION VARIABLES

Alternative solutions to a problem can differ in many respects. Consider the problem of "crossing a river". Possible solutions would include the following: jumping; swimming; flying fox; ford; rowing; launch; ferry; hovercraft; road bridge; rail bridge; tunnel; aircraft; or rocket. Each of these is a valid solution to the broad problem with no consideration given to possible restrictions. The restrictions which would seem to be paramount would include such things as the width, depth, and speed of the stream, and the volume and frequency of crossing traffic. For instance, a small capsule of medicine urgently required on the other side of a flooded creek would obviously demand an entirely different solution to the case of a possible million people crossing daily each way over a big river.

CRITERIA OF SELECTION

The factors influencing the selection of the best solution should be identified during the problem analysis. These tend to change very little from one problem to another, and include such things as:

Suitable materials of manufacture   Aesthetic appeal
Cost of construction   Environmental impact
Durability   Economic viability
Ease of maintenance   Anticipated usage
Safety   Production volume
Reliability

What does change, however, is the relative importance given to each of these criteria. The construction cost of a "one off" bridge designed to last for hundreds of years will be relatively unimportant, whereas the cost of designing and setting up a factory to produce 100 000 washing machines which must be sold, and which are designed to be obsolete in five years, would unfortunately be paramount.

SELECTION OF A SOLUTION

The selection of the *most suitable solution* to the problem is a very difficult part of the exercise. Determining the relative merits, based on the accepted restrictions, of each of the solution variables and the relative importance of the criteria of selection provide the design engineer with a range of possible solutions from which he must make a selection.

"Optimisation" is the design engineer's term for compromise or balancing one criterion against another. The motor vehicle trade provides many excellent examples, one of which is the problem of rusting. The bodies of very many modern vehicles are prone to rusting. Why is this so? It is quite possible to design a car body which will not

SAMPLE PROBLEM ANALYSIS 19/1: A TYPICAL EXERCISE

The design engineer with the total electricity supply authority has been asked to submit a report on the most suitable type of post for supporting the street lighting in a new suburb.

His analysis of the problem may look something like this:

Apparent Problem	Provide lamp posts (maybe?)
Real Problem	Illuminate Newburb!
State A (input)	Dark, hazardous streets of the new suburb (robbery, rape, murder)
State B (output)	Well-lit streets of the safe new suburb
Restrictions	Must be suitable to be installed by the supply authority workmen
Fictitious Restrictions	Usual technique is big light bulbs on wooden poles. (UGH!)

Solution Variables		**Constraints**
Type of illumination	Incandescent bulbs	Power hungry, short lived
	Fluorescent tubes	Installation cost
Usual solutions	Mercury vapour lamps	High capital cost (ugly colours)
	Sodium vapour lamps	High capital cost (ugly colours)
"Think-tank" solutions	Fluorescent road paving	Not developed.
	Arc-lamp suspended from balloon	Impractical
	Mirrors on a satellite	at this stage
	Trained fireflies	
Type of support posts	Cast iron columns	Expensive, hazardous
	Timber poles	High maintenance
	Concrete pillars	Motoring hazards
	Galvanised iron pipes	
	Aluminium extrusions	Cost!
	Moulded plastic products	Cost!
	Fibreglass	Cost!
Location of posts	Near curb	Motoring hazard
	Near building line	Excessive cantilever?
	No posts (lamps on buildings)	Access to buildings?
Criteria of Selection	Safety to motorists	
	Blend with surroundings	
	Cost of installation	
	Ease of installation	
	Maintenance costs	
	Operating costs	

---

rust. The bodies can be made of stainless steel, aluminium, plastic, fibreglass reinforced polymers, or plywood, all of which have been used with varying degrees of success; and yet the motor vehicle manufacturer continues to make car bodies from mild steel which rusts. Why? A compromise or "trade-off" has been made between manufacturing cost and quality of product. If the manufacturer chooses to build a car of A1 quality, the production cost will be so high that very few people will buy the vehicle. The steels chosen for motor bodies have very good deep-drawing characteristics, enabling the shaping of the body to be performed in fewer operations than would otherwise be possible, with consequent savings in cost. Unfortunately one disadvantage of this type of steel is its susceptibility to corrosion. Car designers, rightly or wrongly, have made a compromise between quality and cost.

The car tyre is another example of optimisation. Here we "trade-off" the comfort and silence of soft rubber against a suitable and acceptable amount of wear. If the tyre was designed for negligible wear, then it would become so hard that it would lose its traction on the road. (Imagine a steel tyre—it would provide negligible wear, but also very little acceleration, braking, or cornering ability!)

PROBLEMS IN ANALYSIS

19/2
Use the procedures of analysis previously outlined to justify the method of supporting the street lights in your neighbourhood. Take into consideration their probable date of erection. Suggest an improved street lighting system to cater for the current and probable future needs of the area.

19/3
Critically examine three varieties of modern motor cars. Compare and evaluate the "monster" limousine, the medium-sized family vehicle, and the "mini" personal transport car.

Try to identify the real problem to which each type of vehicle appears to be the current solution. Comment on the success of each type as a suitable solution to the problem as you see it.

Critically compare such properties as the road space occupied by each type and their relative fuel usage.

Assess the total energy involved in their production (materials and construction).

Comment on the need for a prestige vehicle against simple transport.

Discuss the comfort of the ride in each type and consider the effects of the greater inertia of the limousine (for example, on the driver and on passengers).

19/4
There are many different methods available for joining materials together. The *roped joint* as used on Indian bridges provides an example of an early technique, whereas modern methods include the simple pin-joint, as well as bolted, riveted, soldered, brazed, welded and glued joints. Some timber roof trusses have the members joined by *nail plates*, a fairly recent innovation.

Prepare a report on methods now available for joining materials together. This report should include:

(1) A tabulation and comparison of the methods available.
(2) Accounts of *field trips* undertaken to observe joints: for example, to welding works, bridges, other structures.
(3) Sketches of the types of joints considered.
(4) Considerations of joint suitability for particular applications, that is, advantages and limitations of the materials involved. For example,
Where is a rivet chosen in place of a bolt?
Where would a joint be brazed rather than soldered?
When would a joint be glued rather than screwed?
(5) Consideration of forces that can be transmitted by certain types of joints and their strength factors.

19/5

Compare and criticise the varying shapes of beams in common use. Discuss the reasons for the shape of each and the proportions of material in each part of a beam.

In your analysis consider the following questions:

Why is a ceiling joist rectangular?

Why is a rolled steel joist tapered? (The RSJ is also called a *taper flange beam*.)

Why is a universal beam *not* tapered?

What are the advantages of a "built-up beam"?

What are the advantages of a prestressed concrete beam?

Suggest reasons why prestressed concrete bridges are suitable for road transport, and are unsuitable for electric trains.

Offer reasons for the change in choice of materials used in beams in the last fifty years.

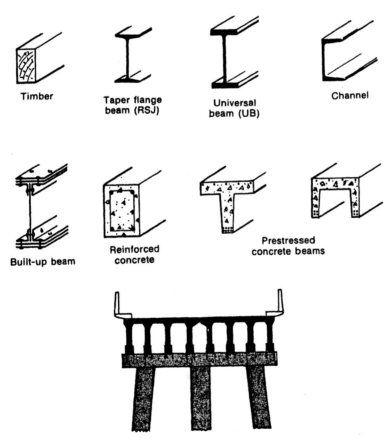

Problem 19/5   A few varieties of beams in common use.

19/6

For many years we have been extending aircraft runways further in order to accommodate aircraft of increasing size and payload capacity. However, some recent developmental work in the aircraft industry has centred around the *vertical take-off and landing aircraft*—the VTOL or "jump-jet".

Investigate the problems of *engine size* and *thrust* associated with large conventional passenger jets—the *jumbo jet*, for example—and with current VTOL aircraft. Suggest reasons why VTOLs are usually small military machines rather than commercial airliners.

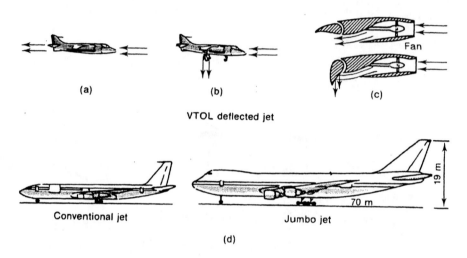

Problem 19/6   The first three diagrams show a typical deflected jet VTOL aircraft: (a) shows the jetstreams in level flight; (b) shows the jets deflected for vertical rise and fall and (c) shows a possible method of deflecting the jet stream. Diagram (d) shows a size comparison of a typical jumbo jet with a passenger jet of more conventional size.

19/7

One aspect of passenger comfort during long trips in trains, buses and cars is centred around the effects of forces and accelerations on both the vehicle and the passenger. One important component of this problem is concerned with the effects of *inertial* forces.

Summarise the effects of such forces on both passengers and vehicles and write a report on the implications for highway and railway engineering. This report should include effects of transition curves and superelevation on the stability of vehicles when rounding bends.

Why is the superelevation of roads and railways always less than the theoretical amount required?

Are there other solutions which could perhaps involve redesigned vehicles?

19/8

Every year an increasing number of people use their motor vehicles to tow trailers, caravans, and boats, often over long distances. Investigate the problem of providing a strong "universal coupling" between a medium-sized family car and a typical caravan or trailer. In this investigation, consider the following problems:

(1) The desirability and cost of providing all modern cars with built-in towing facilities during manufacture.

(2) The appearance and relative locations of towbar and number plate.

(3) The most suitable type of joint between the two vehicles. Consider safety and ease of coupling and uncoupling.

(4) Motor transport regulations concerning the towing of trailers.

(5) Loads that may be imposed on the coupling by the dead weight of the trailer; the acceleration or braking of the car; and by the inertia of the trailer and its load.

(6) The forces and their effects, for example, the shear stresses in the coupling bolts, the shear stresses in the towbar anchor bolts, and the bending stresses in the plate and tubular members of the towbar. (The torsional stresses induced in the towbar are not covered in this text.)

Each of the following projects should receive full analytical treatment comprising investigation and research, analysis as outlined earlier in this chapter, a report of the results giving recommendations and drawings, and a prototype or model to aid in the evaluation of the solution.

(**Note**. When evaluating a model, due care must be taken in assessing the validity of the test results, as it is not always possible to simulate accurately all of the important design features in a model. Consider for instance, a model ship or aircraft being tested in a tank or wind tunnel. If the scale model is 1/10 full size in its linear dimensions, it will have 1/100 of the full surface area, and only 1/1000 of the full volume. These factors must be considered if the tests relate to wetted surface area of the ship, or drag from skin friction of the aircraft.)

19/9

Design and construct a small solar cooker. It should be cheap, light, and portable, easy to use and easy to clean. It should be able to fry an egg and grill a steak, and must be able to boil half a litre of water, in ten minutes or less.

19/10

Design a small hand-operated jib crane to be used for handling heavy slabs of stone in the storage yard of a monumental mason. Your research should include inspections of various types of such cranes and discussions with their operators regarding limitations and possible improvements. (The design of the winch should be a separate exercise.)

Build a scale model from light timber and load it to determine which member will fail. Strengthen that member and repeat the experiments until it is possible to infer suitable proportions for each component.

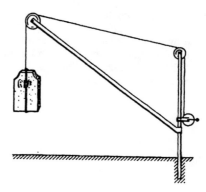

Problem 19/10   Basic design of jib crane. Winch design could be considered later.

19/11

Design a simple truss bridge to cross a small river.

Using this design, build a scale model bridge by gluing together thin timber laths (paddle-pop sticks, or modelling balsa strips will do). Load the model until it collapses. Replace the members which fail with stronger ones and test again. Repeat until it is possible to suggest approximate proportions for the major components of the truss. Some common basic designs of bridges are shown in the diagrams.

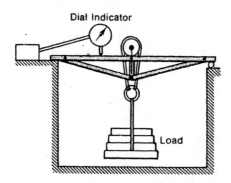

Problem 19/11   Test rig for checking model bridges.

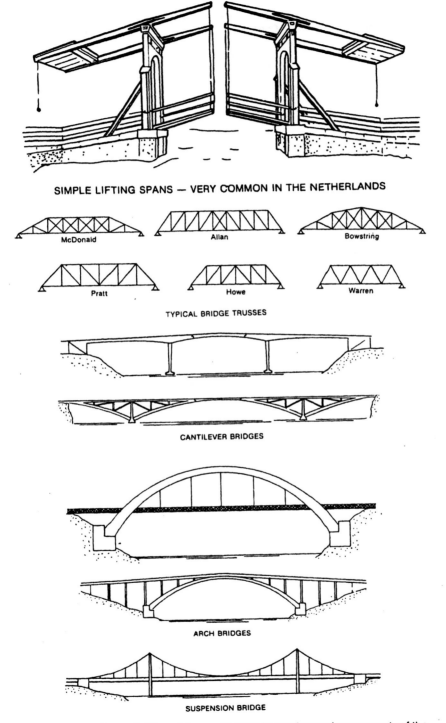

SIMPLE LIFTING SPANS — VERY COMMON IN THE NETHERLANDS

McDonald Allan Bowstring

Pratt Howe Warren

TYPICAL BRIDGE TRUSSES

CANTILEVER BRIDGES

ARCH BRIDGES

SUSPENSION BRIDGE

Problem 19/11 Several different types of bridge in regular use in many parts of the world. Research will provide many variations on these basic designs.

19/12

Develop a mechanised process for trimming the edge of a lawn where it meets a concrete path, or garden edge. Build a prototype and test its effectiveness, making modifications to the original design as necessary. Submit a report.

19/13

Design and build a small electric vehicle to carry two people. The vehicle is to have at least three wheels and be powered by a 12-V car battery and a car starter motor (or any similar motor).

It must be able to drive through a standard internal doorway, and should be small and light enough to be carried in a station waggon. It should be of very simple construction, from readily available, inexpensive materials. The vehicle must be foolproof, safe and simple to operate, and require a minimum of driver instruction. Provide sufficient storage space to carry an optional battery charger, and a small amount of luggage.

Consider the changes required in the above restrictions, and report on modifications needed to your final design in order to make the vehicle suitable for each of the following different uses:
(1) A go-cart for children
(2) Transport for a workshop maintenance crew
(3) (i) Transport to and from the local shops
     (ii) Commuting to and from work
     (iii) An invalid chair
     (iv) A motorised golf buggy

Find out about registration requirements and considerations of safety and stability

19/14

Conduct a feasibility study of the proposal to supplement the ferries on Sydney Harbour with hovercraft.

Build a scale model of an *air cushion vehicle* (ACV) and experiment with the power required to support the vehicle plus loading.

Rate the load-carrying efficiencies of the ACV compared with more conventional forms of transport, and identify the fields in which each is superior.

19/15

Design a vandal-proof public telephone to replace the usual phone box. Consider the advisability and economics of making no charge for calls.

19/16

Improve your school seats. The seats designed by you should be rugged and durable as well as comfortable. Build a prototype seat for use in the classroom, and a prototype fixed seat for use in the playground.

Appendix

Sir Isaac Newton

Galileo died in January, 1642, and on Christmas Day, 1642, in Lincolnshire, a baby was born to the recently widowed wife of a thirty-six-year-old, weak, extravagant man—a frail, undersized baby whose head had to be held for a time in a leather collar. This baby was Isaac Newton.

When he was twelve, Isaac went to the Kings School at nearby Grantham where he was the last in the class, until he had a fight one day. Young Isaac fought the second-last boy in the class, beat him, and decided that if he could beat him with his fists, he could beat him just as well in the school work.

At the age of twenty-two, in 1664, he received his B.A. degree. When, in 1665, the Great Plague caused the Cambridge University to be closed down, Newton returned to his home at Woolsthorpe for about two years until the University opened again—and what a two years they were! The binomial theorem was handed down for future generations of school boys to puzzle over. He had originated investigations into calculus, and spectrum analysis and formulated the laws of gravitation. In 1667, he returned to Cambridge as a fellow of Trinity College and in 1668 constructed the first reflecting telescope.

The works of many great men like Copernicus, Kepler, Galileo, Descartes, Christopher Wren, Edmund Halley and Robert Boyle assisted Newton in his discoveries. Indeed Newton is quoted as saying: "If I have seen further than others it is because I have stood on the shoulders of giants". He could well have added that he also sensed the direction in which to look, had better eyesight and was prepared to spend the time searching the horizons for the as yet uncharted coastline.

There were many general theories but Newton supplied laws which could be proved. He knew that unless there was a binding force acting on a circling body and drawing it towards the centre, this body would fly off at a tangent just like mud spinning off the rim of a car wheel. He knew that there must be such a binding force in the sun. He also knew from Kepler that the time of each revolution around the sun was proportional to the distance. (*The square of the time is equal to the cube of the distance.*) By working on this law of Kepler's he realised that if he could find the force of attraction towards the central body whose magnitude was inversely proportional to the square of the distance the problem was solved. (*If the distance is halved, the effect is four times as great, or the effect gets proportionately less as the distance increases.*)

The legend of the falling apple is generally associated with the laws of gravity. (The tree from which it is said to have fallen was preserved until 1820 when it became so decayed that it had to be cut down but some of the wood was preserved.) This incident is said to have helped Newton to link up the facts: Why did it fall? Why did a stone fall? Why did an arrow fall? If they still fell from the highest point on earth was it not reasonable to assume that they would fall from the moon? Obviously there must be a force exerted by the earth on the apple. Was the moon falling? Why did not the moon fly off at a tangent? It must be falling towards the earth. Why did it not hit the earth? There is no friction in space such as the cricket ball thrown horizontally would encounter on earth, so once it had started circling it would, like the satellites of today, continue circling without the need of additional power and its speed would not diminish as happens with the cricket ball before it falls to the ground.

The force Newton was looking for must be exerted by the sun since the earth was circling the sun but it must also be exerted by Jupiter since four moons at that time were known to be circling Jupiter and it must be exerted by the earth since the moon was circling the earth. Was it possible that in all these cases the force of attraction was inversely proportional to the square of the distance? If he could solve this problem for the moon and the earth would he not have found the key, the universal law of gravitation? Here was a worth-while experiment. The distance of the moon from the earth's centre was known to be sixty times the earth's radius. Assuming that the earth's pull or attraction falls off as the square of the distance, the attraction exerted by the earth on the moon must be 1/60 × 60 of the attraction it will exert on a body at its own surface. (The distance of a body on the earth's surface from the centre of the earth is 1/60 the distance of the moon from the earth's centre.) Now the apple fell at a speed that could be calculated, 16 feet (4.88 metres) per second; therefore, the moon would drop or be pulled a distance of 16/60 × 60 in one second. All he had to do was to calculate the drop and assure himself that it worked out at 16/60 × 60. An examination of Newton's papers had shown that he had worked out the law of universal gravitation "pretty nearly" to satisfy himself in 1666. What did he do with his tremendous scientific achievement? He put it away in a drawer and told nobody about it.

In 1684 most of the members of the Royal Society were trying to solve the problem of gravitation which Newton had solved in principle eighteen years before. Edmund Halley (the astronomer and mathematician after whom the comet is called), Sir Christopher Wren and Robert Hooke in particular were engrossed in this problem. Wren offered a prize to either of them of a book worth forty shillings if they could provide the proof. They both failed but Halley decided that if anyone could prove it it would be Newton. Halley asked Newton: "What will be the curve described by the planets on the supposition that gravity diminishes as the square of the distance?" "An ellipse", said Newton, immediately. "How do you know?" asked Halley, flabbergasted. "Why", said Newton, "I have calculated it". It gives us some picture of Newton's character when we learn that it took him some time to find the calculations and send them to Halley.

The renewal of and reworking over the problems aroused Newton's interest again in the subject and, at Halley's urging, Newton went to Lincolnshire and started to write his book, *Principia Mathematica* (*Mathematical Principles of Natural Philosophy*). Here was an entirely new philosophy presented at a time when the world had just accepted Descartes' vortex theory. The great principle of the *Principia* is the law of universal gravitation:

Every particle of matter in the universe is attracted to every other particle of matter with a force inversely proportional to the square of their distances.

Newton began his book in December 1684 and sent it to the Royal Society in 1686. The majority of this work was, therefore, composed in a period of seventeen months. This was an incomparable achievement, one of the greatest works of all time.

In 1703 Newton was elected President of the Royal Society and re-elected every year until his death twenty-five years later. In 1704 he published his great work, *Opticks*, written more than twenty years previously. In 1705 Queen Anne bestowed the first knighthood ever bestowed on a man of science on Newton. In Newton's later years he was chiefly interested in Religion, Ancient Chronology and Thermometry. He was the first to prove that freezing and boiling take place at a constant temperature. He also displayed an interest in Geology, Electricity and Magnetism and, of course, was the greatest mathematician of his time. When asked by the Queen of Prussia about Newton, Leibnitz, his "calculus rival", answered: "Taking mathematics from the beginning of the world to the time when Newton lived, what he has done is much the better half". It was said that "In passing through his hands, mechanics, optics and astronomy were not merely improved but renovated".

Newton gave to science the ability to make statements about the probable future behaviour of things, statements which are almost certain to be true when put to the test. Newton's works, adequately explaining forces and motions on earth, were to reign supreme until Albert Einstein's *theory of relativity* (1916) showed that some of the unavoidable assumptions made by Newton during his life particularly in regard to space and time could no longer stand up. His ability as a mathematician and experimentalist, the quality and fertility of his mind and the precision of his observation were combined with an instinctive insight into matters which seemed to lead him along pathways of his investigations. He supplied the keys for countless scientific discoveries and laid the basis for the future development of science.

Answers
to Problems

Chapter 1

1/4 1.2 mm; 12.6 kg; 66 s; 56 tonnes; 2 mm; 10 MN.

1/5 10^3 kg; 2×10^3 m; 10^6 N; 6×10^{-3} m; 14.4×10^3 s.

1/8 223.6 km; 26.6° west of north.

1/9 (a) 282.8 km; (b) 45°.

1/10 (a) 56 400; (b) 3 390 000.

Chapter 2

2/5 The newton (N). The kilonewton (kN) and the meganewton (MN). The kgf is not coherent or compatible with SI units.

2/7 9.8 N (10 N); 98 N (100 N); 100 N (102 N); 267.5 N (273 N); 19 600 N (20 000 N).

2/8 (i) $10 \, \text{m/s}^2$; (ii) $9.8 \, \text{m/s}^2$; (iii) $10 \, \text{m/s}^2$; (iv) $10 \, \text{m/s}^2$; (v) precise local g.

2/9 10 N.

2/10 5 kN down at 20° to the horizontal.

2/11 ≈ 30 kg; 300 N.

2/12 125 N. Reactive (compressive).

2/13 220 N.

2/14 (a) While accelerating < 80 kg; (b) At uniform speed = 80 kg; (c) While slowing > 80 kg.

2/15 50 N.

2/16 250 N.

2/17 (i) 50 kN; (ii) 52 kN.

2/18 No. Yes, chipboard!

2/19 (a) $T_{max} \approx 4.9$ kN.
(b) T_{max} occurs near the winding drum when there is a fully loaded car at the ground floor.
(c) (i) 2.9 kN; (ii) 3.3 kN; (iii) 3.6 kN, 3.7 kN.
(d) No. During acceleration and deceleration these tensions can double.

2/20 1.10 m square.

2/21 (i) ≈ 125 kPa (*you* are sitting on the stool).
(ii) ≈ 125 kPa.
(iii) ≈ 5 MPa (40 times as much).

2/22 (i) 15 kN (c); (ii) 306 kPa.

Chapter 3

3/5 1000 N at $\angle 60°$.

3/6 18.8 kN.

3/7 $F_H = 86.6$ kN to the left; $F_V = 50$ kN down.

3/8 $F_x = 10.4$ kN; $F_y = 6$ kN.

3/9 103.6 MN at $\angle 73°$.

3/10 $F_{AB} = 1155$ N; $F_{BC} = 577.4$ N.

3/11 Reaction on vertical wall = 84 N; Reaction on horizontal wall = 130 N.

3/12 73 kN at $\angle 73°$.

3/13 $F_V = 187.5$ N (x2); $F_H = 108$ N.

3/14 No—system is in equilibrium.

3/15 (a) 14 500 MN; (b) One on centreline of the ship and two 10° either side; (c) 14 850 MN.

3/16 $F_x = 2.2$ N; $F_y = 9$ N.

3/17 $R \approx 239$ MN at $\angle 41°$

3/18 5 320 N in each leg.

3/19 $P = 20$ kN; $Q = 51.9$ kN.

3/20 $F_A = 1324$ N; $F_B = 1572$ N.

3/21 $\theta = 120°$; $T = 2.5$ kN.

3/22 Force in jib, 17 kN; force in tie, 10 kN.

3/23 (a) System in equilibrium.
(b) Increase 37.5 kN to 57.5 kN.

3/24 $T_1 = 423$ N; $T_2 = 1086$ N.

3/25 $F_A = 5.5$ kN (c); $F_B = 11.5$ kN (c).

3/26 5.8 m.

3/27 (a) 40 kN; (b) 40 kN; (c) 68 kN, 40 kN.

3/28 Tension in cable, 1125 N; tension in rope, 220 N.

3/29 140 N.

Chapter 4

4/12 200 mm.

4/13 100 N.

4/14 540 N.

4/15 1200 N.

4/16 < 0.5 m from car.

4/17 750 N.

4/18 1.5 kN in each biceps muscle.

4/19 No. $\Sigma F = 0$, $\Sigma M \neq 0$.

4/20 130 N.

4/21 1.25 kN on each side of the pipe.

4/22 (a) 1.2 kN; (b) 250 N; (c) closer to the nail.

4/23 750 mm from Tom.

4/24 1 kN.

4/25 1.2 kN.

4/26 300 N.

4/27 48 kN.

4/28 163 N.

4/29 (a) 40 kN; (b) 50 kN, 40 kN.

4/30 Vertical force on each thrust collar = 5000 N. Horizontal force on each pivot = 2500 N.

4/31 (a) Front pair (with van) 7937 N
(without van) 8571 N

Rear pair (with van) 8173 N
(without van) 6429 N
Van 4445 N

4/32 (a) $R_L = 366$ N, $R_R = 484$ N;
(b) 0.02 MPa.

Chapter 5

5/10 No. $\Sigma M \neq 0$.

5/11 (i) 400 N; (ii) vertically down;
(iii) 2.5 m from left.

5/12 400 MN; 7.5 m to the right of the 200-MN force.

5/13 (i) 50 N upwards;
(ii) 1000 mm (1 m) from the left-hand end of the bar.

5/14 $R_L = 65$ N; $R_R = 95$ N.

5/15 $R_R = 162$ kN $\uparrow$,
$R_L = 271$ kN $\uparrow$, 250 kN $\leftarrow$;
or $R_L = 368$ kN at $\angle 43°$.

5/16 140 N; 275 N at $\angle 60°$

5/17 $R_R = 346$ kN vertically up;
$R_L = 240$ kN at $54°\angle$.

5/18 $R_c = 145.2$ kN $\quad 72.7°$
$R_D = 138.6$ kN $\uparrow$

5/19 $F = 111.2$ N
$R_A = 96$ N $\quad 38.6°$

5/20 200 kN down through the centre of the plate.

5/21 850 N; 780 N at $\angle 5°$.

5/22 3 kN up; 3 m from LH end of beam.

5/23 Front: 2.6 kN each; rear: 2.7 kN each.

5/24 $R_A = 580$ kN $\quad 45°$
$R_B = 560$ kN $\rightarrow$

5/25 (a) $P = 800$ N; $R = 1300$ N.
(b) $P = 200$ N; $R = 500$ N.
(c) $R = 500$ N; $d = 150$ mm.

5/26 $R_B = 810$ kN; $R_C = 190$ kN.

5/27 45 kN; 40 kN at $\angle 11°$.

5/28 (a) $R_L = 189$ kN $\quad 72°$
$R_R = 308$ kN $\uparrow$
(b) $R_L = 224$ kN $\quad 75°$
$R_R = 472$ kN $\uparrow$

5/29 At A: $F_V = 4.7$ kN, $F_H = 3.5$ kN.
At B: $F_V = 54.7$ kN, $F_H = 2.5$ kN.
Tension in $AC = 5.9$ kN.

5/30 328 N; 727 N at $\angle 39°$.

5/31 26.5 kN each; 22 kN each.

Chapter 6

6/6 80 N tension; 20 N shear.

6/7 $F_L = 750$ N; $F_R = 1000$ N.

6/8 180 N m.

6/9 Horizontal load on each hinge = 67 N.
Maximum vertical load on either
hinge = 200 N.

6/10 5 tonnes.

6/11 (a)   180 N

300 N m;   (b) 1200 N.

6/12 (a)   200 N;
400 N m

(b)   (i) 400 N m, (ii) 400 N m.

6/13 (a) $L = 140$ N; $R = 40$ N; 40 N m;
(b) $L = 200$ N; $R = 100$ N; 70 N m;
(c) Hands well apart.

6/14   3 N m, 50 N.

6/15 (a) 2.5 kN ←; (b) 7.5 →, 750 N ↑

6/16 456 N ←, 396 N   1.5°

6/17   130 N
3 N m 67.3°

6/18 (a) 20 N m; (b) 40 N m; (c) 10 N m;
reaction in steering column.

6/19 50 N, 480 mm to the right of B.

6/20   350 kN
200 kN m

6/21 (a) 621.5 N; (b) 11.25 N m.

6/22 (a) 30 kN, 30 kN m; (b) 15 kN m.

6/23 966 N; 710 N.

6/24 $A = 0$; $B = 12.5$ kN →; $C = 0$;
$D = 16$ kN at 39° .

6/25   840 N
1032 N m

6/26 (a) $A = 0.5$ t; $B = 1$ t; $C = 10$ t.
(b) No.
(c) Very little, as total mass of beam
and trolley is only 140 kg.

6/27 Front bolt A loose: $T_B = 675$ N (OK).
Rear bolt B loose: $T_A = 4050$ N. (This
is 6 times as much as the designed
static load.)

Chapter 7

7/5 $F_{CT} \approx 2000$ N; $F_s \approx 1500$ N.

7/6 (i) $R_L = R_R = 500$ kN;
(ii) $F_{AB} = 289$ kN (T),
$F_{BC} = 577$ kN (C).

7/7 7.3 kN (T); 9 kN (C); 10 kN;
30 kN m.

7/8 $F_{AB} \approx 59$ kN (T); $F_{BC} \approx 77$ kN (C).

7/9 $F_A = 1333.3$ kN; $T_{BC} = 953.3$ kN.

7/10 (i) (a) 1196 kN, (b) 1732 kN;
(ii) (a) 1000 kN, (b) 2000 kN.

7/11 (1) $B = 25$ kN ↑; $D = 25$ kN ↑.
(2) $A = 25$ kN ↑; $B = 25$ kN ↑;
$C = 0$.
(3) $A = 33.3$ kN ← ; $D = 50$ kN ↑.
33.3 kN → .
(4) $A = 33.3$ kN ←, indeterminate ↕;
$D = 33.3$ kN →, indeterminate ↕.
(5) $A = 25$ kN ↑; $B = 25$ kN ↑.
(6) Not restrained—will collapse.
(7) $A = 25$ kN ↑; $B = 25$ kN ↑.
(8) Indeterminate.
(9) Not restrained—will collapse.

7/12 289 N; (b) 577 N; (c) 500 N.

7/13 $AD \approx 750$ N (T); $AC \approx 900$ N (C);
$CD = 0$.

7/14 $R_B = 20$ kN; $AC = 22.4$ kN,
$AB = 28.3$ kN, $BC = 20$ kN.

7/15 A and B horizontal = 2310 N;
A vertical = 750 N; B vertical = 0,
resultant = 2.39 kN. Tension in
$C = 10$ kN, in $D = 0$.

7/16 $AB = 25$ kN (T); $BC = 25$ kN (T);
$CD = 50$ kN (C); $DE = 50$ kN (C);
$EA = 50$ kN (C); $BD = 50$ kN (T);
$BE = 50$ kN (T).

7/17 $R_L \approx 10.2$ kN at 11° :
$R_R \approx 7.9$ kN ↑. $AE \approx 13.5$ kN;
$ED \approx 19.5$ kN; $DG \approx 8$ kN;
$GC \approx 13.5$ kN; $EF \approx 2.5$ kN;
$FG \approx 17$ kN; $CF \approx 18.5$ kN.

7/18 (a) $A = 1.82$ kN, $B = 2.36$ kN,
$D = 2.36$ kN; (b) $F_H = 1.67$ kN←,
$F_V = 0.67$ kN ↓ ; (c) $F_H = 1.67$ kN →,
$F_V = 1.67$ kN ↑ .

7/19 (1) $A = 10$ kN ↑, 6.7 kN → ;
$B = 6.7$ kN ← .
(2) Indeterminate.
(3) $A = 5$ kN ↑ ; $C = 5$ kN ↑ .
(4) Indeterminate.
(5) Indeterminate.
(6) $A = 5$ kN ↑ ; $B = 5$ kN ↑
(unstable equilibrium).
(7) $A = 5$ kN ↑ ; $B = 8.33$ kN at
$37°$ ◺ ; $C = 6.64$ kN → .
(8) Indeterminate (one link is
redundant).

7/20 $R_L \approx 1100$ kN, $R_R \approx 1550$ kN;
$AB \approx 1450$ kN, $BC \approx 1580$ kN,
$CD \approx 880$ kN.

7/21 (a) 20 kN; (b) 20 kN at $30°$ ◥ ;
(c) $EB = BG = 23.1$ kN (C),
$AE = GC = 11.5$ kN (T),
$EF = FG = 0$,
$FD = 5.8$ kN (T).

7/22 (a) $A = 5$ kN ↑, 3.75 kN → ;
$B = 3.75$ kN ← .
(b) $A = 11.5$ kN ↑ , 3.75 kN → ;
$T = 7.5$ kN.

(c) $A =$ 5 kN
↯ 30 kN m

7/23 $R_L = 1250$ kN; $R_R = 750$ kN;
$F_{FH} = 1454$ kN (C);
$F_{GH} = 141$ kN (C);
$F_{GI} = 1400$ kN (T).

7/24 $R_L \approx 4.3$ kN at $35.2°$ ◿ ;
$R_R \approx 22$ kN.

7/25 $DE = 36$ kN (C); $HJ = 427$ kN (C).

7/26 $BC = 13.3$ kN (T); $GF = 19$ kN (C).

7/27 $OA \approx 25$ kN; $OB = OC \approx 10.6$ kN.

7/28 (a) 3.1 kN; (b) 5.6 kN; (c) 4.85 kN
each.

Chapter 8

8/11 490 N.

8/12 (a) 0.26; (b) 15.3 N.

8/13 88.2 kN.

8/14 29.4 N.

8/15 0.35.

8/16 200 N.

8/17 (i) 0.64, 25 N; (ii) 0.81, 21.7 N.

8/18 (i) 12.2 kN; (ii) 1.2 kN.

8/19 Wall: 56.6 N ← ;
Path: 204 N ◢ $74°$

8/20 $\theta \approx 40°$.

8/21 Yes. No!

8/22 (i) 36 N; (ii) 180 N.

8/23 23.2 N.

8/24 $\alpha \approx 21.8°$, $T \approx 2.58$ kN.

8/25 82 N.

8/26 (i) Yes; (ii) 180 mm.

8/27 122.5 N (sliding).

8/28 Front ≈ 2.3 kN; Rear ≈ 9.5 kN;
$\mu_s \approx 0.6$.

8/29 (i) 0.6
(ii)

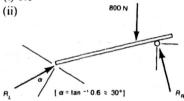

(iii) X ≈ 180 mm.
(iii) NO (since μ_s would need to
equal 1).

Chapter 9

9/16 1.49 m/s..

9/17 2.5 m/s².

9/18 (i) 20 m/s; (ii) 0.67 m/s²; (iii) 600 m.

9/19 17.6 m.

9/20 No.

9/21 (i) 346.4 km/h; (ii) 1.67 km.

9/22 20 s.

9/23 (i) 17.2 m/s; (ii) 7.1 m/s.

9/24 27.6 m/s; 35.6 m.

9/25 5.1 s.

9/26 14.7 m/s; 11 m.

9/27 8820 m; 60 s.

9/28 833 m/s.

9/29 1.2 m.

9/30 347 m west; 510 m above.

9/31 Yes; Yes.

9/32 (i) 4 m/s; (ii) 0; (iii) 2 m/s.

9/33 (a) 1 m; (b) 1.2 m; (c) 17.25 m; (d) 0.5 m/s²; (e) 0.1 m/s².

9/34 0.67 m/s.

9/35 (i) 8 m/s; (ii) 70 s.

9/36 (i) 150 m; (ii) 1.2 m/s².

9/37 (i) 10 m/s²; (ii) 25 m/s²; (iii) − 10 m/s²; (iv) ≈ 55 m.

9/38 (i) 100 m; (ii) 50 m.

9/39 (i) 11° (11° E of N); (ii) 11° (11° W of N).

9/40 (i) 18 knots; (ii) 323°.

9/41 (i) 130°; (ii) 500 nautical mph (500 knots); (iii) 550 nautical mph (550 knots).

Chapter 10

10/17 3 m/s².

10/18 100 kg.

10/19 45 kN.

10/20 50 MN.

10/21 (i) 8.94 s; (ii) 89.4 m/s (≈ 322 km/h); (iii) 8.5 kN.

10/22 2.4 kN; 3 g.

10/23 0.1 m/s².

10/24 (i) 833 N; (ii) 875.5 N.

10/25 40.7 kN.

10/26 39.2 m/s²; 1960 m.

10/27 66.8 kg.

10/28 (i) 2.6 m/s²; (ii) about the same; (iii) Acceleration increases because the mass reduces dramatically. The gravitational attraction of the earth also reduces significantly during the burn.

10/29 8.7 kN.

10/30 27 s.

10/31 31.6 m/s at 18.4° N of E.

10/32 141.9 N; 0.3 m/s².

10/33 (i) $T_A = T_B = T_C = 160.4$ N; (ii) $T_D = 320.8$ N; (iii) 20-kg mass, 1.78 m/s²; 30-kg mass, 0.89 m/s².

10/34 (i) $T_A = T_B = T_C = 179$ N: (ii) $T_D = 306$ N; (iii) 20-kg mass, 0.85 m/s²; 30-kg mass, 0·425 m/s².

10/35 0.125 m/s².

10/36 (i) 0.57 m/s²; (ii) 29.2 s; (iii) 243 m; (iv) 1.55 m/s².

10/37 0.82 m/s²; 26.6 kN, 13.3 kN.

10/38 (i) 3.8 kN; (ii) 2.2 kN.

10/39 5.3 kN.

10/40 84 kN (2 brake shoes).

10/41 15.8 kN.

10/42 (i) 94 N; (ii) 4.5 m/s²; (iii) 13.4 m/s.

10/43 0.5 m/s².

10/44 4.44 m/s.

10/45 0.34.

Chapter 11

11/10 (i) 8×10^3 kg m/s; (ii) 1.2×10^5 kg m/s; (iii) 4.2 kg m/s; (iv) 2.97×10^{10} kg m/s.

11/11 667 m/s.

11/12 2.5 m/s.

11/13 200 N.

11/14 52 km/h.

11/15 32 km/h; 59° E of N.

11/16 $F_v = 917$ N. $F_H = 231$ N.

11/17 3.3 kN.

11/18 700 N.

11/19 0.6 m/s.

11/20 3 MN.

11/21 0.03 km/h.

11/22 450 N.

11/23 (i) − 2 m/s; (ii) 0.3 s; (iii) 0.25 m.

11/24 60.6 kN.

11/25 9.8 kN.

11/26 (a) 12 kN. (b) 5.2 m/s².

11/27 83.5 kN.

11/28 0.77.

Chapter 12

12/7 (i) 1.57 rad; (ii) 0.29 rad; (iii) 2.11 rad; (iv) 5.76 rad.

12/8 (i) 57.3°; (ii) 1088.6° (3.02 revs); (iii) 540° (1.5 revs.)

12/9   (i) 20.9 rad/s; (ii) 73.3 rad/s;
(iii) 150.8 rad/s.

12/10   (i) 7200 rpm; (ii) 5000 rpm;
(iii) 478 rpm.

12/11   0.105 rad/s; 0.147 m/s.

12/12   26.2 m/s.

12/13   20.9 m/s.

12/14   15.7 m/s.

12/15   0.17 rad/s²; 7830 rad.

12/16   70.7 m/s.

12/17   150 rad/s (23.9 rev/s); 159 revs.

12/18   12.3 m/s; 44.3 km/h.

12/19   Hour hand: 145×10^{-6} rad/s;
145×10^{-6} m/s.
Minute hand: 17×10^{-4} rad/s;
23×10^{-4} m/s.

12/20   (i) 0.58 m/s²; (ii) 1.53 rad/s².

12/21   (i) 0.125 rpm; (ii) 80 s.

12/22   12.8 m/s.

12/23   1800 rpm; 60 s.

12/24   7.3 seconds after second gear starts
to rotate.

12/25   (i) 1.57×10^{-2} rad/s;
(ii) 2.62×10^{-4} rad/s²;
(iii) 3.93×10^{-4} rad/s².

12/26   (i) 600 rpm/min.; (ii) 2700 revs;
(iii) 53 s.

Chapter 13

13/13   2961 m/s².

13/14   1645 m/s².

13/15   50 N.

13/16   (i) 7.9 N; (ii) 18.7 rad/s ($\approx$ 3 rev/s).

13/17   41 N.

13/18   645 N.

13/19   -44.3 m/s; 0.2 rad/s.

13/20   428.5 m.

13/21   (i) 6.8 rad/s; (ii) 5.7 N; (iii) 5.1 m/s².

13/22   (i) 71.3 km/h. (ii) tangentially,
(then in a parabola).

13/23   (i) 0.58; (ii) 5.7 m/s².

13/24   32.5 km/h; ~ 60°.

13/25   (i) 85.8 km/h; (ii) 121.4 km/h.

13/26   97.6 mm.

13/27   2.4 m above roadway.

13/28   37.5 N m.

13/29   12.5 N m $\circlearrowright$.

13/30   40 N m.

13/31 $I = 1.6$ kg m²; $k = 0.18$ m.

13/32   (i) 21.6 kg m²; (ii) 540 N m.

13/33   105 seconds.

13/34   (i) 50 rad/s²; (ii) 200 rad/s;
(iii) 63.7 revolutions.

13/35   (i) 2000 kg m/s; (ii) 1047 kg m/s.

13/36   (i) 5 rad/s; (ii) 4 rad/s; (iii) 6 rad/s.

13/37   0.07 kg m².

13/38   8.4 rad/s² $\circlearrowleft$.

Chapter 14

14/15   (i) potential; (ii) kinetic;
(iii) potential; (iv) rotary kinetic.

14/16   (i) 2940 J. (ii) 490 J.

14/17   (i) 20.58 kJ. (ii) 20.58 kJ. (iii) 294 J.

14/18   (i) 4.12 MJ. (ii) 4.12 MJ.

14/19   131.7 MJ.

14/20   (i) 19.6 J; (ii) 19.6 J; (iii) 14.7 J.
Some of the energy deformed both
the ball and the ground; some made
sound and heat.

14/21   (i) 866 J; (ii) 144 J.

14/22   160 N.

14/23   (i) 47 J; (ii) 17.6 J; (iii) 29.4 J.

14/24   29.4 kJ.

14/25   141.2 (sliding) > 48.6 (rolling).

14/26   (i) 960.4 J. (ii) 617.4 J. (iii) 14.7 m/s².

14/27   176.4 MJ.

14/28   (i) 2205 N m. (ii) 7350 N m.
(iii) 9.6 kJ. (iv) 21.2 kJ.

14/29   900 kN.

14/30   3.18 MJ.

14/31   (i) 123.5 kJ; (ii) 2.2 kN.

14/32   7.5 J.

14/33   (i) 5 J; (ii) 5 J.

14/34   4 kJ.

14/35   (i) 118 kN; (ii) 236 MJ.

14/36   2040 m.

14/37   (i) 7 m/s; (ii) 5.5 m/s; (iii) 10.5 kJ;
(iv) 70.1 kN.

14/38 (i) 8 N m; (ii) 1005 J.

14/39 1.56 kN m.

14/40 (i) 30.9 MJ; (ii) 1740 s.

14/41 5.7 kN.

14/42 (i) 26.5 kJ; (ii) 2.19 kN m.

14/43 0.33 kg m².

14/44 (i) 4.28 kJ; (ii) 6.16 kJ;
(iii) 49.08 kJ; (iv) 4.9 kN.

Chapter 15

15/12 4.5 MW.

15/13 1.84 kW.

15/14 (i) 0; (ii) 125 W.

15/15 18.8 kW.

15/16 (i) 118.8 kN; (ii) 117.6 kN;
(iii) 120 kW; (iv) 3.5 MJ.

15/17 294 kilolitres (≈ 294 tonnes).

15/18 > 650 W, say, 750 W.

15/19 367 kW.

15/20 4.1 kW.

15/21 5.90 kW.

15/22 (i) 1176 kilolitres; (ii) 1000 kilolitres.

15/23 955 N.

15/24 (i) 9.63 kW; (ii) 45.37 kW;
(iii) 82.5%.

15/25 8.88 kW.

15/26 (i) 8.125 kW. (ii) 75.4%.

15/27 (i) 36; (ii) 7.

15/28 7.4 kW.

15/29 28.6 kN m.

15/30 (i) 63.4 kJ; (ii) 38 kW.

15/31 > 110 kW.

15/32 16 rpm; 36.

Chapter 16

16/13 (i) 126.7 kN; (ii) 91.5 MPa.

16/14 (i) 114.6 MPa; (ii) 1.5 × 10⁻³;
(iii) 76.4 GPa.

16/15 50 MPa.

16/16 27.9 mm; 445.6 MPa.

16/17 (i) 2.4 kN; (ii) 4 MPa.

16/18 1.25 mm.

16/19 > 50 kg.

16/20 1.67 m.

16/21 12 mm (M12).   > 14.6 mm (M16).

16/22 2.5 mm.

16/23 (i) 80 MPa; (ii) 80.39 MPa.

16/24 (i) 1200 MPa, 300 MPa; (ii) 1.8 mm.

16/25 22%.

16/26 (i) 206 N; (ii) 75 kPa; (iii) 500 kPa;
(iv) 975 kPa; (v) 4.4 MPa each side.

16/27 50 m.

16/28 (i) 29.4 kN; (ii) 18 mm.

16/29 231.25 mm.

16/30 (i) 27.5 MPa; (ii) 33.8 kN.

16/31 (i) 0.14 mm; (ii) 0.36 J;
(iii) 720 J/m³.

16/32 (i) 8.2 kN; (ii) 43.8 MPa.

16/33 Stress in steel, 36.0 MPa; Stress in
concrete, 2.52 MPa.

16/34 49 kN (C); 14.6 kN (T); 3.57 MPa.

Chapter 17

17/7

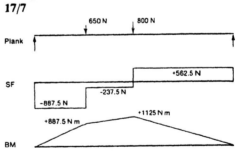

17/8 (i) 7.2 kN; (ii) 7.6 kN;
(iii)

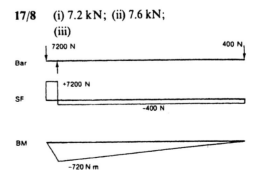

480   *Introduction to Engineering Mechanics*

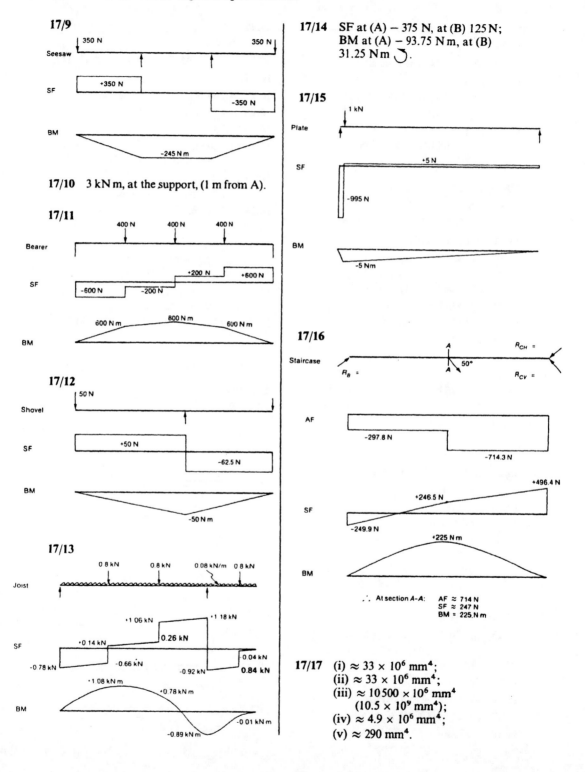

17/9

Seesaw

SF   +350 N   −350 N

BM   −245 N m

17/10   3 kN m, at the support, (1 m from A).

17/11

Bearer   400 N   400 N   400 N

SF   −600 N   −200 N   +200 N   +600 N

BM   600 N m   800 N m   600 N m

17/12

Shovel   50 N

SF   +50 N   −62.5 N

BM   −50 N m

17/13

Joist   0.8 kN   0.8 kN   0.08 kN/m   0.8 kN

SF   +0.14 kN   +1.06 kN   +1.18 kN   0.26 kN   0.04 kN   0.84 kN   −0.78 kN   −0.66 kN   −0.92 kN

BM   +1.08 kN m   +0.78 kN m   −0.01 kN m   −0.89 kN m

17/14   SF at (A) − 375 N, at (B) 125 N; BM at (A) − 93.75 N m, at (B) 31.25 N m ↺.

17/15

Plate   1 kN

SF   +5 N   −995 N

BM   −5 N m

17/16

Staircase $R_B =$ A 50° A $R_{CH} =$ $R_{CV} =$

AF   −297.8 N   −714.3 N

SF   +246.5 N   +496.4 N   −249.9 N

BM   +225 N m

∴ At section *A-A*:   AF ≈ 714 N
SF ≈ 247 N
BM = 225 N m

17/17   (i) ≈ 33 × 10⁶ mm⁴;
(ii) ≈ 33 × 10⁶ mm⁴;
(iii) ≈ 10 500 × 10⁶ mm⁴
(10.5 × 10⁹ mm⁴);
(iv) ≈ 4.9 × 10⁶ mm⁴;
(v) ≈ 290 mm⁴.

17/18   (i)

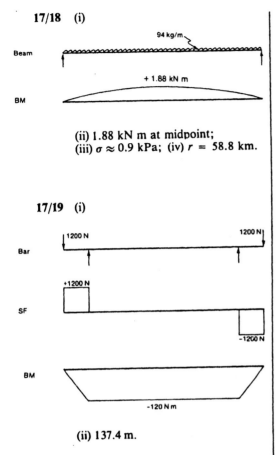

(ii) 1.88 kN m at midpoint;
(iii) $\sigma \approx 0.9$ kPa; (iv) $r = 58.8$ km.

17/19   (i)

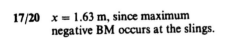

(ii) 137.4 m.

17/20 $x = 1.63$ m, since maximum
negative BM occurs at the slings.

17/21   (i) $F_H = 16.6$ kN,
$F_V = 4.1$ kN,
Tension in $CD = 19.2$ kN;
(ii) $SF_E = 9.6$ kN,
$BM_E = 9.6$ kN m;
(iii) $\sigma = 5.9$ MPa.

17/22   250 MPa.

17/23   1542 N.

17/24

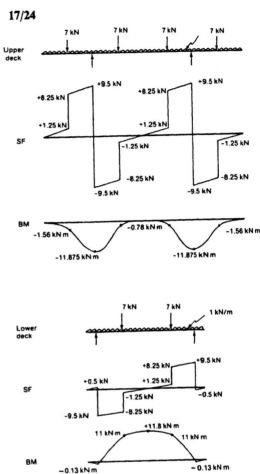

Chapter 18

18/1   50 N; 5:1.

18/2   500 N; 10:1.

18/3   160 N; 6.25:1.

18/4 ≈ 780 N; 6:1.

18/5   250 N.

18/6   400 N.

18/7 ≈ 50 N.

18/8   3:1; Rolling friction is *much* less than
sliding friction.

18/9   3:1; $\not> $ 24-litre bucket.

18/10 $\not< 130$ mm (for a maximum effort
of ≈ 250 N).

18/11 300 mm. No, it's too long. Use a smaller cable drum, or add a simple gear train ($\approx$ 2:1), or better bearings.

18/12 360-mm diameter drum, or 40-litre bucket.

18/13 There is one more rope supporting Tom's mass—the one he is pulling.

18/14 MA = 3:1. 60%. Lighter rope, bigger pulleys, better bearings.

18/15 Tom will slide, because there are 4 ropes pulling Tom and only 3 pulling Dick. 4 metres.

18/16 The three-sheave block, as then 6 ropes will be pulling the car.

18/17 You will be able to lift out the engine.

18/18 (a) >125 N; (b) >500 N.

18/22 $\not< $ 62.5 N.

18/23 (i) 33 m; (ii) 4 kN m; (iii) 75%; (iv) 90 N.

18/24 Cable should be attached to the high point (moment at base of stump is greater; cable will coil under bulldozer if it breaks).

18/25 103 N.

18/26 600 N.

18/27 (i) 83%. (ii) Because of friction. and energy used to bend and straighten the rope.

18/28 1.5 N m.

18/29 0.6 N m.

18/30 A force of 500 N to lift, 700 N to drag, so it is easier to lift it.

18/31 8:1, 80%.

18/32 4.63 kN.

18/33 1920 N.

18/34 (i) 7.59 m/s; (ii) 725 rpm *or* 75.9 rad/s.

18/35 100 rpm, 12:1.

18/36 500 N.

18/37 (i) 86.6 kN each. (ii) Tensile. (iii) 10.8 MPa.

18/38 > 4.66 kN.

18/39 83.7 kN (T), 229 kN (C), 190 kN (T).

18.40 (i) 440:1. (ii) 45 N. (iii) 220:1.

18/41 360 N.

18/42 $\not< $ 135 N.

18/43 17°.

18/44 62 N.

18/45 113.5 kN, 1187.5 kN.

18/46 3.58 kN, SF_A = 1.5 kN, SF_B = 1.85 kN.

18/47 10.7 kN.

Index

Printed in the United States
2101

9 780701 612245